AF738239

Cartesian Empiricisms

STUDIES IN HISTORY AND PHILOSOPHY OF SCIENCE

VOLUME 31

General Editor:

STEPHEN GAUKROGER, *University of Sydney*

Editorial Advisory Board:

RACHEL ANKENY, *University of Adelaide*
PETER ANSTEY, *University of Otago, New Zealand*
STEVEN FRENCH, *University of Leeds*
KOEN VERMEIR, *Centre National de la Recherche Scientifique, Paris*
OFER GAL, *University of Sydney*
NICHOLAS RASMUSSEN, *University of New South Wales*
CLEMENCY MONTELLE, *University of Canterbury*
JOHN SCHUSTER, *University of NSW/University of Sydney*
RICHARD YEO, *Griffith University*

For further volumes:
http://www.springer.com/series/5671

Mihnea Dobre • Tammy Nyden
Editors

Cartesian Empiricisms

Springer

Editors
Mihnea Dobre
Faculty of Philosophy
University of Bucharest
Bucharest, Romania

Tammy Nyden
Philosophy Department
Grinnell College
Grinnell, IA, USA

ISSN 0929-6425
ISBN 978-94-007-7689-0 ISBN 978-94-007-7690-6 (eBook)
DOI 10.1007/978-94-007-7690-6
Springer Dordrecht Heidelberg New York London

Library of Congress Control Number: 2013956125

© Springer Science+Business Media Dordrecht 2013
This work is subject to copyright. All rights are reserved by the Publisher, whether the whole or part of the material is concerned, specifically the rights of translation, reprinting, reuse of illustrations, recitation, broadcasting, reproduction on microfilms or in any other physical way, and transmission or information storage and retrieval, electronic adaptation, computer software, or by similar or dissimilar methodology now known or hereafter developed. Exempted from this legal reservation are brief excerpts in connection with reviews or scholarly analysis or material supplied specifically for the purpose of being entered and executed on a computer system, for exclusive use by the purchaser of the work. Duplication of this publication or parts thereof is permitted only under the provisions of the Copyright Law of the Publisher's location, in its current version, and permission for use must always be obtained from Springer. Permissions for use may be obtained through RightsLink at the Copyright Clearance Center. Violations are liable to prosecution under the respective Copyright Law.
The use of general descriptive names, registered names, trademarks, service marks, etc. in this publication does not imply, even in the absence of a specific statement, that such names are exempt from the relevant protective laws and regulations and therefore free for general use.
While the advice and information in this book are believed to be true and accurate at the date of publication, neither the authors nor the editors nor the publisher can accept any legal responsibility for any errors or omissions that may be made. The publisher makes no warranty, express or implied, with respect to the material contained herein.

Printed on acid-free paper

Springer is part of Springer Science+Business Media (www.springer.com)

Acknowledgments

This volume grew out of discussions during the Bucharest-Princeton Seminar in Early Modern Philosophy, which took place in Bran, Romania, in the summer of 2009. Both of us presented papers on Cartesian experimentation. Among the problems we faced were why seventeenth-century Cartesians would be interested in engaging in experimental practice when they could have just as easily picked up the results from the so-called "experimental philosophers" and what kind of knowledge could Cartesians believe themselves to attain from experiment anyway? In an effort to understand how Cartesian experimentalism fits into seventeenth-century transformations in natural philosophy, we began to collaborate with other scholars working on related problems and authors. In 2010, we organized a symposium on "Cartesian Empiricists" at the HOPOS (History of Philosophy of Science) meeting and a symposium on "Cartesian Physics (as Experimental Philosophy) and its University Receptions" at the ESHS (European Society for the History of Science) conference. These were followed by the symposium "Experimentalism and the Quest for Certainty in Cartesian Natural Philosophy" at the ISIH (International Society for Intellectual History) conference in 2011 and the symposium "Cartesian Physics and its Reception: between Local and Universal" at the ESHS conference in 2012. Our project has also continued to benefit from the annual Bucharest-Princeton seminar. We would like to express our gratitude to all of those who participated in these very stimulating discussions and debates. While it is impossible to name them all, we particularly want to thank Roger Ariew, Delphine Bellis, Patricia Easton, Sorana Corneanu, Daniel Garber, Madalina Giurgea, Dana Jalobeanu, and Epaminondas Vampoulis. We would also like to thank three student research assistants, who helped in the early stages of the project: Ben Aronowicz, Camilia Camacho and Esther Howe.

It was a great pleasure for us to work on this volume and we want to acknowledge the help we received from the following scholars, who read chapters and provided us (and contributors to the volume) with very helpful feedback: Peter Anstey, Roger Ariew, Delphine Bellis, Zvi Biener, Monte Cook, Sorana Corneanu, Sorin Costreie, Joe Cummins, David Cunning, Steffen Ducheyne, Mordechai Feingold, Gary Hatfield, Delphine Kolesnik-Antoine, Paul Lodge, Christoph Lüthy, Gideon

Manning, Sebastian Mateiescu, Trevor McClaughlin, Sophie Roux, Han van Ruler, Eric Schliesser, Paul Schuurman, Laurentiu Staicu, Piet Steenbakkers, Tad Schmaltz, Rienk Vermij, Grigore Vida and Sujeev Wickramasekara.

Our research has been made possible by several supporting grants, including an American Council of Learned Societies Fellowship, an American Philosophical Society Franklin Research Grant, a Scaliger Fellowship from the University of Leiden, a Fellowship-in-Residence at the Obermann Center for Advanced Studies at the University of Iowa, a National Endowment of the Humanities Summer Stipend, generous research support from Grinnell College, a post-doctoral grant from the University of Bucharest and two research grants provided by the Romanian National Research Council (CNCS).

We would like to end with an expression of our sadness at the loss of our colleague Epaminondas Vampoulis during the production of this volume. He is missed.

Contents

List of Contributors

Roger Ariew, University of South Florida, USA

Roger Ariew is Professor and Chair, Department of Philosophy, University of South Florida. He is the author of *Descartes and the Last Scholastics* (Cornell, 1999; 2nd edition, *Descartes among the Scholastics*, Brill, 2011), co-author of the *Historical Dictionary of Descartes and Cartesian Philosophy* (Scarecrow, 2003; 2nd edition forthcoming), and editor and translator of such works as *Descartes, Philosophical Essays* (Hackett, 2000) and *Pascal, Pensées* (Hackett, 2005). He is currently working on a variety of topics, including Descartes' Correspondence and the reception of Descartes' philosophy and science in seventeenth-century France.

Delphine Bellis, Ghent University, Belgium

Delphine Bellis is a postdoctoral researcher at Ghent University. Her research focuses on Descartes and early modern philosophy and physics, in particular optics. Her dissertation was entitled *Le visible et l'invisible dans la pensée cartésienne* (2010). More recently she has been working on Gassendi's philosophy and theory of space. She also currently collaborates on a complete edition of Descartes' correspondence to be published by Oxford University Press.

Mihnea Dobre, Bucharest University, Romania

Mihnea Dobre is a researcher at The Center for the Logic, History and Philosophy of Science (CELFIS), Faculty of Philosophy, University of Bucharest. He is studying the disciplinary shifts in the early modern period. In his dissertation—*Metaphysics and Physics in Cartesian Natural Philosophy: Descartes and Early French Cartesians on the Foundation of Natural Philosophy*—Mihnea Dobre discussed the problems connected with seventeenth-century possibilities of framing a new physics based on Cartesian philosophy. Dobre participated in numerous international conferences and published various articles on topics such as history of philosophy, history of science, and history and philosophy of science.

Patricia Easton, Claremont Graduate University, USA

Patricia Easton is Professor of Philosophy, and Associate Provost for Academic Planning, at Claremont Graduate University. She specializes in the history of modern philosophy, particularly the philosophy of René Descartes and the Cartesians of the seventeenth century. She directed and edited *The Descartes Web Site*, which features seventeenth-century French and English editions of Descartes' work *The Passions of the Soul*, and is the author or editor of three books and several articles and chapters.

Gary Hatfield, University of Pennsylvania, USA

Gary Hatfield is Adam Seybert Professor in Moral and Intellectual Philosophy at the University of Pennsylvania. He has published extensively on the history of modern philosophy and the history and philosophy of psychology from the medieval period into the twentieth century, including *The Natural and the Normative: Theories of Spatial Perception from Kant to Helmholtz* (1990), *Descartes and the Meditations* (2003), and *Perception and Cognition: Essays in the Philosophy of Psychology* (2009). He recently co-edited a volume with Sarah Allred, entitled *Visual Experience: Sensation, Cognition, and Constancy* (2012).

Bernard Joly, Université Lille 3, France

Bernard Joly is an Emeritus Professor at the University of Lille (CNRS UMR 8163 "Savoirs, textes, langage"). He has published *La rationalité de l'alchimie au XVIIe siècle* (Vrin, 1992), *Descartes et la chimie* (Vrin, 2011), *Histoire de l'alchimie* (Vuibert, 2013). His research concerns the history of alchemy and ancient chemistry and their connections with medicine and philosophy in the seventeenth century.

Tammy Nyden, Grinnell College, USA

Tammy Nyden is an Associate Professor of Philosophy at Grinnell College. She has published articles on Spinoza, Dutch Cartesianism and Seventeenth Century Physics and is the author of *Spinoza's Radical Cartesian Mind* (Continuum 2007). Her current research is on the changing views about and practices of experiment in natural philosophy at the University of Leiden in the seventeenth and eighteenth centuries, with particular focus on the similarities and differences among professors represented as Aristotelians, Cartesians and Newtonians.

Sophie Roux, École Normale Supérieure, Paris, France

Sophie Roux is Professor at the École Normale Supérieure, Paris. She was previously appointed at the Max-Planck-Institut für Wissenschaftsgeschichte (Berlin), at the Centre Alexandre-Koyré (Paris) and at the University of Grenoble. Among her publications are *Mechanics and Cosmology* (co-edited with M. Bucciantini and M. Camerota, 2007), *Mechanics and Natural Philosophy before the Scientific Revolution* (co-edited with W. R. Laird, 2008), *Thought Experiments in Historical and Methodological Contexts* (co-edited with K. Ierodiakonou, 2011), *La mathématisation comme problème* (co-edited with H. Chabot, 2011), *L'*Essai de logique *de Mariotte. Archéologie des idées d'un savant ordinaire* (2011), *The Mechanization of Natural Philosophy* (co-edited with D. Garber, 2012), *L'automate. Machine, métaphore, modèle, merveille* (co-edited with A. Gaillard, J.-Y. Goffi and B. Roukhomovsky, 2012). In collaboration with Egidio Festa, she is currently working on a critical edition, translation and commentary of Galileo's *Le Mecaniche*.

Justin E.H. Smith, University of Paris 7-Denis Diderot, France

Justin E.H. Smith is university Professor of the History and Philosophy of Science at the University of Paris 7-Denis Diderot. He is the author of *Divine Machines: Leibniz and the Sciences of Life* (Princeton University Press, 2011), and of the forthcoming *Nature, Human Nature, and Human Difference: Early Modern Philosophy and the Concept of Race* (Princeton University Press). With François Duchesneau, he is the editor and translator of *The Leibniz-Stahl Controversy* in the Yale Leibniz Series.

Wiep van Bunge, Erasmus University, Rotterdam, The Netherlands

Wiep van Bunge is Professor of the History of Philosophy at Erasmus University Rotterdam and the author of numerous papers on early modern philosophy and several books, including *Johannes Bredenburg* (PhD, 1990), *From Stevin to Spinoza* (2001), and *Spinoza Past and Present* (2012). He also (co-)edited a number of books including *De draagbare Spinoza* (1992), *Disguised and Overt Spinozism around 1700* (1996), Balthasar Bekker, *Die bezauberte Welt* (1997), *The Early Enlightenment in the Dutch Republic* (2003), *The Dictionary of Seventeenth and Eighteenth-Century Dutch Philosophers* (2003), *Pierre Bayle* (2008), and *The Continuum Companion to Spinoza* (2011). In addition he is a member of the editorial team publishing *La Correspondance de Pierre Bayle* (1999–).

Koen Vermeir, CNRS, Paris, France

Koen Vermeir is Associate Research Professor (CR1) at *SPHERE, Laboratoire de Philosophie et Histoire des Sciences* (UMR 7219) of the *Centre National de la Recherche Scientifique (CNRS)* and is affiliated to *Sorbonne Paris-Cité*. He has held research positions at the Fund of Scientific Research-Flanders, the Max-Planck-Institute for the History of Science, Harvard University, and the Radcliffe Institute for Advanced Studies and visiting positions at Cambridge University, Cornell University and the ETH Zürich. As a historian of early modern science and philosophy, he has contributed to a wide array of topics, with a special focus on early modern theories of the imagination and on the natural philosophy of preternatural phenomena.

Abbreviations

AT Descartes, René. 1996. *Oeuvres de Descartes*, 2nd ed, ed. C. Adam and P. Tannery. Paris: Vrin.

CSM Descartes, René. 1985. *The philosophical writings of Descartes*, 2 vols. Trans. J. Cottingham, R. Stoothoff, and D. Murdoch. Cambridge: Cambridge University Press.

CSMK Descartes, René. 1991. *The philosophical writings of Descartes*, vol. 3. Trans. J. Cottingham, R. Stoothoff, D. Murdoch and A. Kenny. Cambridge: Cambridge University Press.

OPD Desgabets, Dom Robert. 1983. *Oeuvres philosophiques inedités de Desgabets*. Introduction by Géneviève Rodis-Lewis, text and annotations by Joseph Beaude. 7 opuscules. Paris: CNRS.

Chapter 1
Introduction

Mihnea Dobre and Tammy Nyden

1.1 Cartesian Scholarship in the History of Philosophy and the History of Science

René Descartes and his followers have not received the attention they deserve in the history of experiment. Praised as the founder of modern philosophy and presented as a turning point in the way knowledge, the self, and the world are perceived and analyzed, Descartes is one of the most discussed figures in the history of Western thought. The bulk of attention is directed at his metaphysics and epistemology. One of the great metaphysicians of his time, Descartes is largely portrayed as the Rationalist *par excellence*, for whom the passage from metaphysics to physics raised many difficulties.

While Descartes metaphysics and epistemology deserve an important chapter in any philosophy textbook, in the last 35 years more nuanced views about his work have emerged, giving more attention to his natural philosophy.[1] This was preceded by scattered voices which drew attention to the role Descartes gave to experiment—an interest developed largely by historians of philosophy.[2] This was in

[1] For some examples see: Gaukroger 1980, 2002; Osler 1985, 1994; Hatfield 1985, 1988; Clarke 1989; Nadler 1990; Garber 1992, 2001b, 2002; Grene 1995; Armogathe and Belgioioso 1996; Gabbey 1998; Des Chene 2000; Gaukroger et al. 2000; Roux and Garber 2013 (especially Chap. 3 by Roux 2013); Kolesnik-Antoine 2013.

[2] There is a difference between the English and French literature on this topic. French scholars pointed out Descartes' interest in experiments and empirical knowledge much earlier. See for

M. Dobre (✉)
The Center for the Logic, History and Philosophy of Science (CELFIS),
Faculty of Philosophy, University of Bucharest, Splaiul Independentei 204,
060024 Bucharest, Romania
e-mail: mihneadobre@yahoo.com

T. Nyden
Department of Philosophy, Grinnell College, Grinnell, IA, USA
e-mail: nyden@grinnell.edu

M. Dobre and T. Nyden (eds.), *Cartesian Empiricisms*, Studies in History and Philosophy of Science 31, DOI 10.1007/978-94-007-7690-6_1,
© Springer Science+Business Media Dordrecht 2013

contrast to most twentieth-century studies, which framed Descartes' contributions to the history of science in terms of his mathematics, rejection of Scholastic philosophy and development of particular concepts, such as that of inertia or the laws of motion.[3] Further, focusing on the role of experiment in Descartes' philosophy went against the way early modern thought was framed. Descartes and the Cartesians were overwhelmingly contrasted with John Locke's empiricism and Newtonian experimental method, that is, the discussion was described in terms of mutually exclusive distinctions between Rationalist and Empiricist epistemologies and methodologies. This approach influenced the treatment of Descartes by historians of science, such as Herbert Butterfield and Alexandre Koyré, who in turn influenced a new generation of scholars to greatly limit Descartes' role in the "big picture" of scientific change, traditionally labeled "the scientific revolution."[4] When Butterfield, Koyré, and their followers discussed Descartes, they focused on his deductive methodologies and commitment to a priori ideas, characterizing him and the Cartesians as anti-experiment and a foil for the Newtonian revolution in physics. This is perhaps most striking in Thomas Kuhn's treatment of Cartesian physics as a losing candidate from pre-paradigm science.[5] What is left beyond the imagination of most twentieth-century historiography is that Cartesians could have played a role in the development and acceptance of experimental methodologies and practices. While Descartes' scholars are now rectifying this situation by focusing on new aspects of his philosophy, such as the problem of experimentation and the foundation of knowledge in physics, Descartes' followers (the Cartesians) are waiting to receive the attention they deserve in the histories of both science and philosophy.

This brief survey of the evolution of Cartesian studies and its relation to the scholarship on the history of science is meant to provide a glimpse of a very intricate historiographical problem: what role did Cartesian philosophy play in the introduction, acceptance, and spread of experimental practices and methodologies in late seventeenth-century natural philosophy? The goal of this volume is to encourage discussion of this question by presenting several cases of Cartesian thinkers heavily involved in the practice, pedagogy, and theory of experiment. That is to say, the

---

example Liard 1882; Milhaud 1921; Mouy 1934; Laporte 1945 [1988]. English scholars discovered it quite late—and, in some sense, with great surprise—as it is the case of Clarke 1982, where Descartes is largely presented as an Empiricist.

[3] For some examples see: Burtt 1924, Chap. 4; Keeling 1934, Chap. 5; Scott 1952; Smith 1953, Chap. 8; Blackwell 1966; Aiton 1972; Prendergast 1972–1973, 1975; Williams 1978, Chap. 9; Hatfield 1979.

[4] Butterfield 1949; Koyré 1957. This twentieth-century historiography of science assumed the Rationalist-Empiricist distinction and re-instantiated Newtonian propaganda from the eighteenth century. See Chap. 4 by Van Bunge. See also Cohen 1985; Christie 1990; Shank 2008. On the history of the "big picture" see Cunningham and Williams 1993. The scientific revolution narrative espoused in Butterfield and Koyré was replicated and further developed by their students and other scholars throughout the twentieth century. For examples, see Hall 19074; Kuhn 1957; Gillispie 1960; Dijksterjuis 1961; Westfall 1971. Note that these works were re-released in new editions or versions throughout the twentieth century, holding an impressive influence.

[5] Kuhn 1962 (see 1996 edition, 48, 148).

volume presents studies of figures that might be described as "Cartesian Empiricists," a contradiction in terms according to scientific revolution narratives of seventeenth-century science coupled with the traditional Rationalist–Empiricist distinction in the history of philosophy. The dominance of these narratives explains why contemporary scholars have neglected the figures in this book despite their influence and prestige during the early modern period. The methodology behind this volume combines two elements. First, we study the interplay of two bodies of literature (the history of philosophy and the history of science) and use that interplay to challenge the traditional narratives in both areas of scholarship, thus rediscovering figures and events made invisible by those narratives. We believe this approach leads to a better grasp of the many transformations taking place in the early modern period.[6] Second, we actively seek and study the works of philosophers that escape the Rationalism-Empiricism distinction or any other distinction portraying the early modern period as a battle between two opposing schools of thought. Such distinctions, when taken as essentialist categories, greatly hinder our understanding of the history of philosophy. They do particular damage when they are taken as seriously as the strict Rationalism-Empiricism divide, which has structured both the discipline's curriculum and grand narratives about its history. For example, Thomas Hobbes is often left out of early modern philosophy courses and textbooks, despite his well-established contributions to and influence on the history of philosophy. This neglect results from the fact that his philosophy does not easily align with the Rationalist or Empiricist side. This is all the more true for lesser-known figures, like those discussed in this volume. Many have been neglected *precisely* because they do not fit into the canonical divide. The division becomes so powerful in the historian's imagination, that the very fact that a philosopher's works contains elements of both Rationalism and Empiricism may be taken as evidence that the philosopher is a confused or less-than-systematic thinker and therefore not worthy of study. This warps our understanding of the period as we fail to see that many philosophers, including influential philosophers at the time, carved out conceptual ground between Rationalist and Empiricist extremes. As indicated in the next section, we believe that the time is ripe for a fresh approach.

1.2 The History of the Rationalist-Empiricist Distinction

1.2.1 In the History of Philosophy

Throughout the nineteenth century and for more than half of twentieth century, the difference between Continental Rationalist and British Empiricist thinkers seemed to offer a nice way to portray historical actors as fighting a battle about the origins

[6] For examples of other works in this direction see Barker and Ariew 1991; Hatfield 1985, 1992, 1996a; Garber 1992, 1998, 2001a; Friedman 1992, 1993, 2010.

of ideas. Locke famously contrasted Descartes' philosophy to his "more modest" approach, which rejects innate ideas and takes all knowledge to be derived from sensory experience.[7] Locke's critique of Descartes' theory of ideas was frequently interpreted in terms of an opposition between Empiricism and Rationalism and these epistemologies were seen as undergirding two different methodologies: that conducted by the Royal Society, which valued observation and experiment, and the "Rationalist" method of deducing systems from a priori first principles. It provided a neat narrative of two concurrent early modern philosophies finally merging into the Kantian solution of Transcendental Idealism.[8] This view gradually lost its appeal as more specialized studies on each of the main characters associated with either so-called Continental Rationalism (e.g., Descartes, Spinoza, and Leibniz) or British Empiricism (e.g., Locke, Berkeley, and Hume) exposed crossing issues that made strict categorization untenable.[9] It became the norm to point out the problem of classification,[10] and to stress that the difference between Rationalism and Empiricism is a matter of degree, not kind.[11] Nevertheless, most works continued to use the difference as a central framework for understanding early modern philosophy and continued to focus on roughly the same six philosophers.[12]

In the past three decades historians of philosophy have paid more attention to the historical contexts of the figures and writings that they study, which has resulted in two main historical criticisms of the Rationalist-Empiricist narrative. First, it is anachronistic.[13] The terms Rationalism and Empiricism were not used by early modern philosophers themselves. Rather, they seem to be a projection of post-Kantian notions back onto the period. The second concern is that the single-minded focus on epistemology caused scholars to neglect many issues in early modern writings that were of key importance to early modern philosophers themselves and prevented us from seeing interconnections between these concerns and the epistemologies on which we tend to focus.[14]

---

[7] *Essay Concerning Human Understanding*, Book I in Locke 1975.

[8] See Tennemann 1852; Fischer 1854–1877; Russell 1912; Randall 1940; Cottingham 1984, 1988; Copleston 1958, 1959; Gilson and Langan 1963.

[9] For examples on Descartes, see Clarke 1983. On Leibniz see Francks 1985; Brown 1985. On Spinoza see Brown 1985; Stewart 2007. On Locke see Hanratty 1995; Loeb 1981; Pap 1958; Woolhouse 1971. On Berkeley see Bracken 1974; Hanratty 1995; Ayers 2005. On Hume see Popkin 1959b, 1964; Stewart 1985.

[10] For examples, see Popkin 1959a; Ryle 1960; Matson 1968; Buchdahl 1969; Bracken 1974; Mandelbaum 1976, 1977; Loeb 1981; Norton 1981; Brown 1985; Holland 1983; Ishiguro 1986; Kenny 1986; Sorell 1993; Haakonssen 2006; Fraenkel et al. 2011. There were some exceptions. Ayers 1984, 2005. Perler 1998, for instance, argued for categorizing philosophers as Continental Rationalist or British Empiricists.

[11] For example, von Leyden 1968; Collins 1972, 8.

[12] For example, see Scruton 1982; Cottingham 1984, 1988; Priest 1990; Sorell 1993; Hanratty 1995; Garrett and Barbanell 1997; Perler 1998; Woolhouse 1998; Bennett 2001; Schneewind 2004.

[13] For examples, see Loeb 1981; Cottingham 1988; Lennon 1993; Perler 1998; and Anstey et al. 2010.

[14] For a discussion of an alternative approach, see Corneanu 2011.

1.2.2 In the History of Science

The traditional historiography of science, particularly that of how the modern scientific disciplines separated from philosophy, built on the Rationalist-Empiricist distinction. A direct consequence was the exclusion of "Rationalists" from the histories of science, which focused on how the "Empiricists" developed their views. This approach was also facilitated by the geographical distinction of the historical actors as *Continental* Rationalists and *British* Empiricists. This is particularly evident in the relation between Cartesianism and Newtonianism. When scholars discuss this relation, they tend to emphasize Newton's reaction against Descartes' imagined hypotheses, an approach that seems to go back to Voltaire's writings. In one of the most cited passages on this topic, Voltaire ascribes different world-views to Cartesians and Newtonians:

> A Frenchman who arrives in London finds a great change in philosophy, as in everything else. He left the world full, he finds it empty. In Paris one sees the Universe composed of vortices of subtle matter. In London one sees nothing of this. In Paris it is the pressure of the moon that causes the flux of the sea; in England it is the sea that gravitates toward the moon.[15]

D'Alembert makes similar remarks in his *Discours préliminaire* to the *Encyclopédie*, where Newton is portrayed as creating the much-needed unification of empiricism and mathematics.[16] But even these accounts are more nuanced than they first appear. Take for example another passage from Voltaire, where although he highlights Newton's genius, he gives sufficient credit to Descartes for reforming philosophy:

> Indeed believe, that very few will presume to compare his [Descartes'] philosophy in any respect with that of Sir *Isaac Newton*. The former is an essay, the latter a masterpiece: But then the man who first brought us to the path of truth, was perhaps as great a genius as he who afterwards conducted us through it.
>
> DesCartes gave sight to the blind. These saw the errors of antiquity and of the sciences. The path he struck out is since become boundless. *Rohault*'s little work was during some years a complete system of physicks; but now all the transactions of the several academies in *Europe* put together do not form so much as the beginning of a system....[17]

In the period between Descartes' death in 1650 and the victory of Newtonianism in the mid-eighteenth century, followers of Descartes and Newton shared common elements, which allowed them to exchange ideas and influence each other.[18] Voltaire raises this issue in the second part of this passage when he refers to the Cartesian Rohault as providing a complete textbook on physics. We contend that there are many such exchanges between Cartesians and Newtonians and that their study will illuminate our understanding of Cartesian natural philosophy, as well as the early

---

[15] Voltaire 1741, 89 (it is also cited in *Newtonian Studies*, Chap. 3; see Koyré 1968, 55).

[16] See D'Alembert 1995.

[17] Voltaire 1741, 98.

[18] For a discussion of this uneasy relation between Cartesians and Newtonians, see Shank 2008.

development of Newtonianism. Further, there were many varieties of eclecticism; some were presented as alternatives to Cartesianism or Newtonianism, some brought elements of the two together, but most of them were presented as the *new* science (or natural philosophy). Finally, we point out that there was Cartesian pre-Newtonian experimentation that appears to have had significant influence on and consequences for the eighteenth-century Newtonians. Examples of all three of these cases can be found in this volume. In other words, when the Rationalist-Empiricist distinction and scientific revolution narratives are questioned and we allow ourselves to consider the possibility that Rationalism and Empiricism in general, and Cartesianism and Newtonianism in particular, are not incommensurable and mutually exclusive paradigms, we begin to gain new insights into the early modern period. It is difficult to step away from the traditional narrative and see the philosophy of the period with new eyes. As will become clear in the next section, this is the purpose of the second element of our methodology: to actively seek out cases where the Rationalist-Empiricist distinction fails. In particular, we actively seek out Cartesians that give epistemic and/or methodological value to experiments. We hope this exercise will help us to see new things made invisible by the Rationalist-Empiricist distinction.

1.2.3 A Narrative in Crisis

The Rationalist-Empiricist narrative finds itself in crisis on both historical and philosophical grounds, yet it has not lost its hold on how we conceptualize early modern philosophy and science. As Louis Loeb pointed out in 1981 and remains true today, the Rationalist-Empiricist narrative forms almost all undergraduate curriculums and is upheld by almost all historians as "more right than wrong, even when they have serious reservations about it."[19]

Attempts at alternative narratives have come forward and can roughly be divided into two categories: (1) those that continue to focus on the same six philosophers—Locke, Berkeley, Hume and Descartes, Spinoza, Leibniz—but frame their importance or connections to each other in terms other than Empiricism and Rationalism; and (2) those that replace the Rationalist-Empiricist dichotomy with another binary (e.g., speculative-experimental). Examples of the first category include Gerald Hanratty who argues that when we look at the relationship between Locke, Hume and Berkeley "in the context of the European Enlightenment, the historiographical scheme according to which each made a distinctive contribution to the linear development of the eighteenth-century British empiricism must be revised."[20] In particular,

---

[19] Loeb 1981, 28 and 30. Not unlike Hanratty, who will soon be discussed, Loeb makes the point that even though the Rationalist-Empiricist distinction is "fatally flawed" he is not saying that it is impossible to come up with other criterion that result in the same six canonical figures, placing Descartes, Leibniz, Spinoza in one category and Locke, Berkeley and Hume in another (see pp. 70–71).

[20] Hanratty 1995, 10.

Berkeley's project is better understood as emerging from "his disenchantment with what he saw as a decadent enlightenment project which led inevitably to skeptical, atheistic and materialistic conclusions."[21] In a similar vein, Louis Loeb rejects the opposition of Rationalism and Empiricism and argues that "Continental Metaphysics should be viewed as a philosophical *genre* that emerged and thrived in Europe from 1640 to 1715."[22]

Examples of the second category can be found in the work of Thomas Lennon, as well as the work of Peter Anstey and his former research group at the University of Otago. Lennon interprets the relationship between Locke and Descartes not in terms of an opposition between Rationalist and Empirical epistemologies, but in terms of an ongoing metaphysical debate since the time of Plato, a debate between materialists and those who espouse some version of the forms.[23] Lennon sees this larger debate as connecting to all areas of philosophy and, unlike the Empiricist-Rationalist distinction, his dichotomy allows for the existence of Cartesian empiricists. Peter Anstey, Alberto Vanzo, Kirsten Walsh, and Juan Manuel Gomez at Otago University want to replace the Rationalist-Empiricist dichotomy with the distinction between experimental and speculative philosophy. They argue that their alternative describes actor-categories, that is, terminology used by the very actors it is meant to describe.[24] Peter Anstey explains:

> the experimental/speculative distinction…functioned as a kind of general methodological rubric from the late 1650s until the early decades of the following century and was deeply ingrained in the methodological discourse of many practitioners, promoters and even critics of the new science.[25]

Both of these approaches have been similar to the standard narrative in flavor and result.

Rather than defending the canon or re-conceptualizing oppositions from the traditional historiography, we claim that actively seeking and studying the works of philosophers that escape the categories of traditional historiographies is a useful approach to gaining a fresh perspective on the early modern period. It is likely to bring to light figures, ideas, and tendencies overlooked by traditional approaches. We particularly find any binary used to divide early modern philosophers or mutually exclusive philosophies, including the speculative–experimental distinction,

[21] Hanratty 1995, 11.

[22] Loeb 1981, 363, emphasis is ours.

[23] Lennon 1993.

[24] In their manifesto for the use of this new terminology, the Otago team claim "Philosophers from the early modern period (from Descartes to Hume) are normally divided into Rationalists and Empiricists. Yet this distinction was developed by neo-Kantian philosophers from the late 18th century. In this research project we are exploring the hypothesis that there is a far better way of approaching early modern philosophers. Our central thesis is that the most common and the most important distinction in early modern philosophy is that between Experimental and Speculative Philosophy. This is a distinction that many of the actors actually used, and, we claim, it can explain all that the traditional distinction can explain and more besides" (See Anstey et al. 2010).

[25] See Anstey 2005, 237.

to be problematic. Careful historical investigation of late seventeenth-century natural philosophy reveals a vast array of philosophers and writings that escape such categories, whether they were developed before or after Kant. For example, as Peter Anstey and Alberto Vanzo acknowledge, some philosophers (e.g., the non-Cartesians Robert Hooke and Robert Boyle) are both 'speculative *and* experimental' and present themselves as such:

> By contrast, from the 1660s in England there is an almost monotonous call to avoid the hypotheses and 'castles in the air' of the speculative philosophers. Again and again the methodological writings of the new philosophers pit experimental philosophy against its speculative counterpart. *To be sure, some, like Boyle, argued for the mutual assistance that both might render each other.* But the majority of writers were more inclined to highlight the opposition of experiment to speculation and to warn their readers off the latter.[26]

The Otago group shows the speculative-experimental opposition to be an important element in *English* rhetoric. They concede it does not apply to *all* early modern thinkers, though they contend it applies to most. We do not think this case has yet been made. If what is important in this new distinction is its employment by early modern philosophers themselves—that is, by the use of these words in a methodological context—then it is weakened in all historical cases where the taxonomy of "speculative" and "experimental" is not in use. Such is the case when the focus moves away from England to France or to The Netherlands where such actor categories elude Cartesian philosophers completely. There are different oppositions Cartesians in these countries use, such as that between "reason" and "experience." In several cases (e.g., François Bayle, Jacques Rohault and Burchard de Volder) this taxonomy is coupled with methodological concerns. Further, the taxonomy they do use is not a mutually exclusive binary. Bayle, for example, argues extensively for a joint use of reason and experience in the study of nature.

We do not dispute that terms like Rationalism, Empiricism, speculative and experimental can be useful descriptors in some respects, only that any particular definition of these terms will yield neat divisions of the actors or philosophies of early modern Europe. Any binary will necessary exclude or marginalize eclecticisms and synthesis, which, as this volume demonstrates, are important features of early modern thought.[27]

The philosophers discussed in this volume have one thing in common: they do not easily fit into epistemological and/or methodological distinctions of the standard historiography. Beyond that, they represent a great diversity of attitudes and approaches to reason, experiment, and scientific method. We believe that by studying such figures, fruitful new themes about early modern science will emerge. In bringing these studies of "Cartesian Empiricists" together we hope to show, contrary to the standard historiographical account, that they are not anomalous figures, but both prevalent and important in their own time. In a Kuhnian sense then, we are optimistic that early modern studies has come to a point in which we are able

---

[26] See Anstey and Vanzo 2012, 20 (our emphasis).

[27] For examples of synthesis see Chap. 2 by Ariew; Chap. 5 by Smith; Chap. 10 by Nyden.

to not only notice the "anomalies," but able to develop more useful narratives that can fully incorporate them. This work is a first step towards that end. We do not put forward an alternative narrative ourselves, but only hope to give scholars a new lens with which to examine the historical data, opening the perspective for future narratives.

1.3 On Cartesian Empiricisms

While Descartes' natural philosophy has been getting more attention, the same is not true for how his followers further developed that natural philosophy. We hope this volume begins to fill this gap and offers a fresh view on the complex transformation of Descartes' ideas in the second half of the seventeenth century.

The phrase "Cartesian Empiricism" is not new. In the past two decades, scholars such as Roger Ariew, Patricia Easton, Thomas Lennon, Tad Schmaltz, and Monte Cook have used it to describe French thinkers such as Robert Desgabets (1610–1678), François Bayle (1622–1709), Pierre-Sylvain Régis (1631–1707), Bernard Lamy (1640–1715), and Jacques Du Roure (fl. 1653–1683).[28] *Cartesian Empiricisms* aims to build on their work by introducing additional instances of Cartesians strongly committed to the importance of experimental natural philosophy and furthering our understanding of how Cartesians in the second half of the seventeenth century understood and utilized knowledge from observation, experience, and experiment. By bringing together a number of such figures we hope to show that rather than being limited to a few isolated, eclectic thinkers, it was quite common for Descartes' followers to argue for views that gave experiential knowledge a key role in the Cartesian system.[29] Further, the Cartesians discussed in this volume represent various parts of Europe, complementing the existing scholarship on Cartesians (which is too often restricted to France) and the scholarship on empiricism (which is too often restricted to England).

Both of the terms forming the title of this volume are deeply problematic. The very label "Cartesianism" wrongly suggests *one* shared Cartesian doctrine. As the cases in this volume indicate, there is not a single definition of "Cartesianism" that can satisfy the multitude of directions in which Descartes' philosophy was developed and modified by philosophers who accepted some parts (often not the same parts) of his system in their attempt to explain natural phenomena. This occurred during Descartes' life, as in the case of Henricus Regius, and throughout the second half of the seventeenth century.

[28] See Ariew 2006 on Desgabets, Régis, Lamy, Bayle, and Du Roure; Lennon and Easton 1992 on François Bayle; Easton 2000 on Desgabets; Schmaltz 2002 on Régis and Desgabets; and Cook 2008 on Desgabets.

[29] Note that Roger Ariew makes the stronger claim that empiricism was the predominant view among Cartesians in the second half of the seventeenth century. Ariew 2006, 73.

Descartes himself seemed to reserve the term for his own writings. In his preface-letter to the French edition of the *Principles* (1647)—especially his paragraphs against Henricus Regius he writes:

> I must also beg my readers never to attribute to me any opinion they do not find explicitly stated in my writings. Furthermore, they should not accept any opinion as true—whether in my writings or elsewhere unless they see it to be very clearly deduced from true principles.[30]

This makes it look impossible for Cartesianism to reform itself after Descartes' death. At the end of the seventeenth century, the Dutch divine, Balthasar Bekker expressed this problem with the following puzzlement: "where I follow him, I am praised by you, but where I depart from him, you deem that I err. But to be so thoroughly Cartesian is not Cartesian."[31] No wonder, then, there are many varieties of seventeenth-century approaches to Descartes' philosophy. On the one hand, there were authors who declared themselves "Cartesians" and tried to work within the system provided by the French philosopher, but without being able to contribute to the development of this system; and, on the other hand, there were people working with the principles of the French philosopher, but who allowed for development and change. Even among the first approach a variety of interpretations result depending on which of Descartes' texts the interpreter takes as the most authoritative. After all, Descartes changed some of his views during his life and it was only in the 1640's that he gave his system a metaphysical foundation.[32] Dom Robert Desgabets provides an example of the second type of Cartesian approach in which the entire system is subject to possible development. In a celebrated passage from the *Supplement to the Philosophy of Descartes* (1675), he credited Géraud de Cordemoy, Jacques Rohault, and Louis de La Forge for their contributions to Cartesian physics and claimed that even Descartes' metaphysics required corrections.[33] He took up this task, implying that even Descartes' first principles are not necessarily "the true principles." Desgabets falls into the paradox expressed by Bekker, where the name only applies to him if we allow "Cartesian" to include those who are critical and not mere copyists of their master.

---

[30] AT IXb 20, CSM I 189.

[31] Letter from Balthasar Bekker, S.T.D. and Minister of Amsterdam, to the two honorable pastors D. Joannes Alstius, from Hoornaar, and D. Paulus Steenwinkel, from Schelluinen, concerning their remarks on the first part of his work, *De Betoverde Weereld* (The World Bewitched). Quoted in Thijssen-Schoute 1989, 515: "Al waar ik met hem ga, word ik van U gepresen, maar daar ik van hem wijke, acht gij dat ik doole. Doch dus hard Cartesiaansch te zijn is niet Cartesiaansch." Interestingly, a few decades later Willem 's Gravesande responded to similar charges for not being Newtonian enough because he abandoned Newton's views on living force. His defense is similar to Bekker's: "He only, who in Physics reasons from Phenomena, rejecting all feign'd Hypothes, and pursues this Method inviolably to the best of his Power, endeavours to follow the Steps of Sir Isaac Newton, and very justly declares that he is a Newtonian Philosopher; and not he, who implicitly follows the Opinion of any particular Person." 's Gravesande 1747, xi.

[32] For developments in Descartes' writings see Machamer and McGuire 2009; Schuster 2012. For the metaphysical foundation of Descartes' natural philosophy, see Garber 1992; Gaukroger 2002; de Buzon and Carraud 1994; Hatfield 1985.

[33] See Desgabets 1983–1985, V, 156 (OPD 5, 156). We shall return shortly to this passage.

Changes within Descartes' system of philosophy did take place. Sometimes they were occasioned by external contexts; sometimes they were natural developments within Cartesian ideas; and sometimes they were attempts to reform or expand upon core ideas of the system itself. Among the first case are the various attempts to refute Cartesian ideas based on new observations and experiments. An example in this sense is the report of a curious case of a headless turtle that was observed to move. This report was printed in the *Philosophical Transactions* of June 3, 1667:

> there came a Letter from *Florence*, Written by M. Steno, which has also somewhat perplext the followers of *Des Cartes*. A Tortoise had its head cut off, and yet was found to move its foot three days after. Here was no Communication with the *Conarium* [i.e., the pineal gland]. As this seems to have given a sore blow to the *Cartesian* Doctrine, so the Disciples thereof are here endeavouring to heal the Wound.[34]

Examples of the second case—of natural developments within Cartesian ideas—are seen in seventeenth-century discussions about the beast-machine or Cordemoy's work on speech.[35]

In Robert Desgabets we have already seen an example of the third type of change. In the announcement of the *Supplement*, which seeks to correct errors in the metaphysical core of Descartes' philosophical project, he tells us how he hopes to expand the Cartesian system:

> It is, properly speaking, the topic of Mr. Descartes' Meditations on first philosophy that he always regarded as his masterpiece, and in which I find nevertheless some very important flaws that can only be corrected to the glory of this great philosopher who himself provided the means to give them their ultimate perfection and the remedies to the damage he caused. Thus, in this writing, which I call the first supplement of his philosophy, I take the task to work on such a necessary thing, particularly as I try to correct his own thoughts on things where it seems to me that he departed from the right way of truth; whereas we could call second supplement, the new application of his undoubted principles to phenomena that he did not know, or to truths he did not speak of, and it is on this type of second supplement that Mr. de Cordemoy, Rohault, de La Forge, Clauberg and others have worked in the beautiful writings they've offered to the general public, where one can see in what manner we will be able to extend our knowledge to things equally great and useful.[36]

---

[34] See the *Philosophical Transactions* 1667, 480. For Descartes, the pineal gland is the central place where all the nerves meet; hence, it mediates the capacity of humans and animals to move.

[35] See Cordemoy 1668. An example of the literature generated on the problem of the animal-machines is Pardies 1672. It is worth citing here the attempts to provide mechanical explanations to different bodily functions, including the human body. While these attempts originate in Descartes' manuscript of *L'homme*, a greater influence on the reception of Descartes' ideas is due to Clerselier's edition of 1664. For the importance of this edition, see Zittel 2011.

[36] See Desgabets (1983–1985), V, 156 (OPD 5, 156): "C'est là proprement le sujet des Méditations de M. Descartes touchant la première philosophie qu'il a toujours regardées comme son chef d'œuvre, et où néanmoins je trouve des défauts considérables qu'on ne peut corriger qu'à la gloire de ce grand philosophe qui fournit lui-même les moyens de leur donner leur dernière perfection et les remèdes au mal qu'il a fait. Je me suis donc proposé de travailler à une chose si nécessaire, dans cet ouvrage que j'appelle le premier supplément de sa philosophie, d'autant que je tâche d'y rectifier ses propres pensée dans les choses où il me semble qu'il a quitté le droit chemin qui conduit à la vérité; au lieu qu'on pourrait appeler second supplément, l'application nouvelle que l'on ferait de ses principes incontestables à des phénomènes qu'il n'a pas connus, ou à des vérités dont il n'a

As these different approaches indicate, there is no *one* way to be a Cartesian. By labeling a thinker "Cartesian," we are not making any claims about the philosopher's metaphysical or epistemological positions on any *particular* issue. While all of the Cartesians discussed in this volume affirm key aspects of Descartes' system and see themselves as building on or completing that system, there is a great deal of diversity as to which aspects of the Cartesian system they take as key.

The term "Empiricism" can be equally problematic. As alluded to in the previous section, it has both epistemological and methodological meanings and neither neatly divide all seventeenth-century philosophies or scientific practices. Within epistemology, it refers to the view that all knowledge comes from sensory experience, as opposed to innate ideas. Within the philosophy of science it refers to the view that all theories must be tested with observation, as opposed to a priori reasoning or intuition. In this volume we identify thinkers as Cartesian *Empiricists* if they are Cartesians who give observation, experience, and/or experiment a key role for knowledge acquisition in their natural philosophy.[37] We anticipate some will object that many of the figures discussed in this book are empiricists neither in the epistemological or methodological sense as just defined. This is precisely our point: there are many interesting and important contributors to early modern science for whom the simple distinction between Rationalism and Empiricism does not make sense.

We recognize the rhetorical impact, if not controversy, of pairing of the descriptors "Cartesian" and "Empiricism" and wish to exploit the supposed tension. This tension arises because in the traditional narrative of early modern philosophy—just like we argued above—Descartes' and Cartesian philosophy have been taken to be paradigm cases of Rationalism, an epistemological position characterized as mutually exclusive with Empiricism. This narrative, which still captures the imagination of most non-specialists and is still the basis for organizing most university early modern philosophy courses will be thoroughly challenged. Even though we argue that neither of these two terms fully satisfies the historical cases, we think that by putting together *Cartesian* and *Empiricisms*, we can extract new meanings and give a new life to the decaying historiographical concept of early modern empiricism. We intentionally use the plural Cartesian Empiricisms to indicate that we do not take the figures in this volume to make up a cohesive school of thought, but rather see them as representing a trend in the diversity of views and practices among Descartes' followers throughout Europe.[38] The connection between Cartesian

point parlé; et c'est à cette sorte de second supplément que MM. de Cordemoy, Rohault, de la Forge, Clauberg et autres ont travaillé dans les beaux ouvrages qu'il ont donnés au public, où l'on voit de quelle manière nous pourrons étendre nos connaissances à des choses également belles et utiles."

[37] These are not well-defined categories and each philosopher discussed in this volume will use experiments in their own way. For example of discussions of experience and experiment in the early modern period see Garber 2001b, Chap. 14, 296–328. For a recent account of the multiple transformations in the meaning of the terms associated with empirical practice, such as "observation" or "experiment," see Daston and Lünbeck 2011.

[38] Schmaltz 2002, 11: "Given this variety in opinions among Descartes' followers in France, there is reason to speak not of a single movement, French Cartesianism, but rather a variety of French Cartesianisms."

Empiricist philosophers is not a shared set of core principles, but a family resemblance, where a variety of natural philosophical traits are developed in many different ways. When studied under the category of "Cartesian Empiricisms" the work of these philosophers reflect important early modern trends. We believe these studies will be relevant for both historians of philosophy and science, as well as for anyone interested in the intricate relation between metaphysics, observation, experiment, hypotheses formation, theory, and knowledge of the external world. Again, we do not oppose the use of categories to interpret the insights of early modern philosophers; we simply resist the use of any particular binaries that attempt to locate the central divisions of the time. These will allow us to see some things, but there is much more that they will cause us to miss.[39]

1.4 Chapter Summaries

The first part of this volume examines Cartesian Empiricisms as they arose in various philosophical and local contexts. The second part provides studies on particular natural philosophers and on how they combined elements of Cartesianism with experience and experimentation. The book covers Cartesians in four different countries (France, The Netherlands, Germany and England) and in four "disciplines" (physics, chemistry, psychology, and medicine), although other geographical and disciplinary contexts will come into focus from time to time. We by no means take this list of countries or disciplines to be complete. We only hope to help broaden the dialogue beyond England and France and beyond physics, encouraging future studies in other countries and on other "Cartesian Empiricist" topics.

Part I begins with Roger Ariew's "Censorship, Condemnations, and the Spread of Cartesianism." Ariew examines censures of Cartesian natural philosophy within the French Catholic world and indicates a connection between these censures, the resulting weakening of the commitment to the doctrine of hyperbolic doubt and a tendency for later Cartesians to become more empirical than Descartes himself. With the demise of hyperbolic doubt, the distinction between absolute and moral certainty dissolves and French Cartesians stop privileging knowledge attained through deduction from clear and distinct principles above the knowledge of particulars gained through the senses. Ariew examines the changing status of experience and experiment in close connection to methodological concerns, arguing that as later Cartesians became more empirical, they also tended to use a "limited hypothetical-deductive method."

[39] Gary Hatfield offers a similar argument about the use of "scientific revolution," which is infused with our contemporary notion of "science," a notion that does not map on to seventeenth-century language and thought. Hatfield 1996b, 512: "Yet in learning to focus on the'scientific' facets of seventeenth-century natural philosophy and mathematical science, we learn to slice away the parts of the texts we read or the institutions we study that do not constitute a proper part of the development of science as we now understand it."

Chapter 3 continues to focus on the French context. Sophie Roux asks, "Was there a Cartesian Experimentalism in 1660's France?" and answers through a twofold analysis of the context of the founding of the *Académie des sciences* and the Cartesian conferences of Jacques Rohault. Roux makes a distinction between various types of experimental practice, arguing for a different development within the two communities—on the one hand, the community of *savants* that were opposed to Cartesian philosophy, which slowly moved toward a "radical experimentalism" and formed the *Académie des sciences*; on the other hand, Rohault and his conferences, which are described as "old fashioned." Important in this analysis is how Parisian scientific academies of the 1660s opened the path for the formation of the *Académie*, which is described as a complex process, different than the traditional narrative which aimed to find a continuity between different salons and academies. Rohault—mainly with his late activity—does not fit in the story, as his experimental activity remained unchanged during the decade.

Chapter 4 moves our attention to The Netherlands where both Cartesian and Newtonian physics received their earliest university receptions. In "Dutch Cartesian Empiricism and the Advent of Newtonianism," Wiep van Bunge argues against the received view that an empirical Newtonian natural philosophy rushed in to fill a vacuum left by a failed Rationalist Cartesian natural philosophy in Dutch universities. He argues that Dutch Cartesians, from the very early Henricus Regius and Adriaan Heereboord to the later Burchard de Volder and Christiaan Huygens, challenge the historiography of a Rationalist-Empiricist dichotomy. Further, there is some overlap between the last Cartesians and first Newtonians, including a common effort to hold back the Radical Enlightenment influenced by the philosophy of Spinoza. It was the last Cartesian physicists in Leiden who were the first in The Netherlands to read Newton's *Principia* and while they were not the first Newtonians themselves, in many ways they laid the groundwork for the early Dutch acceptance of Newtonian natural philosophy.

Justin Smith discusses the German reception and development of Cartesian medical philosophy in Chap. 5: "Heat, Action, Perception: Models of Living Beings in German Medical Cartesianism." Smith puts into relief differences among medical Cartesians at Duisburg regarding the distinction of living bodies from non-living matter. Comparing the writings of Johannes Clauberg (1622–1665), Theodor Craanen (1633–1688), and Tobias Andreae (1604–1676), he concludes that a study of these disagreements is instructive to understanding the development of Leibniz's mature philosophy.

The first part ends with Bernard Joly's inquiry in Chap. 6, "Could a Practicing Chemical Philosopher be a Cartesian?" He examines the work of chemists in the French *Académie royale des sciences* in the late seventeenth and early eighteenth centuries, paying particular attention to a particular conception of Cartesianism they formed which emphasized the importance of laboratory experimentation, while maintaining a Cartesian matter theory. Joly warns that they were not "Cartesians" in a traditional sense, but the members of the *Académie royale* debated about the foundations of chemistry and chemistry's relationship to physics and throughout these debates emphasized elements of Descartes' own writings that made room for experimentation. Joly shows that their views do not represent epistemological empiricism as traditionally defined. He does cite an example of a genuine epistemological empiricist in the Académie royale—Etienne-François Geoffroy— but points out that he was neither a Cartesian nor a Newtonian.

Delphine Bellis begins the second part of the volume with a study of Henricus Regius' natural philosophy, "Empiricism Without Metaphysics: Regius' Cartesian Natural Philosophy," which constitutes Chap. 7 of this volume. The Dutch philosopher had a complicated relationship with Descartes, reflected in the variety of scholarly opinions as to what extent he is properly called a "Cartesian." These debates arise from the traditional Rationalist-Empiricist distinction, for Regius appears to offer a combination of rationalist physics and empiricist epistemology.[40] Bellis offers an in depth analysis of Regius' natural philosophy, detailing the importance of imagination and judgment in his epistemology and his commitment to Descartes' explanations and principles of natural philosophy. She argues that by both refuting the Scholastic theory of sense perceptions and rejecting Descartes' theory of innate ideas, Regius managed to leave a space between materialism and empiricism and, as such, can be seen as holding a position "between Hobbes' phenomenism and materialism and Hume's empiricism."[41]

In Chap. 8, "Robert Desgabets on the Physics and Metaphysics of Blood Transfusion," Patricia Easton examines Robert Desgabets' experiments on blood transfusions to illustrate the dual role he gave experimentation: to demonstrate the truth of Cartesian first principles (essences) and to discover which of the many possible worlds God choose to create (existence). Desgabets explained the ability of the mind to be aware of physical objects through Descartes' theory of matter and its motion. The senses differentiate and individuate thought, allowing the soul to know the particularity of a sensible object and "to connect essences (possibles) to actuality."[42]

In Chap. 9, Mihnea Dobre deals with "Rohault's Cartesian Physics" and argues that Rohault's experimentalism should be traced back to the early 1660s. The context of discussion is the Parisian academies of the early 1660s and their connection with the experimental work of the English natural philosophers. Dobre argues that only by placing Rohault's experimental investigations at an earlier date (late 1650s and early 1660s)—which is much closer to the main experimental activities of the so-called "experimental philosophers"—one can get a better grasp of his Cartesian natural philosophy and re-evaluate its eighteenth-century reception within Newtonian milieu, via Samuel Clarke's annotated translation of Rohault's *Traité*. Combining original experimental activities with pedagogical performances in front of his audience, Rohault highlights some of the most important trends in the development of Cartesian natural philosophy.

In Chap. 10, "De Volder's Cartesian Physics and Experimental Pedagogy," Tammy Nyden discusses the role of experiment in Burchard de Volder's physics

[40] See Chap. 7 by Bellis, 142; Verbeek 1993, viii.

[41] Chapter 7 by Bellis, 181.

[42] Chapter 8 by Easton, 199.

pedagogy and natural philosophy. The University of Leiden professor presents a case of pre-Newtonian experimental physics, which combines a Cartesian commitment to a priori reasoning with the demonstration of experiment. For de Volder, experiment demonstrates to students the truth and certainty of theoretical physics. Experiment also provides natural philosophers certain belief that those physical theories are instantiated in God's creation and do not merely reflect the logical possibilities available to the creator. Nyden frames de Volder's pedagogy as a continuation of the University of Leiden's long tradition of teaching through observation and places his natural philosophy within the context of Leiden's eclecticism.

In Chap. 11, "The Cartesian Psychology of Antoine Le Grand," Gary Hatfield discusses the role of experience in the Cartesian psychology of the English philosopher Antoine le Grand. Le Grand elaborated and extended Descartes' treatment of the physiology and psychology of animal and human behavior. He defended Descartes' mechanization of Aristotle's sensitive soul, that is, explaining its functions in terms of size, shape, and motion. Le Grand provided detailed accounts of sensory and motor mechanisms shared by humans and animals: sense perception, memory, and cognitive and appetitive responses to benefits and dangers in the environment. He claims these functions provide "physical certainty," the high standard of certainty required by his Cartesian natural philosophy, which lies between moral certainty and the absolute certainty of metaphysics.

In the final chapter, "Mechanical Philosophy in an Enchanted World," Koen Vermeir examines Balthassar Bekker's *Betoverde Weereld*, arguing that it is not an Enlightenment work, as it is often portrayed, but rather meant to further purify and reform the Protestant religion. In order to purge Protestantism from superstition, that is, from bad religion, Bekker used a combination of Cartesian metaphysics, mechanical philosophy and empiricism to naturalize many phenomena being attributed to the demonic forces. The explanatory power of Cartesian philosophy provided Bekker the tools he needed to combat testimony and empirical evidence of devil phenomena. Bekker used Cartesian epistemological and physical theories to undermine that evidence and to support and privilege empirical evidence against it. While he rejected the idea that the devil could act upon the world, he did not reject all occult phenomena and in fact used Cartesian natural philosophy to try to understand dowsing and some other forms of natural magic.

The second part ends with a brief bio-bibliography on each of the main figures discussed in Chaps. 7, 8, 9, 10, 11, and 12 of the volume.

1.5 Emerging Themes

Many possible themes emerge when all these chapters are studied together. Some, such as the problem of Descartes' theory of ideas and its reinterpretation in Cartesian philosophy, are familiar debates in the scholarly literature. Others are relatively new to the literature and stand out in relief when Cartesian Empiricisms are compared, such as the various attempts by Descartes' followers to reform his theory of certainty; or the role experiment can play in distinguishing the actual world from the possible worlds God could have created. We believe more such themes can be found in this volume and they will provide the basis for future studies of Cartesian Empiricisms. However, we would like to conclude our introduction by focusing on a more general point raised by *Cartesian Empiricisms*: the need to re-evaluate our traditional views concerning the use of historiographical categories in the early modern period. Most of the currently available narratives tend to use—or even frame the problems within—mutually excluding categories, such as the ones discussed above (Rationalism-Empiricism). If such an approach was somehow fruitful in the past, it is becoming less so, especially with contextual studies and the expansion of the canon. Obviously, this historiographical point does not end with Cartesianism or Empiricism, but can be applied to other philosophical categories as well. It is beyond the purposes of our volume to say more regarding this research strategy or its possible future applications. Rather, by focusing on the way Cartesian natural philosophers reflected upon and practiced experiments, we hope to open a new research path for future studies in the history of philosophy and science in the early modern period.

References

Aiton, E.J. 1972. *The Vortex theory of planetary motions*. New York: American Elsevier.

Anstey, Peter. 2005. Experimental versus speculative natural philosophy. In *The science of nature in the seventeenth century: Patterns of change in early modern natural philosophy*, ed. John A. Schuster and Peter R. Anstey, 215–242. Dordrecht: Springer.

Anstey, Peter, and Alberto Vanzo. 2012. The origins of early modern experimental philosophy. *Intellectual History Review* 22(4): 499–518.

Anstey, Peter, Juan Manual Gomez, Alberto Vanzo, and Kirsten Walsh. August 23, 2010 (7:00 a.m.) Experimental Philosophy and the Origins of Empiricism, *Early Modern Experimental Philosophy Blog*. https://blogs.otago.ac.nz/emxphi/2010/08/experimental-philosophy-and-the-origins-of-empiricism/.

Ariew, Roger. 2006. Cartesian empiricism. *Revue Roumaine de Philosophie* 50(1–2): 71–84.

Armogathe, Jean-Robert, and Giulia Belgioioso (eds.). 1996. *Descartes: Principia Philosophiae (1644–1994)*. Napoli: Vivarium.

Ayers, Michael. 1984. Berkeley and Hume: A question of influence. In *Philosophy in history*, ed. Richard Rorty, J.B. Schneewind, and Quentin Skinner, 303–328. Cambridge: Cambridge University Press.

Ayers, Michael. 2005. Was Berkeley an empiricist or a rationalist? In *The Cambridge companion to Berkeley*, ed. Kenneth P. Winkler, 34–62. Cambridge: Cambridge University Press.

Barker, Peter, and Roger Ariew (eds.). 1991. *Revolution and continuity: Essays in the history and philosophy of early modern science*, Studies in philosophy and history of philosophy, vol. 24. Washington, DC: Catholic University of America Press.

Bennett, Jonathan. 2001. *Learning from six philosophers: Descartes, Spinoza, Leibniz, Locke, Berkeley, Hume*. 2 Vols. Oxford: Oxford University Press.

Blackwell, Richard J. 1966. Descartes' laws of motion. *Isis* 57: 220–234.

Bracken, Harry. 1974. *Berkeley*. London: Macmillan.

Brown, Stuart. 1985. Leibniz's break with Cartesian 'Rationalism'. In *Philosophy, its history and historiography*, ed. Alan J. Holland, 195–208. Dordrecht: D. Reidel Publishing Company.

Buchdahl, Gerd. 1969. *Metaphysics and the philosophy of science: The classical origins, Descartes to Kant*. Oxford: Oxford University Press.

Burtt, E.A. 1924. *The metaphysical foundations of modern physical science*. London: Routledge.
Butterfield, Herbert. 1949. *The origins of modern science: 1300–1800*. London: G. Bell and Sons LTD.
Buzon, Frédéric de and Vincent Carraud. 1994. *Descartes et les 'Principia' II. Corps et mouvement*. Paris: Presses Universitaires de France.
Christie, John R.R. 1990. The development of the historiography of science. In *Companion to the history of modern science*, ed. R.C. Olby, G.N. Cantor, J.R.R. Christie, and M.J.S. Hodge, 5–22. London: Routledge.
Clarke, Desmond M. 1982. *Descartes' philosophy of science*. University Park: Pennsylvania State University Press.
Clarke, Desmond M. 1983. Cartesian science in France, 1660–1700. In *Philosophy, its history and historiography*, ed. Alan J. Holland, 165–178. Dordrecht: D. Reidel Publishing Company.
Clarke, Desmond M. 1989. *Occult powers and hypotheses: Cartesian natural philosophy Under Louis XIV.* Oxford: Clarendon Press.
Cohen, I. Bernard. 1985. *Revolution in science*. Cambridge, MA: Belknap Press of Harvard University Press.
Collins, James. 1972. *Interpreting modern philosophy*. Princeton: Princeton University Press.
Cook, Monte. 2008. Desgabets as Cartesian Empiricist. *Journal of the History of Philosophy* 46(4): 501–516.
Copleston, Frederick S.J. 1958. *A history of philosophy. Vol. IV: Descartes to Leibniz*. London: Burns Oates and Washbourne.
Copleston, Frederick S.J. 1959. *A history of philosophy. Vol. V: Hobbes to Hume*. London: Burns Oates and Washbourne.
Cordemoy, de Géraud. 1668. *Discours physique de la parole*. Paris: J. Lambert.
Corneanu, Sorana. 2011. *Regimens of the mind: Boyle, Locke, and the early modern cultura animi tradition*. Chicago: The University of Chicago Press.
Cottingham, John. 1984. *Rationalism*. London: Paladin Books.
Cottingham, John. 1988. *The rationalists*. Oxford: Oxford University Press.
Cunningham, Andrew, and Perry Williams. 1993. De-Centering the 'Big Picture'. *British Journal for the History of Science* 24(4): 407–432.
D'Alembert, Jean. 1995. *Preliminary discourse to the encyclopedia of Diderot*. Chicago: The University of Chicago Press.
Daston, Lorraine, and Elizabeth Lünbeck (eds.). 2011. *Histories of scientific observation*. Chicago/London: The University of Chicago Press.
Des Chene, Dennis. 2000. *Physiologia: Natural philosophy in late Aristotelian and Cartesian thought*. Cornell: Cornell University Press.
Desgabets, Dom Robert. 1983–1985. *Oeuvres philosophiques inedités de Desgabets*. Introduction by Géneviève Rodis-Lewis, Texts and annotations by Joseph Beaude, 7 Vols. Paris: CNRS.
Dijksterhuis, Eduard. 1961 January. *The mechanization of the world picture*. Oxford: Clarendon Press. (Translation of the Dutch *De Mechanisering van het Wereldbeeld*, Amsterdam 1950).
Easton, Patricia. 2000. Desgabets. In *Concise Routledge encylopedia of philosophy*. New York: Routledge. 205.
Fischer, Kuno. 1854–1877. *Geschichte der neueren Philosophie*. 6 Vols. Stuttgart/Mannheim/Heidelberg.
Fraenkel, Carlos, Dario Perinetti, and Justin E.H. Smith (eds.). 2011. *The rationalists: Between tradition and innovation*. Dordrecht: Springer.
Francks, Richard. 1985. Caricatures in the history of philosophy: The case of Spinoza. In *Philosophy, its history and historiography*, ed. Alan J. Holland, 179–194. Dordrecht: D. Reidel Publishing Company.
Friedman, Michael. 1992. *Kant and the exact sciences*. Cambridge: Harvard University Press.
Friedman, Michael. 1993. Remarks on the history of science and the history of philosophy. In *World changes: Thomas Kuhn and the nature of science*, ed. Paul Horwich, 31–54. Cambridge: MIT Press.

Friedman, Michael. 2010. A post-Kuhnian approach to the history and philosophy of science. *The Monist* 93(4): 497–517.

Gabbey, Alan. 1998. Descartes' Physics and Descartes' Mechanics: Chicken and egg? In *Essays on the philosophy of science of René Descartes*, ed. Steven Voss, 311–323. Oxford: Oxford University Press.

Garber, Daniel. 1992. *Descartes' metaphysical physics*. Chicago: University of Chicago Press.

Garber, Daniel. 1998. New doctrines of body and its powers, place and space. In *The Cambridge history of seventeenth-century philosophy: Two volumes*, ed. Daniel Garber, 553–623. Cambridge: Cambridge University Press.

Garber, Daniel. 2001a. Descartes and the scientific revolution: Some Kuhnian reflections. *Perspectives on Science: Historical, Philosophical, Social* 9(4): 405–422.

Garber, Daniel. 2001b. *Descartes embodied*. Cambridge: Cambridge University Press.

Garber, Daniel. 2002. Descartes, mechanics, and the mechanical philosophy. In *Midwest studies in philosophy: Renaissance and early modern philosophy*, vol. 26, ed. Peter French and Howard K. Wettstein, 185–204. Oxford: Oxford University Press.

Garrett, Don, and Edward Barbanell (eds.). 1997. *Encyclopedia of empiricism*. Westport: Greenwood Press.

Gaukroger, Stephen (ed.). 1980. *Descartes: Philosophy, mathematics and physics*. Sussex: Harvester Press.

Gaukroger, Stephen. 2002. *Descartes' system of natural philosophy*. Cambridge: Cambridge University Press.

Gaukroger, Stephen, John Schuster, and John Sutton (eds.). 2000. *Descartes' natural philosophy*. New York: Routledge.

Gillispie, Charles Coulston. 1960. *The edge of objectivity: An essay in the history of scientific ideas*. Princeton: Princeton University Press.

Gilson, Etienne, and Thomas Langan. 1963. *Modern philosophy: Descartes to Kant*. New York: Random House.

Gravesande, Willem J. 's. 1747. *Mathematical elements of natural philosophy confirmed by experiments: Or, an introduction to Sir Isaac Newton's philosophy*. Trans. J. T. Desaguiliers. London: W. Innys.

Grene, Marjorie. 1995. Animal mechanism and the Cartesian vision of nature. In *Physics, philosophy, and the scientific community*, ed. K. Gavroglu, John Stachel, and Marx W. Wartofsky, 189–204. Dordrecht: Kluwer.

Haakonssen, Knud. 2006. The history of eighteenth-century philosophy: History or philosophy? In *The Cambridge history of eighteenth-century philosophy*, ed. Knud Haakonssen, 3–26. Cambridge: Cambridge University Press.

Hall, Alfred Rupert. 1954. *The scientific revolution: 1500–1800, The formation of the modern scientific attitude*. Boston: Beacon Press. (Second edition 1962; published as *The Revolution in Science 1500–1750* by Longman Group Ltd. In 1983.)

Hanratty, Gerald. 1995. *Locke, Hume, Berkeley revisited*. Dublin: Four Courts Press.

Hatfield, Gary. 1979. Force (God) in Descartes' Physics. *Studies in History and Philosophy of Science* 10: 113–140.

Hatfield, Gary. 1985. First philosophy and natural philosophy in Descartes. In *Philosophy: Its history and historiography*, ed. Alan J. Holland, 149–164. Boston: Kluwer.

Hatfield, Gary. 1992. Metaphysics and the new science. In *Reappraisals of the scientific revolution*, ed. D.C. Lindberg and R.S. Westman, 93–166. Cambridge: Cambridge University Press.

Hatfield, Gary. 1996a. The importance of the history of science for philosophy in general. *Synthese* 106(1): 113–138.

Hatfield, Gary. 1996b. Was the scientific revolution really a revolution in Science? In tradition, transmission, transformation. In *Proceedings of two conferences on Pre-Modern Science*, University of Oklahoma, eds. F. Jamil Ragep and Sally P. Ragep and Steven Livesey, 489–523. Leiden: Brill.

Hatfield, Gary. 1988. Science, certainty, and Descartes. In *PSA: Proceedings of the Biennial Meeting of the Philosophy of Science Association 1988*, Vol. 2, eds. A. Fine and J. Leplin, 249–262. East Lansing: Philosophy of Science Association.

Holland, Alan J. (ed.). 1983. *Philosophy, its history and historiography*. Dordrecht: D. Reidel Publishing Company.
Ishiguro, Hidé. 1986. Pre-established harmony versus constant conjunction: A reconsideration of the distinction between rationalism and empiricism. In *Rationalism, empiricism, and idealism*, ed. Anthony Kenny, 61–85. Oxford: Oxford University Press.
Keeling, S.V. 1934. *Descartes*. London: Ernest Benn.
Kenny, Anthony. 1986. *Rationalism, empiricism, and idealism*. Oxford: Clarendon.
Kolesnik–Antoine, Delphine (ed.). 2013. *Qu'est-ce qu'être cartésien?* Lyon: ENS Editions.
Koyré, Alexandre. 1957. *From the closed world to the infinite universe*. Baltimore: Johns Hopkins University Press.
Koyré, Alexandre. 1968. *Newtonian studies*. Chicago: The University of Chicago Press.
Kuhn, Thomas. 1957. *The Copernican revolution: Planetary astronomy in the development of western thought*. Cambridge, MA: Harvard University Press.
Kuhn, Thomas. 1962. *The structure of scientific revolutions*. Chicago: University of Chicago Press.
Laporte, Jean. 1988. *Le rationalisme de Descartes*. Paris: Presses universitaires de France. First published in 1945.
Lennon, Thomas M. 1993. *The battle of the Gods and Giants: The legacies of Descartes and Gassendi 1655–1715*. Princeton: Princeton University Press.
Lennon, Thomas, and Patricia Easton. 1992. *The Cartesian empiricism of François Bayle*. New York: Garland Publishing.
Liard, L. 1882, Du rôle de l'expérience dans la physique cartésienne. In *Descartes*, 111–138. Paris, Germer-Bailliere. Reprinted in Rodis-Lewis, Geneviève, ed. 1987. *La science chez Descartes*, 333–360. New York/London: Garland Publishing Inc.
Locke, John. 1975. *An essay concerning human understanding*. Ed. P. H. Nidditch. Oxford: Clarendon Press.
Loeb, Louis E. 1981. *From Descartes to Hume: Continental metaphysics and the development of modern philosophy*. Ithaca: Cornell University Press.
Machamer, Peter, and J.E. McGuire. 2009. *Descartes' changing mind*. Princeton: Princeton University Press.
Mandelbaum, Maurice. 1976. On the historiography of philosophy. *Philosophy Research Archives* 2: 708–744.
Mandelbaum, Maurice. 1977. The history of philosophy: Some methodological issues. *Journal of Philosophy* 74: 561–572.
Matson, Wallace I. 1968. *A history of philosophy*. New York: American Book Company.
Milhaud, Gaston. 1921. *Descartes savant*. Paris: F. Alcon.
Mouy, Paul. 1934. *Le développement de la physique cartésienne: 1646–1712*. Paris: J. Vrin.
Nadler, Steven. 1990. Deduction, confirmation, and the laws of nature in Descartes's *Principia Philosophiae*. *Journal of the History of Philosophy* 28: 359–383.
Norton, David. 1981. The myth of British empiricism. *History of European Ideas* 1: 331–344.
Osler, Margaret J. 1985. Eternal truths and the laws of nature: The theological foundations of Descartes' philosophy of nature. *Journal of the History of Ideas* 46: 349–362.
Osler, Margeret J. 1994. *Divine Will and the mechanical philosophy: Gassendi and Descartes on contingency and necessity in the created world*. Cambridge: Cambridge University Press.
Pap, Arthur. 1958. *Semantics and necessary truth: An inquiry into the foundations of analytic philosophy*. New Haven: Yale University Press.
Pardies, Ignace Gaston. 1672. *Discours de la connoissance des bestes*. Paris: Sebastien Mabre-Cramoisy.
Perler, Dominik. 1998. *René Descartes*. Münich: Beck.
Philosophical Transactions 1667.
Popkin, Richard. 1959a. Review of *Dominant themes of modern philosophy*, by George Boas. *Journal of Philosophy* 61: 773–778.
Popkin, Richard. 1959b. Did Hume ever read Berkeley? *Journal of Philosophy* 56(12): 535–545.
Popkin, Richard. 1964. So, Hume did read Berkeley. *Journal of Philosophy* 61(24): 773–778.

Prendergast, Thomas L. 1972–1973. Descartes and the relativity of motion. *Modern Schoolman* 50: 64–72.

Prendergast, Thomas L. 1975. Motion, action, and tendency in Descartes' physics. *Journal of the History of Philosophy* 13: 453–462.

Priest, Stephen. 1990. *The British empiricists: Hobbes to Ayer*. London: Penguin.

Randall, J.H. 1940. *The making of the modern mind.* Cambridge: Houghten Mifflin Co.

Roux, Sophie. 2013. An Empire Divided: French natural philosophy (1670–1690). In *The mechanization of natural philosophy*, Boston studies in the philosophy and history of science, vol. 282, ed. Daniel Garber and Sophie Roux, 55–95. Dordrecht: Springer.

Roux, Sophie, and Daniel Garber (eds.). 2013. *The mechanization of natural philosophy*, Boston studies in the philosophy and history of science, vol. 282. Dordrecht: Springer.

Russell, Bertrand. 1912. *The problems of philosophy*. New York: Holt.

Ryle, Gilbert. 1960. Epistemology. In *Encylopedia of western philosophy*, ed. J.O. Urmson. London: Hutchison.

Schmaltz, Tad M. 2002. *Radical Cartesianism: The French reception of Descartes*. Cambridge: Cambridge University Press.

Schneewind, Jerome B. 2004. Teaching new histories of philosophy. In *Proceedings of a Conference, University Center for Human Values*. Princeton: Princeton University Press.

Schuster, John. 2012. *Descartes-Agonistes: Physico-mathematics, method & corpuscular mechanism 1618–33*. Dordrecht: Springer.

Scott, J.F. 1952. *The scientific work of René Descartes (1596–1650)*. London: Taylor & Francis.

Scruton, Roger. 1982. *Kant*. Oxford: Oxford University Press.

Shank, J.B. 2008. *The Newton wars and the beginning of the French enlightenment*. Chicago/London: The University of Chicago Press.

Smith, N.K. 1953. *New studies in the philosophy of Descartes*. London: Macmillan.

Sorell, Tom (ed.). 1993. *The rise of modern philosophy*. Oxford: Oxford University Press.

Stewart, M.A. 1985. Hume and the "Metaphysical Argument *A Priori*". In *Philosophy, its history and historiography*, ed. Alan J. Holland, 243–270. Dordrecht: D. Reidel Publishing Company.

Stewart, Matthew. 2007. *The Courtier and the Heretic: Leibniz, Spinoza and the fate of God in the modern world.* New York: W. W. Norton and Company.

Tennemann, William Gottlieb. 1852. *Manuel of the history of philosophy*, trans. Arthur Johnson, ed, J. R. Morell. London: Henry G. Bohn.

Verbeek, Theo (ed.). 1993. *Descartes et Regius. Autour de l'Explication de l'esprit humain.* Amsterdam: Rodopi.

Voltaire. 1741. *Lettres concerning the English nation*, 2nd ed. London: C. Davis.

von Leyden, W. 1968. *Seventeenth-century metaphysics*. London: Duckworth.

Westfall, Richard S. 1971. *The construction of modern science*. New York: Wiley.

Williams, Bernard. 1978. *Descartes: The project of pure inquiry*. Harmondsworth: Penguin.

Woolhouse, Roger. 1971. *Locke's philosophy of science and knowledge*. Oxford: Blackwell.

Woolhouse, R.S. 1998. *The empiricists*. Oxford: Oxford University Press.

Zittel, Claus. 2011. Conflicting pictures: Illustrating Descartes' *Traite de l'homme*. In *Silent messengers. The circulation of material objects of knowledge in the early modern low countries*, ed. Sven Dupré and Christoph Lüthy, 217–260. Berlin: Lit Verlag.

Part I
Cartesian Natural Philosophy: Receptions and Context

Chapter 2
Censorship, Condemnations, and the Spread of Cartesianism

Roger Ariew

Abstract Descartes and the Cartesians suffered a series of condemnations aimed at several fundamental propositions of corpuscularianism and mechanism, such as the denial of substantial forms and real qualities. Also condemned was the theory of matter and place: extension as the principal attribute of matter, the indefinite extension of the world, and the impossibility of the void. With these objections, came an increased critique of hyperbolic doubt. The rejection of hyperbolic doubt caused Cartesians no longer to distinguish between the absolutely and the morally certain—between that which we cannot doubt and that about which we have no doubt although we could doubt it—and to treat all principles on a par with one another. As a result, Cartesians became more empirical and pursued aggressively a limited hypothetical-deductive method. For example, Huygens describes a hypothetico-deductive method that ends up with high probability, not absolute or moral certainty; in this, Huygens follows a path taken by closer followers of Descartes. This chapter will investigate such issues and their consequences for Cartesianism in the works of Cartesians such as Du Roure and Cordemoy.

2.1 Censorship and Condemnations

I assess two large phenomena in seventeenth-century Cartesianism, first the widespread criticism and condemnation of Cartesian physics and second the multiplication of a more empirical Cartesianism. There is surely a relationship between these two movements. For one, perhaps because of the rampant censorship at the time,

R. Ariew (✉)
Department of Philosophy, University of South Florida, Tampa, FL, USA
e-mail: rariew@usf.edu

M. Dobre and T. Nyden (eds.), *Cartesian Empiricisms*, Studies in History and Philosophy of Science 31, DOI 10.1007/978-94-007-7690-6_2,
© Springer Science+Business Media Dordrecht 2013

many Cartesians, while maintaining Descartes' disposition for mechanistic explanations, aggressively pursued a quasi-hypothetical-deductive method and thus became more empirical. Although I think that the two phenomena are interrelated, I do not think that those relations are obvious and direct; it is clear that some Cartesians adopted positions that were censored, despite their numerous critiques.

The denunciations of Cartesianism were unusually frequent and ferocious. For most of the seventeenth century, the official response to Descartes' philosophy was unfavorable. During his life, Descartes waged fierce battles with his opponents. In the 1640s, he thought himself at war with the Jesuits; he had political problems and received official condemnations by Protestants at Utrecht around 1642 and at Leiden in 1647.[1] The battles intensified after Descartes' death in 1650. There were condemnations by Catholics at Louvain in 1662,[2] culminating with Descartes' works being put on the *Index of Prohibited Books* by the censors of Rome in 1663.[3] The fighting raged in the second half of the seventeenth century: the Jesuits held anti-Cartesian disputations at Clermont College in 1665.[4] There were numerous attacks in print.[5] The Cartesians counter-attacked with satires[6] and learned essays[7] and the anti-Cartesians responded with their own satires.[8] Ultimately, these disputes spilled into the official political arena, the domains of the king, of the Universities, and of the teaching orders: the king issued an edict in 1671, clarifying it in 1675[9]; the faculty of arts at Paris tried to condemn Cartesianism in 1671 and succeeded by 1691[10]; there were skirmishes at Angers and Caen during 1675–1678; the Oratorians prohibited the teaching of Cartesianism in 1678[11]; and the Jesuits formally condemned it as late as 1706.[12]

It would be useful to exhibit the contents of some censures of Cartesian philosophy, together with their aftermath.[13] Perfect for our purposes is the influential 1662

---

[1] Verbeek 1988, 1992. There were plenty of other skirmishes, of course. With the exception of the mentions of Utrecht and Leiden, which are most capably discussed by Verbeek and others, I concentrate on the French Catholic world.

[2] D'Argentré 1728–1736, III, Part. II, 303–304.

[3] Bouillier 1868, I, 446–447.

[4] Prou 1665.

[5] See, for example, Vincent 1677; de la Ville [Louis le Valois] 1680; de la Grange 1682; Huet 1689.

[6] See the "arret burlesque" in Boileau 1747, III, 150–153.

[7] *Plusieurs raisons pour empecher la censure ou la condemnation de la philosophie de Descartes* in Boileau 1747, III, 117–141 (reprinted in Cousin 1866, III, 303–317). See also Bayle 1684.

[8] Daniel 1690; M. G. de L'A. [Pierre Daniel Huet] 1689; Daniel 1693.

[9] Bouillier 1868, I, 469; Babin 1679, 6–7.

[10] D'Argentré 1728–1736, III, Part. I, 149.

[11] *Concordat entre les Jesuites et les Peres de l'Oratoire, Actes de la Sixiéme Assemblée, September 1678*, in Bayle 1684, 11–12.

[12] de Rochemonteix 1889, IV, 89–93; see Ariew et al. 1998, 254–260.

[13] There are obviously institutional, social, and other factors behind the censures; such discussions exceed the scope of this essay.

condemnation at Louvain, which focuses primarily on Cartesian physics (taken broadly). The Louvain condemnation lists five difficulties with Cartesian doctrine: (1) Descartes' definition of substance; (2) his rejection of substantial forms or real accidents; (3) extension considered as the essential attribute of matter; (4) the indefinite extension of the world; and (5) the rejection of the possibility of a plurality of worlds.[14] These five difficulties did not all originate with the censors of Louvain, but they were, in fact, repeated again and again during the seventeenth century.

2.1.1 *The Definition of Substance*

Descartes defines substance as a thing existing in such a way as to depend on no other thing, needing only the concurrence of God in order to exist; corporeal substance and mind (or thinking substance) are thus understood as falling under this common concept.[15] The authorities of Louvain inferred from the definition that there would not be any substantial forms, except for rational soul. And, in particular, there would not be any substantial forms in animals and plants. The issue was picked up in the textbooks. Oratorians required their professors to teach that "in each natural body there is a substantial form really distinct from matter."[16] The Jesuits condemned the proposition "There are no substantial forms of bodies in matter."[17] Moreover, one can find numerous discussions of the Cartesian definition of matter and body and repeated criticisms of the consequence that animals are machines lacking sensation and knowledge. These spanned such diverse thinkers as the Franciscan Claudius Frassen and the Oratorian Jean-Baptiste de la Grange, among others.[18] Ultimately, the Jesuits condemned the proposition that "Animals are mere automata deprived of all knowledge and sensation."[19]

[14] Ariew 1994, 3; all translations in this essay are mine, unless otherwise indicated. Armogathe and Carraud 2003 shows that the Louvain condemnations were the catalyst for Descartes' works being put on the *Index*.

[15] *Principles* I 51–52. AT VIIIa 24–25.

[16] Ariew 1994, 4; trans. in Ariew et al. 1998, 256–257: "Qu'en chaque corps naturel il y a une forme substantielle réellement distinguée de la matière."

[17] Ariew 1994, 6; trans. in Ariew et al. 1998, 260: "Nullae sunt formae substantiales corporeae à materia distinctae."

[18] See, for example, Frassen 1668: on the rejection of Descartes' definition of matter and body, 30; on the Cartesian doctrine of animal-machines, 646. Cf. also de la Grange 1682, I, 13: on animals having no reason.

[19] Ariew 1994, 6; trans. in Ariew et al. 1998, 259: "Belluae sunt mera automata omni cognition ac sensu carentia." The trial at Angers produced a variant on this theme. Fromentier was harshly accused of holding the immortality of animal souls, 41.

2.1.2 The Rejection of Substantial Forms or Real Accidents

Descartes said: "it is completely contradictory that there should be real accidents, since whatever is real can exist separately from any other subject."[20] The Louvain objection was that, as a consequence, there would be a problem with the accidents of bread and wine remaining without subject in the Eucharist. This was the most frequently repeated criticism of Cartesian philosophy in the Catholic world, beginning as early as Antoine Arnauld's *Fourth Set of Objections*.[21] The Jesuits of Clermont College argued in 1665 that the Cartesian hypothesis must be distasteful to theology because "there is no necessity to allow a substantial form in man, which favors the impious and dissolute, [and] there can be no conversion of bread and wine in the Eucharist into the blood and body of Christ, nor can it be determined what is destroyed in that conversion, which favors heretics."[22] Oratorians required their professors to teach "there are real and absolute accidents inherent in their subjects, which can supernaturally be without any subjects."[23] And at Angers, the Oratorians Eugene Fromentier, Bernard Lamy, and Cyprien de Villecroze were removed from their teaching positions for having taught a number of Cartesian doctrines, one of them being that "There are no species or real accidents in the Eucharist."[24] The Jesuits condemned the proposition "There are no absolute accidents."[25] Most textbooks contained discussions of the doctrine.[26] But the discussions about the Eucharist shifted from the denial of substantial forms and real accidents to the principle that quantity or extension is corporeal subsistence—that is, to the third Louvain objection.

2.1.3 Extension as an Essential Attribute of Matter

The authorities at Louvain found offensive Descartes' principle that the extension of bodies constitutes their essential and natural attribute.[27] Oratorians required one to teach

---

[20] *Replies* VI 7. AT VII 434–435: "omnino repugnat dari accidentia realia, quia quicquid est reale, potest separatim ab omni alio subjecto existere."

[21] AT VII 217–218. The general issue of the Eucharist is clearly one that would concern a Catholic or some Protestants reconcilers like Leibniz. Still it is surely the most frequently repeated criticism of Descartes in general.

[22] *To Boyle*, July 4, 1665, in Oldenburg 1965–1986, II, 431–432; trans. 435: "Nihil esse necessarium, Substantialem in homine formam, admittere; quod impiis et disciplinae solutioris amantibus favet….Nullam fieri in Eucharistia conversionem panis et vini in ipsum Christi Corpus sanguinem, nec assignari, quid in illa conversion destruatur, posse quod favet haereticis."

[23] Ariew 1994, 4; trans. in Ariew et al. 1998, 256–257: "Qu'il y a des accidens réels et absolus inherens à leurs sujets, surnaturellement être sans aucun sujet."

[24] Babin 1679, 39, 44: "Il n'y a point d'especes ny d'accidents réels dans l'Eucharistie."

[25] Ariew 1994, 6; trans. in Ariew et al. 1998, 260: "Nulla sunt accidentia absoluta."

[26] De la Grange 1682, I, 3; see also 109–135.

[27] *Replies* VI. AT VII 442 and *Principles* I 53. AT VIIIa 25.

"That actual and external extension is not the essence of matter."[28] The Oratorian censors objected to Father Lamy's definition of extension as the essence of body and to his rejection of substantial forms.[29] The University of Paris condemned the proposition that "The matter of bodies is nothing other than their extension and one cannot be without the other,"[30] and the Jesuits echoed with a prohibition of "The essence of matter or of body consists in its actual and external extension."[31] Textbooks were filled with such statements, again with a naturalistic explanation of the Eucharist being the issue at stake.[32]

2.1.4 *The Indefinite Extension of the World*

Louvain objected to Descartes' assertion that "we recognize that this world, that is, the whole universe of corporeal substance, has no limits to its extension."[33] The Oratorians of Angers are said to have wrongly taught that "the world is infinite in its extension, a principle which is no less dangerous than the first principle [about the Eucharist]"[34]; the censor comments: "it is true that the Cartesians do not make use of the word *infinite* but only *indefinite*, which is the same thing and merely adds a single syllable to what we say about the *infinite*."[35] And the Jesuits condemned the proposition that "In itself, the extension of the world is indefinite."[36] The issue was also given full play in Scholastic texts.[37]

---

[28] Ariew 1994, 4; trans. in Ariew et al. 1998, 256: "Que l'extension actuelle et extérieure n'est pas de l'essence de la matière."

[29] See Lamy's propositions 4 and 8 in Babin 1679, 37 (also in Girbal 1964, 156–157), with the censor's replies, propositions 1 and 5. Babin 1679, 43–45; Girbal 1964, 158–161.

[30] Ariew 1994, 5; trans. in Ariew et al. 1998, 257: "La Matière des corps n'est rien autre chose que leur étendue, et l'une ne peut être sans l'autre."

[31] Ariew 1994, 6; trans. in Ariew et al. 1998, 259: "Essentia materiae seu corporis consistit in extensione externa et actuali."

[32] For example, Duhamel 1692, 189–201. The classic discussion of this issue is Armogathe 1977.

[33] *Principles* II 21. AT VIIIa 52: "Cognoscimus præterea hunc mundum, sive substantiæ corporeæ universitatem, nullos extensionis suæ fines habere."

[34] Babin 1679, 40: "Le monde soit infiny dans son étenduë, il n'est pas moins dangereux que le premier."

[35] Babin 1679, 40: "Il est vray que les Carthesiens ne veulent pas se servir de ce mot d'Infiny, qui serait trop odieux, mais seulement de celuy d'Indefiny qui est la même chose, et qui n'ajoûte qu'une seule syllabe à tout ce que nous disons de l'Infiny."

[36] Ariew 1994, 6; trans. in Ariew et al. 1998, 259: "Mundi extensio indefinite est in se ipsa."

[37] De la Grange 1682, I, Chap. 28: "de la nature du lieu et du vide [393] que le monde est infini, qu'il n'y a point d'espaces vuide au de-la des cieux, et que plusieurs mondes sont impossibles"; Vincent 1677, 69: "an mundus sit indefinite extensus"; Duhamel 1705, 5, 16: "Cartesio possibilis non est alter mundus, quia noster mundus est infinite, vel, ut loquitur, indefinite extensus."

2.1.5 The Plurality of Worlds

The authorities at Louvain objected to Descartes' principle: "if there were an infinite number of worlds, the matter of which they were composed would have to be identical; hence, there cannot in fact be a plurality of worlds, but only one."[38] Oratorians affirmed that one must teach "that there is no repugnance in God's creating several worlds at the same time,"[39] and the Jesuits condemned the proposition that: "There can be only one world."[40] Typically, the argument found in the textbooks was that Descartes was infringing on God's omnipotence, as de la Grange made clear:

> For who would believe that Descartes teaches only the truth and what is known clearly by natural light, when he tells us in Part II of his *Principles*, article 22, that several worlds are impossible? Can anything more novel and more shocking to reason be uttered? Ever since people have attempted to reason about God's works, possibly there has not been one who has dared to teach this doctrine, or even who has been of that opinion. In fact, there is nothing that seems more clear and natural to us than to assert that God, having produced this world, can still produce another.[41]

Cartesianism was censured not only for doctrinal reasons, but also on pragmatic and pedagogical grounds. It was often asserted that being taught Cartesian philosophy would leave one unprepared for the higher faculties of theology, law, and medicine. During a 1665 disputation the Jesuits of Clermont College summarized some of these difficulties:

> To say no more, the Cartesian hypothesis must be distasteful to mathematics, philosophy, and theology. *To philosophy* because it overthrows all its principles and ideas which commonsense has accepted for centuries; *to mathematics*, because it is applied to the explanation of natural things, which are of another kind, not without great disturbance of order; *to theology*, because it seems to follow from the hypothesis that too much is attributed to the fortuitous concourse of corpuscles, which favors the atheist.[42]

---

[38] *Principles* II 22. AT VIIIa 52: "Si mundi essent infiniti, non posse non illos omnes ex una et eadem materia constare; nec proinde plures, sed unum tantum, esse posse."

[39] Ariew 1994, 4; trans. in Ariew et al. 1998, 257: "Qu'il n'y a aucune répugnance que Dieu puisse produire plusieurs mondes à même temps."

[40] Ariew 1994, 6; trans. in Ariew et al. 1998, 259: "Mundus existere non potest nisi unicus."

[41] De la Grange 1682, I, 6: "Car qui croiroit que Descartes n'enseigne que la verité, et ce qui est connu clairement par la lumiere naturelle, lors qu'il nous dit dans l'article 22. de la seconde Partie de ses Principes, *que plusieurs mondes sont impossibles*. Peut-on dire quelque chose de plus nouveau, et qui choque davantage la raison? Depuis que les hommes se mélent de raisonner sur les Ouvrages de Dieu, il n'y en a possible pas eu un, qui ait osé enseigner cette doctrine, ou mesme qui ait esté de ce sentiment. En effet, il n'y a rien qui nous paroisse plus clair et plus naturel, que de dire que Dieu ayant produit ce monde, peut bien encore en produire un autre." See also Vincent 1677, 75; Duhamel 1705, V, 16.

[42] *To Boyle*, July 4,1665, Oldenburg 1965–1986, II, 431–432; trans. 435: "Ne plura dicam, necesse est, ut et mathematicae, et Philosophiae et Theologiae displiceat Hypothesis Cartesiana. *Philosophiae*, cujus omnia principia notionesque, multis abhinc seculis communi consensione receptas, evertit: *Mathematicae*, quad ad res naturales, quae sunt alterius generis, explicandas, non sine magna perturbatione ordinis traducit: *Theologiae*, quatenus ex hoc hypothesi videtur esse consequens, Nimium aliquanto tribui corpusculis fortuito concurrentibus; quod favet Atheis."

The Jesuit summary is broken down into three categories: the first, a complaint already issued at Utrecht, is the rejection of any novel philosophy. Descartes had previously attempted to defend himself against that charge by arguing (unsuccessfully, it seems) that his philosophy was not novel, but the oldest of all philosophies, since he only accepted principles that had been generally admitted by all philosophers.[43] The second refers to the Scholastic teaching about the objects of mathematics and natural philosophy (or science), usually discussed under the topic of the classification of the sciences. Given that mathematics is an abstraction from natural things, the application of mathematics to natural things would be a "disturbance of order." The Jesuit claim is that mathematical sciences should be subalternated to physics and not vice-versa, as they seem to be with Descartes. Finally, in the third, Cartesian philosophy is unfairly linked with atomism and the standard complaint against atomism is issued against it.[44]

2.2 Descartes, the Cartesians, and Atomism

This was not the first time Descartes was criticized for his "corpuscles." To some of his critics, Descartes' matter theory looked very much like atomism. As early as 1637, responding to the publication of the *Discourse on Method*, Libertus Fromondus sent Descartes a work against Epicureans and atomists he had written earlier[45] and provided him with his objections to what he saw as Descartes' over-reliance on atomistic and mechanical principles. Concerning Descartes' account of body in the *Meteors*, Fromondus commented: "This composition of bodies made up of parts with different shapes…by which they cohere among themselves as if by little hooks, seems excessively crass and mechanical."[46] In 1679, the Oratorian censors criticized the corpuscularianism of both the Oratorian Fromentier and that of Descartes,[47] even though they recognized that both philosophers formally rejected atomism:

> The opinion of Epicurus and Democritus, that the world has been formed by the fortunate encounter of atoms and small bodies flying about from all parts, has been treated as extravagant and impious. One wants to believe that Descartes and his followers do not teach that the universe was made by chance and without God's providence, but, at bottom, what they say is not different than what Democritus and Epicurus advance, since Descartes only wants God to have created all matter, divided it into almost equal parts, agitated these parts

---

[43] AT VII 580–581, 596. See Ariew 1994. See also de la Grange 1682, I, 1–2.

[44] The summary of the disputation continues with the two complaints referred to previously concerning the rejection of substantial forms and real accidents.

[45] Fromondus' *Labyrinthus, sive de compositione continui* from 1631.

[46] AT I 406: "Compositio deinde illa corporum ex partibus diversarum figurarum…quibus invicem tanquam uncinis cohærescant, nimis crassa et mechanica videtur."

[47] Babin 1679, 36 and 36n.

in various directions, each to its own proper center, and several around a common center; after that, God can remain at rest....Is there something more odious in Epicurus' opinion not found in Descartes' hypothesis?[48]

We should emphasize that the difficulty with atomism as potentially harmful to the Catholic faith was also the legal basis for the king's prohibition of the teaching of Cartesianism. In 1671, François de Harlay, the archbishop of Paris, announced a verbal decree from king Louis XIV requiring that "no other doctrine be taught in the universities than the one set forth by the rules and statutes of the university, and that nothing [of these other doctrines] be put into theses."[49] The king prohibited "certain opinions the faculty of theology once censured, whose teaching or publication was prohibited by the Parliament," which, as he put it could, "bring some confusion in the explanation of our mysteries."[50] The reference in the decree to "certain opinions the faculty of theology once censored" was an allusion to a condemnation of 14 anti-Aristotelian propositions some 50 years earlier. In 1624, the Sorbonne had censored various opinions disseminated by some alchemical atomists.[51] The faculty had objected to some propositions put forward by the atomists that attacked Peripatetic matter theory, arguing that the prime matter of the Peripatetics is utterly fictitious, and their substantial forms are no less absurdly defended.[52] The faculty had also censored propositions physical alterations happening through the introduction or destruction of an accidental entity, because, they said, it attacked the holy sacrament of the Eucharist.[53] Thus, the "confusion in the explanation of our mysteries" in the king's 1671 edict also alluded to the 1624 event, a condemnation of atomists attacking Aristotelian matter theory. The king's exhortation—"to bring it about that no other doctrine than the one set forth by the rules and statutes of the University is taught in the Universities"—recalled the subsequent *arret* issued by

---

[48] Babin 1679, 41: "L'opinion d'Epicure et de Democrite, qui vouloient que le Monde se fût formé par la rencontre heureuse des Atomes et des petits corps qui voltigent de toutes parts, a été traittée d'extravagante et d'impie. On veut bien croire que des Carthes et ses Partisans n'enseignent pas que l'Univers ait été fait par hazard et sans la Providence de Dieu: mais au fond ce qu'ils disent n'est pas different de ce qu'avancent Democrite et Epicure, car des Carthes veut seulement que Dieu ait fait toutre la matiere, qu'il l'ait divisée en de petites parties à peu pres égales, qu'il les ait agitées en divers sens chacun en son propre centre, et plusieurs d'elles au tour d'un centre commun; apres quoy Dieu peut demeurer en repos....Y a-il quelque chose de plus odieux dans le sentiment d'Epicure, qui ne se trouve point dans l'hypothese de des Carthes?"

[49] Jean Duhamel, *Quaedam recentiorum philosophorum, ac praesertim Cartesii, propositiones damnatae ac prohibitae* in Duhamel 1705, V, 17–18: "L'on n'enseigne point dans les universités d'autre doctrine que celle qui est portée par les règlements et les statuts de l'Université, et que l'on n'en mette rien dans les thèses." See also Schmaltz 2002, 29–34.

[50] Duhamel 1705, V, 17–18: "Certaines opinions que la Faculté de Théologie avait censurées autrefois et que le parlement avait défendu d'enseigner ni de publier...[qui pourraient] porter quelque confusion dans l'explication de nos mystères."

[51] I.e., Antoine Villon, Etienne de Clave and Jean Bitauld. See Garber 2002.

[52] *Positiones Publicae*, in de Launoy 1653, 128–129.

[53] See De Launoy 1653, 132.

the Court du Parlement.[54] That legal document prohibited "all persons, under pain of death, from either holding or teaching any maxims against the ancient and approved authors, nor hold any dispute that those approved by the aforementioned doctors of the Faculty of Theology."[55] Although Louis did not mention Cartesianism explicitly, it was clearly the "other doctrine" against which the 1671 decree was directed. In any case, he clarified his intent by 1675, specifically naming those who taught "the opinions and thoughts of Descartes" as ones who "might bring disorder to our Kingdom which it would be good to prevent." Louis ordered they be prevented from continuing their lessons in any way whatsoever.[56]

The linkage between atomism and heresy or near heresy was very common in the seventeenth century. Even Mersenne, in his 1624 *L'impiété des Deistes*, complained about the adherence of Gorlaeus and Hill to "Epicureanism" and to the doctrine "that inside bodies there are atoms which have quantity and figure." According to Mersenne, "ultimately, they are all heretics, which is why we should not be surprised that they agree, being all as thick as thieves."[57] Gassendi, in 1624, accepting the seemingly innocuous doctrine that "the essence of quantity is nothing but its external extension,"[58] felt compelled to point out that his doctrine had negative consequences for the sacrament of the Eucharist and to take steps to reaffirm his orthodoxy: "To continue, let us now turn our attention to the famous difficulty concerning the essence of quantity. Our philosophers explain it so well that nothing could be more obscure, though nothing would seem to be more obvious than quantity. However, I must confess that the mystery of the Eucharist, as our faith conceives it, may cause some difficulty in this matter."[59]

So Descartes was properly cautious about atomism. He went as far as to deny formally the possibility of physical atoms and voids. At the end of *Principles*, Part IV, Descartes reflected generally on the method he used in his physics. He claimed that he has not used any principle not accepted by all the philosophers of every age, including Aristotle. Descartes adds: "He considered the shapes, motions and sizes of bodies and examined the necessary results of their mutual interaction in accordance

[54] Duplessis d'Argentré 1728–1736, II, Part. II, 147.

[55] Duplessis d'Argentré 1728–1736, II, Part. II, 147: "toutes personnes à peine de la vie tenir, ni enseigner aucunes maximes contre les Autheurs anciens et approuvez, ni faire aucunes disputes que celles qui seront approuvés par les Docteurs de ladite Faculté de Theologie."

[56] Louis Phelypeaux, in Babin 1679, 6–7: "les opinions et les sentimens de Des Carthes…[qui] pourrait causer en notre Royaume quelque desordre qu'il est bon de prevenir."

[57] Mersenne 1624, 237–239: "qu'il y a des atomes dedans les corps, qui ont quantité, et figure, etc.;…au bout de conte ils sont tous Heretiques, c'est pourquoy il ne faut pas s'estonner s'ils s'accordent comme larrons en foire."

[58] Gassendi 1624, II, Exer. 3, Art 10: *Quantitatis essentiam esse extensionem externam.*

[59] Gassendi 1624, II, Exer. 3, Art. 10; also Art 11: *Species Eucharisticas non item fore Fides nos Orthodoxa docet*: "Sequitur ut celebris quoque difficultas seligatur de Quantitatis essentia, quae sic explicatur a nostris, ut cum nulla res esse videatur quantitate evidentior, nulla jam dici possit obscurior. Et fateor quidem Eucharisticum, quod Fides nostra tuetur, Mysterium aliquam posse huic materiae difficultatem conciliare."

with the laws of mechanics, which are confirmed by reliable everyday experience."[60] According to Descartes, no one has doubted that bodies move and have various sizes and shapes, that their various different motions correspond "to these differences in size and shape; and that when bodies collide bigger bodies are divided into many smaller ones and change their shapes."[61] The difference between his approach and that of others is that he considers "that in each body there are many particles so small that they are not perceived with any of our senses."[62] Descartes then dealt with the question of the nature of these insensible bodies. Among other things, he rejected the Democritean "suppositions" that the minute bodies are indivisible and that there are voids around them, calling these suppositions inconsistent. In the French edition of the *Principles*, he added that he rejected "all of Democritus' suppositions" with the one exception of "the consideration of shapes, sizes, and motions."[63]

Ironically, Descartes' arguments for the impossibility of void were also condemned by Scholastics as overly strong. According to Descartes, it is a contradiction to suppose that there is such a thing as a vacuum.[64] Thinking of a vessel, its concave shape, and the extension that must be contained in this concavity, Descartes asserted: "it would be as contradictory of us to conceive of a mountain without a valley, as to conceive of this concavity without the extension contained in it, or of this extension without an extended substance."[65] In fact, he argued that if God were to remove the body contained in that vessel and did not allow anything else to take its place, the sides of the vessel would thereby become contiguous. Textbook authors such as Jean Duhamel took on Descartes' actual argument: "God can absolutely destroy the bodies presently between the heavens and earth, having produced them and conserving them freely.... God could put a third body between them without displacing them...and, as a consequence, heaven and earth would not be touching truly and effectively."[66]

---

[60] *Principles* IV 200. AT VIIIa 323–324: "Nempe figuras et motus et magnitudines corporum consideravi, atque secundum leges Mechanicæ, certis et quotidianis experimentis confirmatas."

[61] *Principles* IV 200. AT VIIIa 323–324: "Pro quarum diversitate ipsorum etiam motus varientur, atque ex mutua collisione, quæ majuscula sunt in multa minora dividantur, et figuras mutant."

[62] *Principles* IV 201. AT VIIIa 324–325: "At multas in singulis corporibus particulas considero, quæ nullo sensu percipiuntur."

[63] *Principles* IV 202. AT VIIIa 325: "Tout ce que ce dernier [Démocrite] a supposé...[sauf] la considération des figures, des grandeurs et des mouvements."

[64] *Principles* II 16–18. AT VIIIa 49–50.

[65] *Principles* II 18. AT VIIIa 50: "Non magis repugnet nos concipere montem sine valle, quam intelligere istam cavitatem absque extensione in ea contenta, vel hanc extensionem absque substantia quæ sit extensa."

[66] Duhamel 1692, Chap. 4, Si le vide des philosophes est impossible, 202. Also Duhamel 1705, III, 203, *vacuum divinitus possibile est*: "Dieu peut absolument détruire les corps qui sont presentement entre ciel et la terre, les ayant produits, et les conservant librement....Dieu pourrait mettre un troisième corps sans les déplacer...et par conséquent le ciel et la terre ne se toucheroient pas veritablement et effectivement." Cf. also Frassen 1668, 372: *Cartesius contendit, non solum nullum vacuum existere; sec nec etiam divinitus esse possibile*; de la Grange 1682, Chap. 30, 410–417: *si le vide est possible*; Vincent 1677, 63: *de vacuo philosophico*.

Descartes' views on the void were officially condemned: Oratorians affirmed "That the void is not impossible"[67]; Jesuits censured the proposition that "The compenetration of bodies properly speaking and place void of all bodies imply a contradiction."[68]

The Dominican Antoine Goudin spent almost 30 pages of his *Philosophia* arguing against Cartesian principles.[69] He disputed, for example, Descartes' conservation of quantity of motion based on God's immutability. According to Goudin, God can, without inconsistency, augment, diminish, or vary the motions he has given to bodies. Goudin also argued against Descartes' corpuscles as a first principle. For Goudin the core of his argument was that Descartes' corpuscles can no more explain the variety of animals with sensation and life than can atoms. In fact, though he knew fully well and cited the *Principia* passages against Democritus, Goudin began his discussion of Descartes' principles by referring his reader to his previous criticism on the principles of the atomists; as he said, "Since these principles [of Descartes] do not differ from those of the atomists in their principal points, they are refuted by the reasons we have just given."[70]

In his prior disputation against the atomists,[71] Goudin argued that there are no atoms, and even if atoms are accepted *per impossibile*, they cannot provide any foundation as a first principle. His line of reasoning was that, however small a body, it is always divisible; thus, there are no indivisible bodies, that is, no atoms. He considered the reply that atoms are so small that nature cannot abide a smaller body—they are divisible mathematically, or only by an operation of the mind, but that they are indivisible naturally and in reality. Goudin replied that atoms are different from one another—they have different shapes from one another, one longer, one larger. Nature therefore allows things smaller than some atoms. He asked rhetorically: what would prevent the branch or hook of an atom to be broken into two atoms, since there are such smaller proportions in nature? So there are no atoms; but even if there were atoms, Goudin asserted, they cannot be the principle of all things because they are not sufficient in themselves to explain the generation of sensitive and animate life out of their combinations alone—witness the exception made for humans and the insuperable difficulties with accounts of animals as machines without sensation. Ultimately, atoms and their combination cannot explain differences in kind.

In the same fashion as Goudin, some Cartesians reacted negatively to what they thought was the inadequacy of Descartes' account of bodies. Take, for example,

---

[67] Ariew 1994, 4; trans. in Ariew et al. 1998, 257: "Que le vuide n'est pas impossible."

[68] Ariew 1994, 6; trans. in Ariew et al. 1998, 259: "Penetratio corporum propriè dicta, et locus omni corpore vacuus involvunt contradictionem."

[69] Goudin 1726 [1668], II, Art. 4, 16–44.

[70] Goudin 1726 [1668], II, Art. 4, 16: "Et Hujus Principia ferè satis impugnata sint articulo praecendenti, cum in praecipius capitibus non different à Principiis Atomistarum."

[71] Goudin 1726 [1668], II, Art. 3, 10–16.

Géraud de Cordemoy whose fame in part rested on his attempts to extend Cartesian philosophy to the fields of language and communication and his advocacy of Cartesian orthodoxy, such as his defense of the doctrine of animal-machines and the consistency of Cartesianism with Genesis; above all, Cordemoy is known for the views he propounded in the 1666 *Le discernement du corps et de l'ame*, which expounded Cartesian physics. In the work, Cordemoy offered a variation of Cartesian mechanical philosophy—everything in the physical world is explained in terms of the size, shape, and motion of particles—but one that required atoms and the void. He rejected the indefinite division of body and the Cartesian identification of space and extension. He distinguished body and matter, matter being an assemblage of bodies, and claimed that bodies as such were impenetrable and could not be physically divided or destroyed. These views were intended as an answer to his criticism of the Cartesian principle of individuation of bodies as shared motion. According to the principle, a body at rest between other bodies would have to constitute a single body with the other bodies, even though we have a clear and natural idea of a body at rest between other bodies.[72] Cordemoy proposed that shape, rather than motion, distinguishes the indivisible atoms. Thus Cordemoy's atomism arose from an attempt to provide a principle of individuation for bodies within a broader Cartesian framework.

In 1685, Leibniz commented upon Cordemoy's atomist solution to the Cartesian problem of individuation; although he appreciated Cordemoy's criticism and elaboration of Cartesianism, he thought that Cordemoy had not gone far enough with his solution. As Leibniz said:

> These are difficulties for Cordemoy himself: let us suppose two triangular atoms come into contact and compose a perfect square, and that they rest next to each other in this way, and let there be another corporeal substance or atom, a square one equal to the other two. I ask, in what respect do these two extended things differ? Certainly no difference can be conceived in them as they are now, unless we suppose something in bodies besides extension; rather they are distinguished solely by memory of their former condition and there is nothing of this kind in bodies.[73]

Leibniz proposed instead a return to a kind of hylemorphism and to an individuating form or *haecceity* as a principle of individuation.

The case of Cordemoy provides an example of a Cartesian adopting a position that was censored and severely criticized numerous times—of a Cartesian going against the stream, as it were. Cordemoy did write his treatise before the king's edict in 1671, but there were plenty of indications before then that atomism would be considered incompatible with the Catholic faith.

---

[72] Cordemoy 1666, 11–12.

[73] Leibniz 1923–, VI.4, 1799: trans. in Leibniz 2001, 279: "Sunt quae ipsum Cordemoium premant, ponamus duas atomos triangulares se tangere et componere quadratum perfectum, et ita juxta se quiescere, detur alia substantia corporea seu Atomus quadrata aequalis composito ex his duabus, quaero in quo differant haec duo extensa; certe nulla in ipsis ut nunc sunt concipi potest diversitas nisi ponamus aliquid in corporibus praeter extensionem, sed sola memoria pristinorum discernuntur qualis in corporibus nulla est."

2.3 Descartes and the Cartesians: Hypotheses and Doubt

Another of the pragmatic reasons for dispensing with Cartesianism concerned Cartesian doubt, which was very often the target of criticism. Most of the *Seventh Set of Objections* by the Jesuit Pierre Bourdin was directed against it. According to Bourdin: the method is faulty in its principles; in the implements it uses; because it is deficient; by failing to reach its goal; by being excessive; through negligence; willfully, etc.[74] The Oratorian censors objected to Cartesian doubt,[75] contending that: "To say that we must doubt all things is a principle that tends toward atheism and upsets the foundations of the highest of mysteries.... This principle manifestly entails atheism or at least the heresy of the Manicheans, who accepted a good and an evil principle for all creatures."[76] In 1691, the University of Paris formally condemned 11 propositions of Descartes, including the following three:

1. One must rid oneself of all kinds of prejudices and doubt everything before being certain of any knowledge.
2. One must doubt whether there is a God until one has a clear and distinct knowledge of it.
3. We do not know whether God did not create us such that we are always deceived in the very things that appear the clearest.[77]

In 1706, the Jesuits officially censured five Cartesian/Malebranchian propositions concerning doubt:

1. The human mind can and must doubt everything except that it thinks and consequently that it exists.
2. Of the remainder, one can have certain and reasoned knowledge only after having known clearly and distinctly that God exists, that he is supremely good, infallible, and incapable of inducing our minds into error.
3. Before having knowledge of the existence of God, each person could and should always remain in doubt about whether the nature with which one has been created is not such that it is mistaken about the judgments that appear most certain and evident to it.
4. Our minds, to the extent that they are finite, cannot know anything certain about the infinite; consequently, we should never make it the object of our discussions.
5. Beyond divine faith, no one can be certain that bodies exist—not even one's own body.[78]

---

[74] AT VII 527–536. See Ariew and Grene 1995.

[75] Babin 1679, 35 and 35n.

[76] Babin 1679, 36, 40–41: "Dire qu'il faut douter de toutes les chose, c'est un principe qui tend à l'atheisme et renverse les fondemens des plus hauts mysteres.... Ce principe insinuë manifestement l'Atheisme ou du moins l'heresie des Manichéens qui admettoient un bon et un mauvais Principe de toutes les creatures."

[77] Ariew 1994, 5; trans in Ariew et al. 1998, 258: "I. Il faut se défaire de toutes sortes de préjugés, & douter de tout avant que de s'assurer d'aucune connaissance. II. Il faut douter s'il y a un Dieu, jusqu'à ce qu'on en ait une claire connaissance. III. Nous ignorons si Dieu ne nous a pas voulu créer de telle sorte que nous soyons toujours trompe dans les choses mêmes qui paraissent les plus claires."

[78] Ariew 1994, 5; trans in Ariew et al. 1998, 258: "1. Mens humana de omnibus dubitare potest ac debet, praeterquam quod cogitet. 2. Reliqua non prius nobis certa et explorata esse possunt, quam clare innotuerit Deum existere, summeque bonum esse, non fallacem qui mentem nostram inducere in errorem velit. 3. Ante certam notitiam Divinae existentiae dubitare semper quisque posset ac

Arnauld had already indicated in the *Fourth Set of Objections* that "he fears that some people may take offense at [Descartes'] rather wide-open style of philosophizing in which everything is called into doubt." He thought this style "dangerous to people of ordinary intelligence" and recommended that the *Meditations* "be bolstered with a brief preface in which the author indicates that these things are not being seriously doubted at all."[79] The view can be captured nicely by the Jesuit René Rapin who asserted: "In truth, Descartes teaches one to doubt too much, and that is not a good model for minds who are naturally credulous."[80] Though not in exactly the same way, all of the critics reject doubt as a path to certainty. Duhamel's rejection of the method of doubt is exemplary:

> The Cartesians pretend to distinguish themselves from the Pyrrhonists in that they do not want to doubt for the sake of doubting, but to be certain, after a sufficient examination, of things about which they have doubts; instead the Pyrrhonists doubt for the sake of doubting, without ever being certain of anything.
>
> But it is clear that, once one doubts everything seriously and effectively, it is impossible to be certain of anything, whatever examination one might conduct, because, if one could be certain of something after such a serious doubt, it would be only by the evidence of the thing, since there is no other rule of human certainty other than the evidence of the thing, according to the Cartesians; now we suppose that they seriously doubt the most evident things, even their own thought and their own existence, and that consequently, it is clear that, after such a general and serious doubt, it would be impossible to be certain of anything, whatever examination one might conduct.
>
> That is why the Cartesians are to be distinguished from the Pyrrhonists in that they do not reason soundly when they say that after a general doubt one can be certain of something, whereas the Pyrrhonists reason soundly and in conformity with their principles when they say that we cannot be certain of anything after having doubted everything.[81]

---

deberet, an non talis naturae conditus fuerit ut in omni judicio suo fallatur, etiam in iis quae certissima et evidentissima ipsi apparent. 4. Mens nostra eo quod finite sit, nihil certi scire potest de infinito, poindeque a nobis disputari de illo nunquam debet. 5. Non nisi per fidem divinam certo cognoscere quisquam potest quod aliqua extent corpora, ne suum quidem."

[79] AT VII 214–215: "Vereor ne quosdam offendat liberior hæc Philosophandi ratio, qua omnia revocantur in dubium….mediocribus ingeniis hanc viam esse periculosam….haud scio an aliqua præfatiuncula hæc Meditatio præmuniri debeat, qua significetur de iis rebus serio non dubitari."

[80] Rapin 1725, 366: "A la vérité, il [Descartes] enseigne trop à douter: et ce n'est pas un bon modèle à des esprits naturellement credules: mais enfin il est plus original que les autres."

[81] Duhamel 1692, Chap. 4; see also Chap. 1–3: "Les Cartésiens pretendent se distinguer des Pyrroniens en ce qu'ils ne veulent pas *douter pour douter, mais après un examen suffisant s'assurer des choses dont ils ont douté*, au lieu que les Pyronniens doutent pour douter, sans jamais s'assûrer de rien. Mais il est clair que si l'on doute une fois serieusement et effectivement de tout, il est impossible de s'assûrer d'aucunes choses, quelque examen qu'on en fasse, parce que si aprés un doute serieux on pouvoit s'assûrer de quelque chose, ce ne seroit que par l'évidence de la chose; puisqu'il n'y a point d'autre regle dela certitude humaine, que l'evidence de la chose, selon les Cartésiens; or on suppose qu'ils doutent serieusement des choses les plus évidentes, même de leur propre pensées et de leur propre existence, et par consequent il est clair qu'aprés un doute general et serieux il seroit impossible de s'assûrer d'aucune chose, quelque examen qu'on en fist. C'est pourquoy les Cartésiens sont distinguez des Pyrroniens en ce qu'ils ne raisonnent pas consequemment, lors qu'ils disent qu'aprés un doute general on peut s'assûrer de quelque chose, au lieu que les Pyrroniens raisonnent consequement et conformément à leurs principes, lors qu'ils disent qu'on ne peut s'assûrer de rien aprés avoir douté de tout."

Thus, from 1641 to 1706 various critics of Descartes mounted an attack on the method of doubt (that is, on "serious and effective" or hyperbolic doubt). It can be shown that this criticism had an effect, with Cartesians toning down or reinterpreting hyperbolic doubt. But to understand how this can result in Cartesian empiricism we must make a detour into the issue of Descartes' treatment of hypotheses and his correlate notion of moral certainty.

As Descartes fully understood, the new matter theory explained the behavior of sensible bodies by reference to imperceptible particles. So the question arose, how can we arrive at the knowledge of the shapes, sizes, and motions of these particles? The answer involves the epistemic status of hypotheses, but the role of hypotheses in Descartes' philosophy was not clear, or it seemed to have undergone some change, and the Cartesians did not seem to have accepted Descartes' view fully. Descartes' usual view was that his hypotheses could be grounded in non-hypothetical, self-evident first principles about general things, that he had or could provide such a derivation.[82] By Part IV of the *Principles* he knew that such a demonstration would be futile. Descartes' opinion in the *Principles* is that his hypothetical principles about particular things are not absolutely, but merely morally certain, meaning that there is at least some logical connection and coherence in them, such that his physics would have to be rejected and taken only as a fiction, or else it all has to be accepted, and not be rejected until another is found more capable of explaining all the phenomena of nature.

In the *Principles*, Descartes describes the method he has used with respect to his hypotheses or suppositions. He has first considered in general all the clear and distinct notions the understanding can contain with regard to material things—those of shapes, sizes, and motions—and the rules in accordance with which these three things can be modified by each other—that is, the principles of geometry and mechanics. So he has concluded that all the knowledge people have of the natural world must be derived from these notions. Next he has deduced the principal differences between the bodies that are imperceptible by the senses merely because of their small size and the observable effects that would result from their various interactions. Then, when he has observed just such effects as perceived by the senses, he has concluded that they in fact arose from such an interaction of bodies that cannot be perceived—"especially since it seemed impossible to think up any other explanation for them."[83] His legitimation for this seemingly abductive procedure is an analogy: "those who are experienced in dealing with machinery," like a clock, "can take a particular machine whose function they know and, by looking at some of its parts, easily form a conjecture about the design of the other parts, which they cannot see."[84] Descartes then extends his analogy about such machines as clocks to make clear the limitations of the explanations of phenomena referring to corpuscles

[82] For a different reading of Descartes and hypotheses, see Chap. 11 by Hatfield.

[83] *Principles* IV 203. AT VIIIa 326: "Præsertim cum nullus alius ipsas explicandi modus excogitari posse videbatur."

[84] *Principles* IV 203. AT VIIIa 326: "Qui in considerandis automatis sunt exercitati, cum alicujus machinæ usum sciunt et nonnullas ejus partes aspiciunt, facile ex istis, quo modo aliæ quas non vident sint factæ, conjiciunt."

our senses do not perceive. Two clocks identical on the outside may indicate the time equally well but use different operating mechanisms. So also God could have produced the phenomena we perceive in innumerably different ways. As a result, the causes postulated by Descartes to explain some effects may correspond to the phenomena manifested by nature, but may not be the ones by which God produced those effects: "With regard to the things that cannot be perceived by the senses, it is enough to explain their possible nature, even though their actual nature may be different."[85] These explanations, according to Descartes, are only *morally* certain, that is, they suffice for the conduct of life, although, given the absolute power of God, they *can* be doubted. In the French edition of the *Principles*, Descartes adds: "Thus those who have never been in Rome have no doubt that it is a town in Italy, even though it could be the case that everyone who has told them this has been deceiving them."[86] In this way Descartes distinguishes between two kinds of certainty, one he calls moral, and another he calls absolute. The situation is different with absolute certainty, which, according to Descartes, we possess for mathematical demonstrations, the knowledge that material things exist, and the evidence of all clear reasoning that is carried on about them: "Absolute certainty arises when we believe that it is wholly impossible that something should be otherwise than we judge it to be."[87] So absolute certainty accrues to metaphysical principles that have passed the test of hyperbolic doubt and to the general physical principles that can be derived from them. Moral certainty accrues to the physical principles about particular things that cannot be perceived. We do not have real doubts about these principles, but they fail the test of hyperbolic doubt, because we understand that God could have brought about things in some other way.

Descartes uses another example to illustrate moral certainty. He refers to a code-breaker who has decoded a message and who is certain of his solution, but who understands that another solution might be possible. He states: "even though his knowledge may be based merely on a conjecture, and that it may be the case that the person who wrote the message…encoded quite a different meaning in the message; but it would be so difficult for this to happen, that it does not seem credible."[88] Descartes adds in the French edition: "especially if the message contains many words."[89] He concludes, cashing in his analogy:

[85] *Principles* IV 204. AT VIIIa 327: "Sufficere si de insensibilibus qualia esse possint, explicuerim, etsi forte non talia sint."

[86] *Principles* IV 205. AT IXb 323: "Ainsi ceux qui n'ont jamais été à Rome ne doutent point que ce ne soit une ville en Italie, bien qu'il se pourrait faire que tous ceux desquels ils l'ont appris les aient trompés."

[87] *Principles* IV 206. AT IXb 324: "L'autre sorte de certitude est lorsque nous pensons qu'il n'est aucunement possible que la chose soit autre que nous la jugeons."

[88] *Principles* IV 205. AT VIIIa 328: "Etsi hoc sola conjectura cognoscat, et fieri forsan possit…alium in ea sensum occulta verit: hoc enim tam difficulter potest contingere, ut non credibile videatur."

[89] *Principles* IV 205. AT IXb 323: "Principalement lorsque le chiffre contient beaucoup de mots."

> Now if people look at all the many properties relating to magnetism, fire, and the fabric of the entire world, which I have deduced in this book from just a few principles, then, even if they think that my assumption of these principles was arbitrary and groundless, they will still perhaps acknowledge that it would hardly have been possible for so many items to fit into a coherent pattern if the original principles had been false.[90]

It is tempting to think that moral certainty for Descartes is merely high probability, because of the examples of the code-breaker who decodes a message and the person who is told about Rome. Are we not more secure in our decoding, given that we have broken a larger code than a smaller one? Is it not relevant that we are told about Rome from many sources as opposed to a few? Still, moral certainty and high probability are usually distinguished.[91] And despite Descartes' examples, his moral certainty does not admit of any degree. Moral certainty suffices for the conduct of life, but not in the sense that it is a good rule of thumb or something highly probable; it is genuine certainty within its own sphere. If something is morally certain we lack any reason to doubt it, though we could doubt it if we considered God's absolute power. Descartes' two kinds of certainty are thus dependent on our being able to construct hyperbolic reasons for doubt: Absolute certainty, the certainty attaching to his metaphysical principles and the principles about general things he deduces from them, passes that criterion, whereas moral certainty, the certainty attaching to physical principles about particular things, fails it.

Once one understands Descartes' peculiar notion of moral certainty and the role it plays in his system, it is easy to see what can become of it in the hands of followers who might discard some aspects of the method of doubt. The rejection of hyperbolic doubt caused some Cartesians no longer to distinguish between the absolutely and the morally certain in the fashion of Descartes—that is, between that which we cannot doubt and that about which we have no doubt although we could doubt it—and thus to treat principles on a par with one another. As a result, many Cartesians became more empirical and pursued a limited hypothetical-deductive method.[92]

We can see the method of doubt being toned down in Pierre-Sylvain Régis' reply to Pierre Daniel Huet's critique of Cartesian philosophy. Huet rejects the method of doubt because he is a skeptic and would rather just continue in doubt:

> Both he and they [Descartes and the skeptics] saw that we must doubt; but he stopped doubting when it was most necessary to doubt, namely at a principle which is not any less uncertain than all the other things that led him to doubt. They continue to doubt this

[90] *Principles* IV 205. AT VIIIa 328: "Sed qui advertent quam multa de magnete, de igne, de totius Mundi fabrica, ex paucis quibusdam principiis hic deducta sint, quamvis ista principia tantum casu et sine ratione a me assumpta esse putarent, forte tamen agnoscent, vix potuisse contingere, ut tam multa simul cohærerent, si falsa essent."

[91] See Ariew 2010.

[92] The empirical hypothetical-deductive nature of Cartesian science is well established. See Mouy 1934, esp. 147, 165–166, concerning Pierre-Sylvain Régis. See also Clarke 1989; Ariew 2006. For Jacques Rohault's empiricism, see Chap. 9 by Dobre.

principle and believe that they have many reasons to doubt it. Descartes could not have reproached them if he knew their reason, which is that nothing appears clear enough to them to be admitted as true.[93]

Régis in his reply asserts that Descartes has not abandoned his promise to doubt everything when he accepts something as true after having examined it.[94] He claims that Descartes never accepted the general rule to hold everything as false, but merely resolved to consider as false whatever appears doubtful. He distinguishes between real doubt, arising from the nature of things, and a feigned, methodological doubt—what Descartes called hypothetical, hyperbolic and metaphysical doubt—arising from his resolution to doubt.[95] In keeping with this interpretation of Descartes, he asserts that Descartes only held the rules of logic as false "hypothetically" in order to examine them. He asks rhetorically: "who can prevent Descartes from holding them as true, if they have appeared to him as such, after he has examined them?"[96]

One of the first Cartesians was Jacques Du Roure, who belonged to the group centering about Descartes' literary executor Claude Clerselier.[97] Du Roure is the first to have published a complete textbook of Cartesian philosophy, *La Philosophie*

---

[93] Huet 1689, Chap. I, Art. 14: "Hi enim et ille [Cartesius et Scepticorum] viderunt esse dubitandum; at dubitare ille tum desiit, cum erat maxime dubitandum; in hoc videlicet principio, quod minus incertum est ac reliqua omnia quibus adductus erat ad dubitandum;…hi dubitare pergunt in eodem illo principio, de quo vel maxime dubitandum esse vident: hautquaquam certe dubitantes ut dubitent; quod ipsis minime insimulasset Cartesius, si rationes eorum diligentius perspexisset; sed ideo dubitantes, quod nihil ipsis satis liquido, satisve certo percipi posse videatur." This is basically in agreement with Duhamel' position (as above).

[94] Régis 1691, I, Art. 5.

[95] Régis 1691, I, Art. 1. Although Régis uses the word *hyperbolic* to describe Descartes' methodological doubt, it is clear that his notion of doubt is radically different from that of Descartes.

[96] Régis 1691, I, Art. 6: "Or qui le peut empêcher, quand il les a examinées, de les tenir pour vrayes, si elles lui ont paru telles?"

[97] Among Du Roure's bona fides for being a Cartesian is the fact that he knows the Cartesian texts very well and that he is acquainted with a number of Descartes' letters before their publication in Clerselier's edition of Descartes correspondence (see Ariew 2012). Du Roure also met the Cartesian Johann Clauberg when the latter was in Paris and ultimately published a work in Clauberg's compilation of Cartesian papers in Dutch translation: *Cartesiaanse reden-konst: met het onderscheid tusschen de Cartesiaanse en schoolse philosophie* (Amsterdam, 1683). The verso of Du Roure's title page from his *Philosophie* tells the story very well; while he appreciates Gassendi and Hobbes and quotes them at times (Hobbes, in particular is crucial to his "La Morale Demontrée"), his admiration for Descartes knows no bounds: "On peut opposer Hobbes, Gassendi et Descartes à tous ceux dont l'Italie et la Grece se glorifient.…Ceux qui voudront se donner la peine de lire cette Philosophie, y trouveront plusieurs sentiments de ces trois sçavants Philosophes; mais principalement de Descartes. C'est pourquoy je veux faire voir par les témoignages suivants, combien il est estimé. (One can oppose Hobbes, Gassendi, and Descartes against all those whom are glorified by Rome and Greece.… Those who would take the trouble to read this philosophy will find numerous opinions of these three wise philosophers, but principally those of Descartes. This is why I want to show the extent he is esteemed by the following testimony)." The six subsequent paragraphs are superlative praise for Descartes, including: "Descartes est le premier Philosophe de tous les temps (Descartes is the premier philosopher of all time)."

divisée en toutes ses parties (1654),[98] and subsequently the somewhat less-Cartesian *Abrégé de la vraye philosophie* (1665). In Du Roure's case, the parts of philosophy included natural theology and the usual elements of the curriculum: metaphysics, logic, ethics, and physics. Du Roure begins his Physics with a discussion of moral certainty, clearly weakening the notion to mere probability and introducing physical certainty as a third element. There is a full exposition of his grades of certainty in his *Abrégé*:

> There are three kinds of evidence. One is Moral, when the contrary does not happen ordinarily. The other is Physical, when the contrary never happens. The last is Metaphysical, or absolute, when the contrary can never happen. Of these three Propositions: (1) This man will die before he reaches one hundred years old, (2) he will die, (3) he can die, the first is Morally evident and certain, the second Physically, and the last Absolutely.[99]

This probabilistic and empirical epistemology is affirmed by Du Roure even in his *Logic*; the latter starts with an examination of method, by which Du Roure means primarily analysis and synthesis; he continues by discussing experience, including the following statements he takes to be true: "All our knowledge comes from experience [that is, the senses]….And whoever makes use of reason more than experience or reflections on experiences often falls into error."[100] So with Du Roure we have somebody who falls into the Cartesian camp, though he defends a view that might be thought at variance with orthodox Cartesianism, displaying an epistemology that looks more like Gassendist empiricism.

Of course, not all Cartesians followed the same path in their espousal of probabilism and a hypothetico-deductive method in physics. Here is a typical paragraph supporting a hypothetico-deductive method ending up with high probability, not absolute or moral certainty (though it sounds very much like a considerably weakened moral certainty in the fashion of Du Roure). It is from the Preface to the second edition of Christiaan Huygens' *Traité de la Lumière* (1690):

> One finds in this work these kinds of demonstrations that do not produce as great a certainty as those of Geometry, and that even differ much from geometrical demonstrations, given that geometers prove their propositions by certain and incontestable principles, while here principles are verified by conclusions derivable from them; the nature of these things does not allow any other treatment. It is always possible, however, to attain in this way a degree of probability, which very often is little short of complete evidence. This is the case when

[98] The full title of the work is: *La Philosophie divisée en toutes ses parties, établie sur des principes évidents et expliquée en tables et par discours, ou particuliers, ou tirés des anciens et des nouveaux auteurs, et principalement des péripatéticiens et de Descartes*.

[99] Du Roure 1665, Discours General, Les Siences, No. 7–8: "7. Il y a trois sortes d'Evidence. L'une Morale, quand le contraire n'arrive pas ordinairement. L'autre Physique, quand le contraire n'arrive iamais. La derniere Metaphysique, ou Absoluë, quand le contraire ne peut arriver. 8. De ces trois Propositions: Cet Homme mourra devant cent ans, il mourra, il peut mourir : La premiére ét (sic) certaine et évidente Moralement, la deuxiéme Physiquement, la derniére Absolument." Even Du Roure's example of absolute certainty, "this man can die," seems less than absolute.

[100] Du Roure 1665, Logique, Sect. 20: "Toutes nos connoissances viennent de l'experience.… Et quiconque fait plus de raisonnemens que d'experiences ou de reflexions sur elles, tombe souvent dans l'erreur."

> things demonstrated by these assumed principles correspond perfectly to the phenomena that experiment has brought under observation—especially when there are a great number of them, and further, principally, when one can devise and predict new phenomena that should follow from the hypotheses one uses, and one finds that the effect corresponds to our expectations. But if all these proofs of probability are encountered in what I propose to treat, as it seems to me they are, this should be a very strong confirmation of the success of my inquiry, and it is scarcely possible that the facts are not just about as I represent them.[101]

Huygens, who is not an orthodox Cartesian nor a Catholic, has his own reasons for adopting a hypothetical-deductive method leading to high probability; these do not have to be the same as what would motivate Cartesians in a Catholic country. But he is here following a path taken by many followers of Descartes, such as Du Roure, and Régis, among others.

References

Ariew, Roger. 1994. Quelques condamnations du cartésianisme: 1662–1706. *Bulletin cartésien XXII, Archives de Philosophie* 57: 1–6.

Ariew, Roger. 2006. Cartesian Empiricism. *Revue roumaine de philosophie* 50: 71–85.

Ariew, Roger. 2010. The new matter theory and its epistemology: Descartes (and Late Scholastics) on hypotheses and moral certainty. In *Vanishing matter and the laws of nature: Descartes and Beyond*, ed. D. Jalobeanu and P. Anstey, 31–46. London: Routledge.

Ariew, Roger. 2012. Descartes' correspondence before Clerselier: Du Roure's *La Philosophie*. *Journal of Early Modern Studies* 1(1): 43–64.

Ariew, Roger, and Marjorie Grene (eds.). 1995. *Descartes and his contemporaries*. Chicago: University of Chicago Press.

Ariew, Roger, John Cottingham, and Tom Sorell, (eds. and trans.). 1998. *Cambridge texts in context: Descartes' meditations*. Cambridge: Cambridge University Press.

Armogathe, Jean-Robert. 1977. *Theologia cartesiana: l'explication physique de l'Eucharistie chez Descartes et dom Desgabets*. The Hague: M. Nijhoff.

Armogathe, Jean-Robert, and Vincent Carraud. 2003. The first condemnation of Descartes's *Oeuvres*: Some unpublished documents from the Vatican archives. *Oxford Studies in Early Modern Philosophy* 1: 67–110.

[101] Huygens 1690, Preface (unpaginated) 2–3: "On y vera de ces sortes de demonstrations, qui ne produisent pas une certitude aussi grande que celles de Geometrie, et qui mesme en different beaucoup, puisque au lieu que les Geometres prouvent leurs Propositions par des Principes certains et incontestables, icy les Principes se verifient par les conclusions qu'on en tire; la nature de ces choses ne souffrant pas que cela se fasse autrement. Il est possible toutefois d'y arriver à un degré de vraisemblance, qui bien souvent ne cede guere à une evidence entiere. Sçavoir lors que les choses, qu'on a demontrées par ces Principes supposez, se raportent parfaitement aux phenomenes que l'experience a fait remarquer; sur tout quand il y a un grand nombre, et encore principalement quand on se forme et prevoit des phenomenes nouveaux, qui doivent suivre des hypotheses qu'on employe, et qu'on trouve qu'en cela l'effect repond à notre attente. Que si toutes ces preuves de la vraisemblance se rencontrent dans ce que je me suis proposé de traiter, comme il me semble qu'elles sont, ce doit estre une bien grande confirmation du succês de ma recherche, et il se peut malaisement que les choses ne soient à peu pres comme je les represente."

Babin, François. 1679. *Journal ou relation fidele de tout ce qui s'est passé dans l'université d'Angers au sujet de la philosophie de Des Carthes en l'execution des ordres du Roy pendant les années 1675, 1676, 1677, et 1678*. Angers.

Bayle, Pierre. 1684. *Recueil de quelques pièces curieuses concernant la philosophie de Monsieur Descartes*. Amsterdam: Chez H. Desbordes.

Boileau. 1747. *Oeuvres de Boileau*, ed. Saint-Marc. Paris.

Bouillier, Francisque. 1868. *Histoire de la Philosophie cartésienne*, 3rd ed. Paris: C. Delgrave.

Clarke, Desmond M. 1989. *Occult powers and hypotheses: Cartesian natural philosophy under Louis XIV*. Oxford: Clarendon.

Cordemoy, Géraud de. 1666. *Le Discernement du Corps et de l'Ame en Six Discours. Pour servir à l'éclaircissement de la Physique*. Paris: F. Lambert.

Cousin, Victor. 1866. *Fragments Philosophiques pour servir à l'histoire de la philosophie*. Paris: Didier.

Daniel, Gabriel. 1693. *Nouvelles difficultés proposées par un péripatéticien à l'auteur du "Voyage du monde de Descartes"*. Paris: Vve de S. Benard.

Daniel, Gabriel. 1690. *Voyage du monde de Descartes*. Paris: Vve de S. Bernard.

d'Argentré, Charles Duplessis. 1728–1736. *Collectio judiciorum de novis erroribus*. Paris: Cailleau.

de la Grange, Jean-Baptiste. 1682. *Les principes de la philosophie contre les nouveaux philosophes, Descartes, Rohault, Regius, Gassendi, le P. Maignan, etc*. Paris: J. Couterot.

de L'A, M. G. [Pierre Daniel Huet]. 1692. Nouveaux mémoires pour servir à l'histoire du cartésianisme [n.p.].

de la Ville, Louis [Louis le Valois]. 1680. *Sentimens de Monsieur Descartes touchant l'essence et es proprietez du corps opposez à la Doctrine de l'Eglise, et conforme aux erreurs de Calvin sur le sujet de l'Eucharistie*. Paris: E. Michalet.

de Launoy, Jean. 1653. *De varia Aristotelis in Academia parisiensi fortuna*. Paris: E. Martini.

de Rochemonteix, Camille. 1889. *Un collège des Jesuites au XIIe et XIIIe siècle: le collège Henri IV de la Flèche*. Le Mans: Leguicheux.

Descartes, René. 1996. *Oeuvres de Descartes*, 2nd ed, ed. C. Adam and P. Tannery. Paris: Vrin

Du Roure, Jacques. 1654. *La Philosophie divisée en toutes ses parties*. Paris: Chez Francois Clouzier.

Du Roure, Jacques. 1665. *Abrégé de la vraye philosophie*. Paris: chez l'auteur.

Duhamel, Jean. 1692. *Reflexions critiques sur le système cartesien de la philosophie de mr. Régis*. Paris: Edme Couterot.

Duhamel, Jean. 1705. *Philosophia universalis, sive commentarius in universam Aristotelis philosophiam, ad usum scholarum comparatam*. Paris: apun viduam C. Thiboust et P. Esclassan.

Frassen, Claude. 1668. *Philosophia Academica, quam ex selectissimis Aristotelis et Doctoris Subtilis Scoti rationibus....* Paris: Apud E. Couterot.

Garber, Daniel. 2002. Defending Aristotle/Defending Society in early seventeenth century Paris. In *Wissensideale und Wissenskulturen in der frühen Neuzeit*, ed. C. Zittel and W. Detel, 135–160. Berlin: Akademie-Verlag.

Gassendi, Pierre. 1624. *Exercitationes Paradoxicae Adversus Aristoteleos*. Gratianopoli: Verdier.

Girbal, François. 1964. *Bernard Lamy (1640–1715)*. Paris: PUF.

Goudin, Antoine. 1726 [1668]. *Philosophia juxta inconcussa tutissimaque Divi Thomae dogmata*. Coloniae Agrippinae: T. von Collen and J. Huisch.

Huet, Pierre Daniel. 1689. *Censura philosophiae cartesianae*, 2nd ed. Paris: Apud Danielem Horthemels. 1695.

Huygens, Christiaan. 1690. *Traité de la Lumière*. Leiden: van der Aa.

Leibniz, G.W. 1923. *Sämtliche Schriften und Briefe*. Berlin: Akademie-Verlag.

Leibniz, G. W. 2001. *The Labyrinth of the continuum: Writings on the continuum problem, 1672–1686*, ed. and trans. Richard Arthur. New Haven: Yale University Press.

Mersenne, Marin. 1624. *L'impiété des Déistes, Athées, et Libertins de ce temps, combattue, et renversée de point en point par raisons tirée de la Philosophie, et de la Théologie*. Paris: P. Bilaine.

Mouy, Paul. 1934. *Le développement de la physique cartésienne, 1646–1712*. Paris: Vrin.
Oldenburg, Henry. 1965–1986. *The correspondence of Henry Oldenburg*, 13 vols, ed. A. Rupert and Marie Boas Hall. Madison/London: University of Wisconsin Press.
Prou, Ludovicus. 1665. *De Hypothesi Cartesiana positiones physico mathematica*. Clermont.
Rapin, René. 1725. *Reflexions sur la philosophie, Oeuvres*. La Haye: P. Gosse.
Régis, Pierre-Sylvain. 1691. *Réponse au livre qui a pour titre P. Danielis Huetii...Censura Philosophiae Cartesianae*. Paris: Chez J. Cusson.
Roux, Sophie. 1998. Le scepticisme et les hypothèses de la physique. *Revue de synthèse* 4: 211–255.
Roux, Sophie. 2000. Descartes atomiste? In *Atomismo e continuo nel XVII secolo*, ed. R. Gatto and E. Festa, 211–274. Naples: Vivarium.
Roux, Sophie. 2012. An empire divided: French natural philosophy (1670–1690). In *The mechanization of natural philosophy*, ed. D. Garber and S. Roux, 55–98. Dordrecht: Springer.
Schmaltz, Tad. 2002. *Radical Cartesianism*. Cambridge: Cambridge University Press.
Verbeek, Theo. 1992. *Descartes and the Dutch. Early reactions to Cartesian philosophy 1637–1650*. Carbondale: University of Southern Illinois Press.
Verbeek, Theo (ed. and trans.). 1988. *La Querelle d'Utrecht*. Paris: Les impressions nouvelles.
Vincent, Jean. 1677. *Discussio peripatetica in qua philosophiae cartesianae principia*. Toulouse: Ex officina Colomeriana.

Chapter 3
Was There a Cartesian Experimentalism in 1660s France?

Sophie Roux

> *A true philosophy topic for an essay in the baccalaureate... must deal with a problem and not a doctrine*
> *Empiricism is not* the *theory of experience, but* one *possible theory of experience....*
>
> Georges Canguilhem

Abstract In order to determine if there existed an experimentalist Cartesianism in France in the 1660s, I concentrate on Jacques Rohault and address the three following questions. (1) Is there a difference in the way Descartes and Rohault deal with experiments? I state that there is no doctrinal difference between them: the experiments they carry out are of the same order; they attribute the same epistemological functions to them; they share the same ontology. The main difference between them is that, unlike Descartes, Rohault made experiments a means of popularization of the Cartesian philosophy. (2) How does Rohault treat experiments in his *Mercredis*? Studying quite closely the evolution that led to the greater priority attributed to experiments in the scientific circles that prefigure the *Académie des sciences*, I show that, in 1660s France, the treatment Rohault give to experiments in his *Mercredis* is exceeded by the radical experimentalism of the other French learned societies. (3) Did this radical experimentalism bring out a transformation of Cartesianism? I establish that, while the first criticism to Descartes concerns his dogmatic pretentions, there emerges in the last 30 years of the seventeenth century what has since become a historiographic cliché, the idea that Cartesians neglected experiments in favor of hypotheses and speculation.

S. Roux (✉)
École Normale Supérieure, Paris, France
e-mail: sophie.roux@ens.fr

M. Dobre and T. Nyden (eds.), *Cartesian Empiricisms*, Studies in History and Philosophy of Science 31, DOI 10.1007/978-94-007-7690-6_3,
© Springer Science+Business Media Dordrecht 2013

3.1 Introduction[1]

The title of this volume and the title of this chapter seem to constitute a philosophical paradox and a historiographic provocation. If Descartes was a rationalist, and all those called "Cartesians" were faithful to him, then the existence of Empiricist Cartesians would be a philosophical paradox. If the most remarkable characteristic of modern science has been its commitment to experimentation, or more precisely, the commitment to experimentation that was proper to the Royal Society, then it would be historiographic provocation to speak of an experimentalist Cartesianism.[2] But before going any further, we should indicate what we mean by "experimentalism" and "Cartesianism": a discussion that failed to set out the scope of these categories would wade through the mire of uncertainty.

The subject of this chapter is not empiricism as such, but what I would call experimentalism. Both empiricism and experimentalism are philosophical categories that apply to doctrines rather than to practices. It's not enough to have experiences or to carry out experiments to be an empiricist or an experimentalist: there must be an explicit doctrine derived from them. But these two categories must in turn be distinguished from each other: empiricism relates to doctrines on the origin of our knowledge, while experimentalism describes doctrines derived from the constitution of the natural sciences. Doctrines are called empiricist when they defend the thesis that holds that all our knowledge absolutely comes from experience, and they are most often opposed to doctrines described as rationalist. Nonetheless, the category of empiricism includes a wide range of various doctrines, in particular depending on what one means by "knowledge" (an idea, a proposition, know-how, etc.) and by "experience" (a dream, a quotidian observation, a laboratory measurement, etc.). Experimentalist doctrines are those that support the thesis that experiments, or a certain kind of experiment, have a certain function in the development of the natural sciences. Here, too, this is a vast category, with associated doctrines varying depending on the experiences taken into account and the function attributed to them. It is common, for example, to oppose immediate sensory observations, experiences mediated by instruments, and data produced by sophisticated devices such as computers. It is also common to distinguish the function held by experiments that allow for the development of hypotheses and the function held by experiments that test hypotheses.

Empiricism and experimentalism are distinct categories. It is possible to support a doctrine arising from one of these categories without supporting any doctrine arising from the other category. Clearly one can be an experimentalist without being an empiricist: it's enough to recognize that certain experiments in the natural sciences have a function without having to support the notion that all our knowledge absolutely arises from experience. This is the case, for example, if one defends the idea that,

---

[1] Unless otherwise indicated, translations are mine. I thank Mihnea Dobre and Tammy Nyden for their challenging questions and careful editing.

[2] See Chap. 1 by Dobre and Nyden on the history of this issue.

alongside experiments, mathematic reasoning, whether relying on innate ideas or arising from the construction of concepts in pure intuition, also has a role in the natural sciences. But it does seem that any empiricist, insomuch as she defends a theory of all knowledge, is an experimentalist: a doctrine of knowledge in general applies in the particular case of knowledge of natural sciences. Nonetheless, we can avoid this conclusion by playing—it's true, not very fairly—on the many meanings of the words "experiment" or "experience." For example, we can consider that the senses, commensurate with things, are the source of all knowledge that we have of things, and even go so far as to say that they allow us to know things perfectly, and yet not support the version of experimentalism by which the only experiences worthy of interest are those provided by observation and measuring devices. On the contrary, it is precisely because we consider that our senses can measure things that we can judge that observation and measuring devices are superfluous, or even deceptive.

Although empiricism and experimentalism are categories that can include a wide range of doctrines, they are still fairly easy to define. But what about Cartesianism? We could adopt a strong definition of Cartesianism, according to which to be a Cartesian, you must support the same set of theses as Descartes. But if so, either this set of theses coincides exactly with the set of theses found in the Cartesian corpus, in which case the only Cartesians are those like the fictional twentieth-century French writer Pierre Ménard, who rewrote *Quixote* word for word, or else this set of theses is a subset of theses one more or less explicitly favors. In these conditions it is not surprising that certain historians fall back on a weaker definition of Cartesianism by which any author influenced by Descartes is a Cartesian. Thus, almost every author from the second half of the seventeenth century becomes one way or another a Cartesian, for they all had read Descartes. The situation quickly becomes aporetic, and it must be said, irremediably, for they all did not read the same works by Descartes, nor did they understand them in the same way. We consistently fail to achieve a view from nowhere of Cartesianism, and even more to provide an essential definition. To escape this aporia, I have proposed in other articles what I call a "polemic conception of Cartesianism."[3]

To summarize, and to simplify, the historian who has an essentialist conception of Cartesianism is seeking an essential definition of this category. The historian who has a polemic conception of Cartesianism, while admitting from the outset that this definition does not really exist, attempts to tease out the polemics and controversies in which a Cartesian configuration can at some point be seen. While the essentialist historian studies works to evaluate their faithfulness to the works of Descartes and rate them according to their distance from them, the polemic historian pays particular attention to the controversies in which Cartesians and anti-Cartesians explicitly opposed each other, as well as to the internal quarrels in which Cartesians tried to define what constituted their identity, for even those who claimed to be Cartesians were not in agreement as to what this claim implied. The essentialist wants to

[3] Roux 2012, 2013.

isolate the intrinsic meanings of philosophemes, while the polemist takes into account their historical variations.[4]

A certain historical relativism goes with the polemic conception of Cartesianism: even if one limits oneself to France, 1670s Cartesianism is not the same as 1750s Cartesianism, and that of the 1840s is not that of the 1930s.[5] This historical relativism does not however lead to an absolute nominalism in which each text would embody its own conception of Cartesianism. Rather, there are particularly significant moments in the history of philosophy in which a particular conception of Cartesianism succeeded in becoming stable beyond the particular moment of its inception. As I have shown in one of my papers illustrating the polemical conception of Cartesianism, such a case emerges from the controversy between the "new philosophers" and the "old philosophers" that occurs in France in the years 1670–1690. It is essential for the polemical historian to note this type of moment.

If a category like that of Cartesianism is relative to the historical status of the perpetual battlefield that is philosophy, one can wonder if the same is not true for the category of experimentalism. In a word, it seems to me that despite the formal similarity of these "-isms," there is less historical variability in the case of the category of experimentalism than in the case of Cartesianism, but that it still exists, at least in the manner we implement in the history of philosophy. Historical variability is less strong in the case of experimentalism than in the case of Cartesianism: while experimentalism can be defined outside of history, as I did in the beginning, Cartesianism cannot be defined without arbitrary choices, as I have just recalled. Yet historical variations exist: in order for Jeanne, a historian of philosophy, to affirm that philosopher Suzanne was an experimentalist, Suzanne's doctrine must correspond to the general definition that Jeanne gave to what it means to be an experimentalist, but in addition, Jeanne must feel it relevant to characterize Suzanne as an experimentalist. But it seems to me that this relevance is relative: it is relevant to say that Suzanne is an experimentalist rather than a feminist, it is relevant to say the Suzanne is an experimentalist rather than saying that Mathilde is an experimentalist. In other terms, while one can provide an absolute definition of experimentalism, it will be more or less absolute depending on to what this experimentalism is compared.

By recapitulating what I've stated so far, I can provide details on what I will deal with in this chapter. On one hand, I have distinguished experimentalism and empiricism: in this chapter, I'll speak only of experimentalism, and not of empiricism.[6]

---

[4]For a study of a polemic among Cartesians, see Moreau 1999. For a study of a controversy between proponents and opponents of Cartesianism, see Roux 2012. In the present chapter, I'll return to a point briefly touched on in this article, that is the way in which the Cartesians appear opposed to experimentalism in this controversy, see Roux 2012, 84–87. The present chapter uses some of the ideas present here and there in Roux 1998.

[5]Azouvi 2002.

[6]For some comments on the refusal of empiricism among most Cartesians, see Clarke 1989, 43–70. There were however some exceptions, for example Dom Robert Desgabets, Henricus Regius or Pierre-Sylvain Régis. On Desgabets and Régis, see Schmaltz 2002; on Régis see Chap. 6 by Joly; on Regius, see Chap. 7 by Bellis; on Desgabets, see Chap. 8 by Easton.

On the other hand, Cartesianism, and to a certain extent experimentalism, are relative categories: I will have to pay particularly attention to the point of view I adopt to judge the experimentalism of this or that philosopher. These premises stated, the problem is to determine if, as the title of this chapter suggests, there exists an experimentalist Cartesianism in France in the 1660s. I hope to resolve this problem by discussing a particular case, that of Jacques Rohault (1618–1672), the French Cartesian the most often cited for his experimental commitment and his experimentalist doctrine.[7] If I succeed in showing that this Cartesian given as an example of an experimentalist was not truly one, or in any case, not in an exceptional fashion, then it seems likely that in 1660s France there is no experimentalist Cartesianism, or at the very least, that describing Cartesianism as experimentalist is not the most relevant description. As one can see, inasmuch as relevance is relative, my discussion will not deal with Rohault alone, but with Rohault as compared with other philosophers. More precisely, I'll deal in turn with the three following questions:

1. Is there a difference in the way, according to their texts, Descartes and Rohault treat experiments? I will state that there is no doctrinal difference between them: the experiments they carry out are qualitatively and quantitatively of the same order; they attribute the same epistemological functions to experiments; they share the same ontology. The main difference between them is that, Rohault, unlike Descartes, seems to have made experiments a means of communication (popularization or propaganda) of the Cartesian philosophy.
2. How can we consider the way in which Rohault treats experiments in his *Mercredis*? It is no longer a matter of comparing Rohault to Descartes, but rather of situating him in the field of natural philosophy of his day. My thesis is that, in 1660s France, the treatment Rohault give to experiments in his *Mercredis* is exceeded by the practices of the other learned societies. To establish this thesis, I will study quite closely the evolution that led to the greater priority attributed to experiments in the scientific circles that prefigure the *Académie des sciences*.
3. Did this break bring out a transformation of the category of Cartesianism? My response to this question is affirmative. While the criticism of Samuel Sorbière (1615–1670) and Jean Chapelain (1595–1674) with respect to Descartes was based on his dogmatic pretentions, there emerges in the last 30 years of the seventeenth century, among all sorts of adversaries of Cartesianism, what has since become a historiographic cliché, the idea that Descartes and the Cartesians neglected experiments in favor of hypotheses and speculation.

[7] Already at the end of the seventeenth century, Rohault was the Cartesian who could be saved as an experimenter and experimentalist; in this regard see Leibniz to Nicaise published in *Journal des savants* cited below n132. See also Savérien 1783, xxviii–xxx, lv–lvi; Mouy 1934; Blay, "Introduction," in Rohault 2009, xxix; Chap. 9 by Dobre in this volume. Clarke 1989, 202–211, proposes a more nuanced, and in my opinion more exact, discussion, if only because coming from a systematic comparison of Malebranche and Rohault, he gives a relative appreciation of Rohault's experimentalism. The different articles by Trevor McClaughlin devoted to Rohault (in particular McClaughlin 1977, 1996, 2000) must be read, but aside from the fact that they repeat themselves, they do not in my opinion go into enough details of the texts.

3.2 Descartes and Rohault

Historians who want to characterize the Scientific Revolution in terms of scientific experiments must respond to a massive objection: there were plenty of experiments before this period. To overcome this objection, they must define what is special about experiments during the Scientific Revolution. They have held that these experiments were special because they allowed for testing a hypothesis in an interventionist fashion, were founded on the measurement of quantities, were proper experiments in the Baconian sciences, were singular experiments attested to by detailed reports by trustworthy witnesses, or supposed the creation of some form of scientific community.[8]

Whatever type of experiments that one may claim to be proper to the Scientific Revolution, we can find them in the Cartesian corpus. In his first writings, Descartes attacks "philosophers who neglect experiments and believe that truth must come from their own brain, like Minerva from the head of Jupiter," and pays homage to Francis Bacon, a fairly strong one, given his reticence to recognize any interest there might be in reading an author other than himself.[9] The reader of the *Discourse on Method* is invited to go see with his own eyes the dissection of a heart, and the experiments by which Descartes, largely inspired by Harvey, intends to show the circulation of blood are manifestly implementing a hypothetico-deductive scheme.[10] The explanation of the origin of the rainbow found in the *Meteors* supposes the use of a device that is, if not complex, at least carefully designed (a prism, a glass flask) and that requires delicate measurement of angles.[11] His meticulous and detailed observation of sleet and snowflakes in Amsterdam in the evening of February 4, 1635 compares well to the texts of Robert Boyle, which have been described as perfect examples of a style proper to the Royal Society.[12] Lastly, certain passages show that Descartes was well aware of the material conditions that must be met for a scientific community to be able to work effectively. I'm thinking neither of the project for an Academy written for the Queen of Sweden, which in the end only assigns turns to speak, nor to the plan to establish within the Royal College rooms, professors, and celebrations aimed specifically to the various types of trades.[13] I'm thinking instead of the care given by Descartes at times to the way in which a

[8] See for example Koyré 1953; Kuhn 1976; Shapin and Schaffer 1985; Dear 1995.

[9] *Regulae ad directionem ingenii*, Regula V, AT X 380. References to Bacon are to be found in Descartes to Mersenne, January 1630, December 23, 1630, May 10, 1632, AT I 109, 195–196, 251.

[10] *Discours de la méthode*, Cinquième Partie, AT VI 46–55. For comments, see Des Chene 2001, 19–25.

[11] *Météores*, Discours Huitième, AT VI 325–344. For comments, see Garber 2001a, 94–104; Zittel 2009, 202–206, *passim*.

[12] *Météores*, Discours Sixième, AT VI 298–308. For comments, see Zittel 2009, 219–225. For the idea that such reports are typical for the Royal Society, see Shapin and Schaffer 1985, 60–65; Dear 1995, *passim*.

[13] These two projects, published in AT XI 659–660 and 663–665, were known to Baillet 1691, II, 433–434 and 663–665.

community of observers must be set up, for example, when the goal is to compare barometric measurements in various places and times, and that to do so, he sent Mersenne a duplicate of the graduated scale he used himself.[14] To be honest, the only experiments we don't find in Descartes are those carried out by a community of scientists assigned to establish facts.[15] And it still remains to be proven that this practice actually existed other than as an ideal, and that it was indeed important in the Scientific Revolution, whatever one means by that.

If Descartes is not considered to leave room for experiments, it is not because of his experimental practice described in his most scientific writings. Rather, it is because of the doctrine of experience he supports in his more philosophical writings. This doctrine is opposed to empiricism, and is only moderately experimentalist. On one hand, with regard to empiricism, the *Meditations* create a distance from sensory experiences, not because they are strictly speaking false, but because reason must decipher them to gather objective physical meaning.[16] On the other hand, and more important with regard to experimentalism, if the question of knowing the place of experimentation in Cartesian science has always been a topic of discussion, we can nonetheless state that Descartes attributes to experience and experiments a function that is not negligible, but one that is nonetheless secondary.[17]

As indicated by texts cited countless times, experiments have an important epistemological function for Descartes: they allow us to identify the way God chose to produce a given phenomenon from among all the manners possible; and similarly to chose, among all the possible manners we have to conceive a way to explain the phenomenon, the explanation that corresponds to the actual created world.[18] But this epistemological function is secondary, in two ways. First, Descartes feels that the true work of physics is not to carry out experiments to establish facts, but rather to explain the facts according to a few general principles. When mathematical physics intervenes, as for the rainbow, at least part of this explanation can take the deductive form in the sense that we understand it today. But in all other cases, and there are many of them in Cartesian physics, the explanation relies on exhibiting a hypothetical causal chain of motions of corpuscles. This is of course the case of the rainbow itself, when the issue is explaining the nature of colors. In these conditions, it's not surprising that, in a secondary position, there are certain fundamental propositions that experience and experiments cannot refute, that is, propositions that establish the ontological paradigm according to which all phenomena can and must be explained in terms of motions of corpuscles. This is the case for the proposition that the

[14] Descartes to Mersenne, December 13, 1647, AT V 99.

[15] On this point, see Garber 2001b, who holds that this is true for all natural philosophers prior to the Royal Society.

[16] I developed this point in Roux 2011, 178–180.

[17] The bibliography is large but useless, because it is very repetitive. The discussions that are the most reliable, because they are more nuanced, although not exactly in the same way, seem to me those of Clarke 1982 and Garber 2001a. Homage must also be paid to the studies "Descartes expérimentateur" and "Descartes et Bacon" published in Milhaud 1921.

[18] *Discours de la méthode*, Discours Sixième, AT VI 64–65; *Principia philosophiae* II 204, AT VIII 327.

essence of matter is its extension, or for the three laws of motion that determine all phenomena of nature and determine them as phenomena of nature. It is only after this paradigm was established, according to Descartes once and for all, and in a totally demonstrative manner, that experiments are used to allow for a choice among several possible causal chains.

Are things any different for Rohault? This has been the position often defended, as if though, failing to totally save Descartes, it was important to at least preserve one of his disciples. In this light, while Descartes may not have been totally modern, from Descartes to Rohault a great step would be taken to a more substantial experimentalism and to a more clearly accepted modernity. Contrary to Descartes, Rohault would have been able, in his *System of Natural Philosophy*, to recognize the necessary alliance between experience and reasoning, distinguishing three forms of experience, and recognizing the importance of hypothetico-deductive reasoning in physics. Indeed, Rohault does note that there exist two symmetrical errors that are among the causes of the lack of progress in physics, the first being inattention to experiments, the second being exclusive promotion of experiments, to the detriment of reasoning:

> For they who fall into the first of these Errors, hinder themselves of the best Means of finding out new Discoveries, and of confirming their own Arguments [*raisonnements*] likewise; And they who fall into the second, by depriving themselves of the Liberty of drawing Conclusions, hinder the Knowledge of a large Train of Truths, which may many Times be deduced from one single Experiment. Wherefore it cannot but be very advantagious to mix Experiments and Arguments [*raisonnement*] together.[19]

Indeed again, Rohault notes, as would any good commentator systematizing the remarks of his master, that, alongside the sensory observations we all make without particular intent, and the specific knowledge that men of the field acquire by experimental practice, there is a third type of experiment,

> those which are made in Consequence of some *Reasoning* [*celles que le raisonnement previent*] in order to discover whether *it* was just or not. As when after having considered the ordinary Effects of any particular Subject, and formed a true Idea of the *Nature* of it, that is *of That in it which makes it capable of producing those Effects*; we come to know by our Reasoning, that if what we believe concerning the *Nature* of it be true, it must necessarily be, that, by disposing it after a certain Manner, a new Effect will be produced, which we did not before think of; and in Order to see if this Reasoning holds good, we dispose the Subject in such a manner as we believe it ought to be disposed in Order to produce such an Effect.[20]

But just saying that one needs experiments and reasoning does not really make for a substantial thesis in epistemology: as Rohault himself recalls a page later, this was also the position of Aristotle. And the application made by Rohault of the hypothetico-deductive scheme is very special. If one judges from the ordinary effects of a "subject," "the true Idea of the Nature," nature cannot be a hypothesis

---

[19] Rohault 1987, The Author Preface, I, unpaginated.

[20] Rohault 1987, The Author Preface, I, unpaginated; 1681, Préface, unpaginated. The verb "prévenir" used transitively did not have the same meaning in the seventeenth century as it does today: "prévenir," according to Furetière's *Dictionnaire*, is "to be the first to do the same thing, to win in races; *celui qui prévient* arrives the first at the goal, wins the prize."

that could possibly be refuted by a later experiment. The aim in fact is merely to reformulate rough sensory experience in terms of the Cartesian ontology. To see this, one can look, for example, at Chap. 12 of the first part of the *System of Natural Philosophy*, devoted to explaining the effects that Aristotelians would attribute to *horror vacui* by the weight of air and the presence of a subtle matter.[21]

This is certainly a chapter rich in experiments, experiments with syringes, siphons, glass tubes, an experiment with a carp bladder, the experiment known as the vacuum in the vacuum, the Puy-de-Dôme experiment, this time carried out in the clock towers of Notre-Dame.[22] It is not important that Rohault was not the first to imagine or carry out these experiments; the real problem, in fact, is in the explanations he uses.[23] For him, explaining means to expose which of the general principles of Cartesian physics are compatible with these experiments (e.g., a body can only be placed in motion by a body that touches it; all motion is in circles, that is to say a closed curve; vacuums are impossible; there is a subtle matter; the existence of pores allows for its circulation, etc.). Rohault presents these physical principles as "a Foundation which cannot be contested."[24] These uncontestable foundations are the very same principles of general physics of Descartes. It is even, so to say, their epistemological characterization that they are uncontestable: in particular, they cannot be refuted by an experiment.

Consider, for example, the analysis given of the operation of a syringe. It is obvious, writes Rohault, that because one end is open, the piston cannot be drawn back without a circular motion of air. This is obvious because of the general principle that in a full world, all motion is circular. But what if the end is closed? Either the syringe has pores, and the motion will take place, or the syringe does not have pores and there will be no motion. And since motion does take place, there must therefore be pores in the glass of the syringe.[25] The hypothetico-deductive process corresponding to the third type of experiment in the Preface is present, but under the general supposition that the world is full. Thus the epistemological framework used by Rohault is in fact as follows: given, on one hand, the general principles of Cartesian philosophy, and on the other, the fact that the piston of a syringe can be drawn, it means that there are pores in the glass. The experiment intervenes not to

---

21 See for example McClaughlin 2000, 336n52. In the following paragraph, I detail the comments presented in Roux 2011, 128–134.

22 Rohault 1987, I, Chap. XII, 56–78.

23 McClaughlin 1977, 227–228; 1996, 471–475, 480–481 identify the various sources of Rohault's experiments. We sometimes read that Rohault helped Florin Périer edit Pascal 1663, but I don't see an argument for this. Nonetheless, the two remarks that make up the Avertissement, unpaginated, of this edition show that the editor knew the work of Rohault; likewise the presentation entitled *Nouvelles expériences faites en Angleterre, expliquées par les principes establis dans les deux Traitez precedens de l'Equilibre des Liqueurs, & de la Pesanteur de la masse de l'Air*, shows that he knew the work of Boyle.

24 Rohault 1987, I, Chap. XII, Sect. 5, 57; 1681, I, 80 for the French.

25 Rohault 1987, I, Chap. XII, Sect. 6–9, 57–58. This very chapter is analysed in this volume by Mihnea Dobre as well, albeit with quite different conclusions. See Chap. 9 by Dobre.

allow for a choice between the Cartesian ontology and another. Instead, the general principles of this ontology being given, it is simply about allowing a choice between two of their possible instances. This is precisely the epistemological situation described by Descartes when he writes that experiments allow us to choose an explanation among the various explanations possible for a phenomenon.[26]

One could object that this kind of situation is common in physics: far from the basic ontological choices being questioned by experiments, these choices must be considered as given in order to formulate theories, among which the experiment will allow a choice to be made. This is a legitimate objection. But even a theory that has in the background predetermined ontological principles must offer something new with respect to the experimental situation that it is supposed to explain, for example, by allowing for quantitative predictions for what will be the case in analogous experimental situations. What happens in Rohault's physics is simply a translation or a change of language: to the description of experiment in terms of objects perceived by the senses is added the description of a matter that would by its properties be able to produce the experiment that is perceived by the senses. The problem of the translation proposed by Rohault is that the second description offers nothing more than the first. If we continue to read this chapter, we will reach the presentation of the weight of air: it is not an experimental demonstration, as it was for Torricelli or Pascal, by the prediction of the height reached by mercury in a tube, or the heights it would reach at different altitudes. Rather, it is deduced from the principle that a body can only be moved by a body touching it: as air is the only body touching the piston, "we must think that it is the Air that causes this surprizing Motion; for, considering that the Air always contains in it a great Quantity of the Particles of Water, and other terrestrial Bodies;…we shall…assert; that the grosser Air is heavy, and consequently that by its weight, the Sucker is forced into the Syringe."[27] There is absolutely no quantitative prediction that could potentially invalidate a hypothesis, but rather the consequences of the principle that a body can be moved only by a body touching it. Thus, contrary to what the Preface affirms, and also to what Rohault's mentor Claude Clerselier (1614–1684) will claim, reasoning cannot "prévenir," that is to say anticipate, experience and experiments. It does not allow us to anticipate what it will be.[28] As we will see towards the end of the second part of this chapter, when we examine Huygens' judgment of Rohault, it happened that Rohault would refuse to take into account experiments that contradicted what he considered to be established. When we realize that most of the experiments presented in this chapter come from the writings of Pascal, who, in addition to his very firm position on the importance of experiments in physics, had practiced quantitative physics and was opposed to formulating hypotheses on subtle matter,

---

[26] See the texts whose references are given above, n17.

[27] Rohault 1987, I, Chap. XII, Sect. 10, 59. Rohault does mention what happens in the case of mercury in Sect. 23, 64, but absolutely not as a crucial experiment.

[28] On the use by Clerselier of the affirmation that the reasonings of Rohault anticipate ("préviennent") experiments, and the way that Rochon mocks this affirmation, see below in the third part of this chapter.

we can see that Rohault presents, with regard to experimentalism, no progress toward a stronger and more substantive doctrine, but rather a regression toward a weaker and less substantial doctrine.

Rohault's doctrine of experiments does not come from an epistemological broadening that might have left more room for experiments than had Descartes. As we have just shown, Rohault's epistemology is not only, as has written Trevor McClaughlin, "compatible" with that of Descartes, it is fundamentally identical to it.[29] This epistemological identity does not however mean that there are no differences between the *System of Natural Philosophy* and the *Principles of Philosophy*, or between Rohault and Descartes.

First, the *System of Natural Philosophy*, without excluding considerations that arise from metaphysics or general physics, does not always assign them the same function as do the *Principles of Philosophy*. Thus, while Descartes states the idea that God conserves the motion he created, according to Rohault, the conservation of motion is not based on the immutability of the action of God.[30] While Rohault, like Descartes, argues that everything we can think of can be done by God (in order to establish the infinite divisibility of matter), he adds that we can see smaller and smaller animals depending on the power of our microscopes, or the division of gold into finer and finer leaves.[31] Rohault also often insists that he wants to reason "as a physicist." Thus he declares on several occasions that, reasoning as a physicist, he does not include the omnipotence of God, by which, of course, the created things could have been different than they are.[32] Although it is a matter of relatively undecidable questions, I think that this distancing of metaphysics can be read not only in intellectual terms, but in institutional terms as well. In the context of the great battle of Cartesianism, Rohault had adopted a classical strategy to retain a bit of autonomy for physics by sharing tasks between physicists and metaphysicians: by claiming not to enter the domain of the metaphysician, he can expect that the metaphysician will return the favor.

Secondly, Rohault has indubitably set up, with his *Mercredis*, a social scheme with no equivalent in Descartes. Descartes never considered that public experiments would constitute a serious weapon in his great battle against Aristotelians. While he had opened the possibility of a new audience for philosophy, for example when he declared that he wrote the *Discourse on Method* in French so that "even women can understand something,"[33] as we see in his correspondence, except for a few princesses, it was in fact primarily written for and commented by professional philosophers, theologians, and professors. Rohault, on the contrary, gave his lectures,

---

[29] *Pace* McClaughlin 1996, 478.

[30] Rohault 1681, I, Chap. XI, Sect. 5, 71.

[31] Rohault 1681, I, Chap. IX, Sect. 9–12, 56–60 and Chap. XXI, Sect. 2–3, 160–161. On the addition of these empirical facts, see Roux 2006, 127, where I note that in the preface, Rohault considers this question as too metaphysical.

[32] Rohault 1987, I, Chap. V, Sect. 12–13, 34–35; Chap. VII, Sect. 9, 41; Chap. VIII, Sect. 2, 45–46; Chap. IX, Sect. 2, 51 and Sect. 12, 60.

[33] Descartes to Vatier, February 22, 1638, AT I 560.

according to Clerselier, before "people of all stations and conditions, prelates, abbots, courtesans, doctors, physicians, philosophers, surveyors, regents, schoolboys, provincials, foreigners, artisans, in a word, people of all ages, sex and profession."[34] Other sources attest to the fact that a certain number of persons of quality, young students (for example Pierre-Sylvain Régis) and women (for example Madame de Bonneveaux or Madame de Guerderville) attended these lectures. It also seems fair to note, as has McClaughlin, that Rohault's family relations gave him an exceptional familiarity with artisans.[35]

Given this, if Rohault is different from Descartes, it is not because he supported a different epistemology of physics. Rather, it is because, in a different institutional situation, he systematized social practice tied to the communication of philosophy that had at been best dimly seen by Descartes. Hence the idea that, to seize any novelty of Rohault, it is appropriate to not only read his *System of Natural Philosophy*, but to take seriously the very practice of his lectures. To do so, one must compare them to comparable enterprises from the same period. As I will now show, Rohault flourishes at the moment when something that can be called a radical experimentalism began to develop in France; that is to say, a doctrine by which the veritable work of physicists was mainly to make observations and to carry out experiments in closed social spaces like the Royal Society or the Académie des sciences. Some have held that Rohault participated in the establishment of this radical experimentalism, or that he was influenced by it.[36] Instead, it seems to me that compared with this radical experimentalism, Rohault's Cartesian experimentalism switched to become "arrière-garde," something old fashioned.

3.3 The Académie Montmor, the Compagnie des Sciences et des Arts, Rohault's Mercredis

The history of learned societies in seventeenth-century France has often been written in a retrospective mode. The aim has been to seek out in these societies the first hints of an institutionally decisive event, the founding of the *Académie des sciences*. Thus, the first histories of the *Académie des sciences* sought in these earlier learned societies the proof of seniority for the French compared with the English and their Royal Society.[37] Later histories have tended to project institutional phenomena that would only apply to the *Académie des sciences* of the eighteenth century back onto the relatively informal societies of the seventeenth century, which were merely a

---

[34] Clerselier 1682, unpaginated.

[35] McClaughlin 1996, 475–476.

[36] McClaughlin 1996, 478, 2000, 341–342.

[37] Cassini 1693, 26; Duhamel 1698, 7–9; Fontenelle 1733, 4–5. See on the contrary, and correctly, Brown 1934, 91–105.

certain number of vectors of scientific exchanges among many others. (In particular, it seems to me that scientific work was carried out and sanctioned much more in correspondence than in these societies, which is why the following discussion bears no judgment as to the importance of these societies for the progress of science.) Whether in the earliest histories or in those written by our contemporaries, this retrospective style has imposed what I would call a continuist genealogy, according to which an uninterrupted line connects one society to the next, and all of them to the *Académie des sciences*.[38] The meetings held by Mersenne would thus have begat the *Académie Le Pailleur* (Pascal's *Academia parisiensis*), which begat the *Académie Montmor*, which begat the *Compagnie des sciences et des arts* (the name of a project from 1664, also known as the *Académie Thévenot*), which in turn begat the *Académie des sciences*. In this series of begettings, Théophraste Renaudot's Conférences and the *Académie Bourdelot* are often left aside, as are other specialized circles like that of the mathematician Claude Mylon (1615–1662), a choice that creates its own problems.

Yet the main problem with this continuist genealogy is that it produces between these societies a difference that is both too great and too small. Too great, because this genealogy leads us to think that each of these societies had a well-defined identity and individuality. But not only is it true that some scientists, for example Gilles Personne de Roberval (1602–1675), Ismaël Boulliau (1605–1694), or Pierre Petit (1598–1677), worked throughout the century in various societies, but that at any given time, the same individuals were simultaneously members of multiple societies, even when they embodied opposing parties: Adrien Auzout (1620–1689), who was anti-Cartesian, attended Rohault's *Mercredis*; Pierre Michon Bourdelot (1610–1685), at the time physician attached to the Condé family and moderator of the eponymous *Académie*, gave talks at the *Académie Montmor*; Géraud de Cordemoy, a Cartesian attorney, joined some times at Melchisédech Thévenot when he brought together the anti-Cartesian experimentalists.[39] But the difference produced by the continuist genealogy is also too great in that it neglects the differences of type of institution or orientation among some of these societies. By difference of type of institution, I mean, for example, the difference between informal meetings held occasionally and societies with strict rules, whether for the days they met, the terms of admission of participants, or the way sessions were held, with for example, the designation of a moderator and a secretary. As for the discontinuity of orientation, this refers to the general goals set out by these societies and the intellectual means they afforded themselves to achieve them. Discontinuity of type of institution and discontinuity of orientation are not always linked. I will thus be able to show that there is a discontinuity of orientation between the *Académie Montmor* in its early years and

[38] Brown 1934 is the pioneering work, on which all others rely. See also Mesnard 1963; Taton 1966; Hahn 1971; Hirschfield 1981. For a pertinent critique of the manner in which relations between learned societies and the *Académie des sciences* were conceived, see Mazauric 2007.

[39] Huygens 1888–1950, XXII, 535, 540, 543–544, 554; Borch 1983, III, 423, 435; IV, 173. On Petit and Auzout, see below, n75.

in its later years, with the development of a radical experimentalism that began in the *Académie Montmor* and continued in the *Compagnie des sciences et des arts*.

On the whole these questions have not been given much consideration in the secondary literature since the seminal work of Harcourt Brown, with the exception perhaps of works on the *Académie Bourdelot*, which I will set aside for the very reason that its functioning, at least as a means of aristocratic patronage, is fairly well known.[40] To evaluate Rohault, not as the author of the *System of Natural Philosophy*, but rather as the organizer of his *Mercredis*, I must get at the root and explore primary sources. I'll begin by presenting the early days of the *Académie Montmor*. I'll then show that it experienced the development of a radical experimentalism as of 1661. This will allow me to then evaluate Rohault's *Mercredis*.

The *Académie Montmor* operated, as far as we can tell, and with some long interruptions, from the end of 1657 to June 1664.[41] Its name comes from the man who was both its patron and its moderator, Henri-Louis Habert de Montmor (1600–1679), Conseiller du Roi and Maître des Requêtes, a member of *Académie française* from almost the earliest days of that institution. As a patron, Montmor protected both Gassendi and Descartes. On one hand, he offered Descartes a house in the country and, as we know from Adrien Baillet, attended Descartes' funeral in 1667. On the other, he hosted Pierre Gassendi in the last 2 years of his life, served as the executor of his estate, and helped to publish his *Œuvres complètes* (Lyon, 1658). Despite this, Montmor's *Académie* was considered to be Cartesian. Sorbière indeed wrote that Montmor endeavored to make Descartes' physics a Latin poem; Chapelain wrote that he created his *Académie* solely to establish the doctrine of Descartes and to allow for its beautiful dreaming; Dom Robert Desgabets (1610–1678) noted, in a letter he sent to Jean-Baptiste Denis (1643–1704) on July 28, 1667, that 10 years earlier, "his esteem for the philosophy of Monsieur Descartes led him to be very faithful to the assemblies held by Monsieur Montmor."[42] It is appropriate in these conditions not to speculate on its exact doctrinal orientations, but rather to consider two documents of what one might call an institutional nature, since they were written by the *Académie's* secretary, Samuel Sorbière, as well as the testimony of travel journals and correspondence.

---

[40] Pintard 1951; Béguin 1999, 362–379.

[41] Brown 1934, 68–74, discusses the informal meetings that took place at Montmor's house before 1657. The meetings were interrupted by Roberval's insult to Montmor, then by political affairs between December 1658 and August 1659 (Boulliau to Huygens, December 6, 1658 and Chapelain to Huygens, August 20, 1659, in Huygens 1888–1950, II, 287, 468; Oldenburg to Saporta, July 11, 1659, in Oldenburg 1965–1973, I, 294–295), then from May to October 1661 because of the illness of Madame de Montmor (Chapelain to Huet, September 26, 1661 and Chapelain to Huygens, October 16, 1661, in Chapelain 1880–1883, II, 153, 159). On the end of Montmor Académy see below, n70.

[42] Baillet 1691, II, 442, 462; Bougerel 1737, 372–373, 434–436; Sorbière to Montmor, August 22, 1657, in Sorbière 1660, 371; Chapelain to Heinsius, September 22, 1667, to Bernier, February 16 and April 26, 1669, in Chapelain 1880–1883, II, 530, 622, 640; Chapelain 1662, 52; Denis 1668, 2–3.

The first of these documents, the *Académie's Règlement* (1657), shows that the list of participants and the conduct of the sessions were strictly regulated. The circle of Montmorians, "made up of people curious about nature, medicine, mathematics, liberal arts, and mechanics" was limited to its founding members and those that two thirds of those present allowed to join, even if it was planned that they would "carry out correspondence with scientists from France and abroad."[43] The high point of each meeting was two speeches prepared in advance on a subject chosen by the President. Once these speeches were read, each member present could give his comments "each in turn, and in a few words."[44] The fact that this academy was specialized in the study of nature was indicated by the description of the people admitted, but also by the introductory affirmation: "the goal of these conferences will not be the vain exercise of the mind and useless subtleties, but rather always the clearest knowledge of the works of God and the advancement of the commodities of life, in the arts and sciences that serve to best establish them."[45] We will note, however, that while the 1657 *Règlement* gives a few instructions as to the conduct of exchanges—in sum, that one must not speak to say nothing, nor speak without having the floor—it says absolutely nothing about carrying out experiments in the study of natural things. The point of the conferences was quite literally to meet to confer, or more accurately, to listen to speeches written in advance.

It is therefore not surprising that in the list of topics proposed for 1659 sent by Oldenburg to Boyle nothing seems to be able to or have to rely on experiments that would have taken place during the meetings:

> the source of the truth of opinions now in fashion. The explanation of the principles of Descartes, the insufficiency of motion and shape to explain the phenomena of nature (and enterprise to be proven by an Aristotelian). After, on the brain, on nutrition, on the use of the liver and spleen, on memory, on fire, on the influence of the stars, if the fixed stars are suns, if the Earth is alive, on the generation of gold, if all our knowledge is dependent on the senses.[46]

This remark is even more true for the speeches Sorbière himself made in 1658–1659, on fevers and the cold, on motion, on rarefaction and condensation, on the idea that our limited knowledge of the natural sciences should not discourage us from studying them, on the truth of our knowledge of nature, on the source of the diversity our opinions on a given subject.[47] Although Sorbière insists that the aim is to avoid "metaphysical thoughts," or insists on the need to "mix into our discussions all that

[43] Sorbière and Du Prat 1657, 634, Art. VII–IX.

[44] Sorbière and Du Prat 1657, 633, Art. II–VI. In his letter to Hobbes dated 1 February 1658, in Sorbière 1660, 632, Sorbière indicates that this scheme drew the opposition from those who did not want to have to write speeches.

[45] Sorbière and Du Prat 1657, 633, Art. I.

[46] Oldenburg to Saporta, July 11, 1659, in Oldenburg 1965–1973, I, 294–295. The question of knowing if the Earth is alive was dealt with by Chapelain, see BNF, Ms. 12847, mentioned in Collas 1912, 331.

[47] Sorbière 1660, 60–64, 181–189, 190–193, 194–202, 694–700, 701–704, 712–714.

we know that is useful and curious in the Arts of the Sciences and that applied immediately to the commodities of life,"[48] when we read these speeches, we would readily describe them as metaphysical, in the sense that this term has sometimes taken to mean those theories that no specific experiment could confirm or contradict. And there was also metaphysics in the contemporary meaning, as in the session where they examined the manner in which secondary and primary causes work together to produce the phenomena of our world.[49] In addition, of course, the sessions were devoted to reading scientific letters, but even when these referred to experiments, they did not imply carrying out any: the intent was to report on experiments done elsewhere. It was, for example, to the *Académie Montmor* that Huygens communicated in 1658, by means of a letter to Chapelain, his "System of Saturn," that is, the fact that Saturn is surrounded by a thin flat ring; it was before the Montmorians that Clerselier read a letter in which Descartes is purported to have responded to the attacks of Roberval, a letter that Clerselier would admit a dozen years later having written himself to refute him "more gallantly and with more authority"; it is at least in part before the *Académie Montmor* that were read and commented the letters Fermat sent to Marin Cureau de la Chambre (1594–1669), Clerselier, and Rohault to contest the explanation of refraction proposed by Descartes.[50] On these lines, it was to the *Académie Montmor* that Jean-Baptiste Denis presented for the first time his theories on the transfusion of blood, but we have no evidence that he did so other than in the form of a speech.[51] The least we can conclude is that experimental practices were not at the heart of the first *Académie Montmor*.

Given this conclusion, the second document, the speech given by Sorbière in 1663 to the Académie, and which he then sent to Colbert, may appear surprising. Sorbière insists on the place experiments had held in the *Académie Montmor*, contrary to what its *Règlement* and a look at the subjects dealt with have led us to conclude: "we left each the freedom to bring his experiments, we exhorted the most industrious to experiment, we preferred them to any other discussion."[52] Later, Sorbière details these experiments:

> We have even seen with pleasure Monsieur Rohault come here with his set of magnets, and Monsieur Pecquet put on his ceremonial garb to carry out according to his method his dissections. Monsieur Petit played his artillery with gunpowder and fulminating gold. Monsieur Thévenot showed his tubes designed to examine the ascension of water, which rises on its own to its own level. Monsieur de Monconys brought an enchanted horse that the Devil curried, according to its grooms, and that he had bought to show in our presence the falseness of this opinion.[53]

---

[48] Sorbière 1660, 695, 100.

[49] Richard Jones to Boyle, March 20, 1660, in Boyle 2001, I, 405–406.

[50] Chapelain to Huygens, May 10, 1658, in Huygens 1888–1950, II, 173–176. Clerselier 1667, unpaginated. Clerselier to Fermat, May 13, 1662, in Descartes 1667, 284–286, *passim*.

[51] Denis 1668, 2–3.

[52] Denis 1668, 161.

[53] Denis 1668, 216–217.

This description certainly corresponds to the interests of the various parties, with Petit already in the days of Mersenne carrying out experiments to see if a cannon ball would fall, whatever the height from which it was dropped. Monconys often tried to refute all sorts of beliefs. It is attested elsewhere that Jean Pecquet (1622–1674), author of the *Experimenta nova anatomica* (1661), did dissections, that Rohault demonstrated magnetic phenomena, that Monconys gave a talk on the ascension of water, and that Thévenot showed a spirit level.[54] Beyond the doubts one may have as to the intellectual gains from seeing Petit's artillery or Monconys' horse, what is remarkable is the way in which Sorbière insists on these experiments, which were not mentioned at all in the 1657 *Règlement.*

To understand Sorbière's move, we must place this 1663 speech in its historical context. At a time when the *Académie des sciences* was in gestation, his intent was to prepare the future by drawing lessons from the past, as well as to show the *Académie Montmor* in the most favorable light, as if as the mother of all the learned societies specialized in the study of natural things, it had full legitimacy to determine what the *Académie des sciences* should be.[55] But even in this context, Sorbière disqualifies the radical experimentalists, meaning by that those who gave the highest, or even the exclusive, priority to experiments. He criticized them in particular for having made the most of the troubles of the *Académie Montmor*, and of then signing its death warrant. In addition to the people who fomented discord to be able in the end to position themselves as the arbiters among scientists,

> There is another type of person, who in the general collapse of this assembly wanted to rally and take control, using a very obvious pretext, but with a plan whose execution was impossible for us. They preached only for experiments and demanded that we meet only to do them, or that we speak only spontaneously. They said that we need to take care only to act properly, and that there is no need to reason on a topic before doing some experiment, which would supply enough material for discussion without any other meditation.[56]

The 1663 speech is a strategic rewriting of the past. Faced with the radical experimentalism that, as I will show, developed as of 1661, the aim is to affirm that the *Académie Montmor* carried out experiments as much as any private learned society could reasonably do, and called on the public authorities to do more. For "to endeavor to create an academy of physics" only can be done by "kings, rich sovereigns, or a few wise and wealthy republics."[57]

---

[54] One of the experiments of Pecquet is reported in Oldenburg's letter to Saporta, August 27, 1659, in Oldenburg 1965–1973, I, 308: "Only Monsr Pecquet brought an experience of his of the winds engendered in the body of man wch was odde, vid. yt he had known a man, who, wherewoever he touched him on his body, gave from him much wind by his mouth, even when he touched him on his tigh or his feeth," but about Pecquet's dissections, see especially Sorbière 1660, 22–59. We find the *Discours sur l'ascension de l'eau sur un niveau, en un tuyau étroit, récité par Mr. de Montconys, chez Mr. de Montmor* in Montconys 1665–1666, III. Thévenot's spirit level is mentioned in Thévenot 1681, 10–12 and in Thévenot to Huygens, in Huygens 1888–1950, IV, 18–19.

[55] Sorbière 1663, 160.

[56] Sorbière 1663, 162.

[57] Sorbière 1663, 160, 216.

If there was any doubt that Sorbière is rewriting history and that the *Académie Montmor* was a society where speeches and not experiments took first place, we can conclude by looking at some third-party testimony. Whether he's writing to Michaelis, Hartlib, or Boyle, Oldenburg gives the same verdict on the state of natural philosophy at the *Académie Montmor*: "in Paris there are many men who promise a great deal but few who give"; "I wish only, these discourses may not rather tend to speculation and shew of wit, yn usellnes to the life of man, wch latter I much doubt off, considering the nature of most of ye French, and indeed of most of men, yt lover rather to praise yn to worke"; "French natural philosophers are more discursive yn active or experimentall"—which, according to Oldenburg, confirms the Italian proverb, "Le parole sono femine, le fatti maschii."[58]

One could say that Oldenburg, who was not yet secretary of the Royal Society, but was already well established in England and tied to Boyle by a relationship of patronage (it was in his capacity as tutor to Richard Jones, Boyle's nephew, that he was traveling in France), was merely projecting well-established prejudices on the people he met in France. But the judgment of French astronomer Ismaël Boulliau in a letter to Huygens from July 1661 was scarcely any different:

> If one could persuade our gentlemen of Paris, who are rich, curious about beautiful things, and desirous of immortal glory, to make some expense to acquire knowledge by experiments, one might hope for something. But they want to acquire what is the best and most beautiful with no pain, other than that given to making beautiful speeches and philosophizing in the clouds, with no application, and with no expense. You have seen it by your own experience, and I will say no more in writing.[59]

In that last sentence, Boulliau alludes to Huygen's experience during his second stay in Paris. The brief notes Huygens took in his travel journal about the sessions of the *Académie Montmor*, which he attended very faithfully from November 1660 to February 1661, constitute another conclusive document with regard to the content of these sessions: "Des Argues made a speech on whether the mathematical point truly exists," "Monsieur de Neuré read his paper on the causes of thunder," "Rohaut read the experiments of water rising in small tubes," "Rohault explained the little tubes," "a dispute between Rohault and Auzout," "Bourdelot spoke of gout.... Pequet against Bourdelot," "Monsieur de la Potterie spoke of the elementary fire under the sky of the Moon," "Bourdelot again spoke of gout and did so very well," "Monsieur Pecquet spoke of the generation of the chicken in the egg and was booed."[60] From these notes and from the testimony of Oldenburg and Boulliau, we see both that the sessions of the *Académie Montmor* were acts of speech (talking, reading, explaining, making speeches), and that despite the worthy resolutions of the 1657 *Règlement*, the atmosphere was not totally polite, and the speeches almost always ended in arguments. Ten years later, Desgabets still recalls the

58 Oldenburg to Michaelis, April 26, 1659, to Hartlib, July 30, 1659, and to Boyle, July 23, 1659, in Oldenburg 1965–1973, I, resp. 240, 260, 287.

59 Boulliau to Huygens, July 11, 1661, in Huygens 1888–1950, III, 293.

60 Huygens 1888–1950, XXII, 535, 537, 539, 540, 543, 544, 546, 553, 554, 560.

"the slight mocking [*petites railleries*]" that accompanied his reading of his *Discours de la communication ou transfusion du sang*, which led him to understand that "some held this thought to be ridiculous."[61]

The contrast is great with the radical experimentalism that developed from 1661 and until the foundation of the *Académie des sciences* in 1666. In a general fashion, by radical experimentalism I mean the doctrine according to which the true work of those who study things of nature is nearly exclusively to carry out experiments in a socially closed space. This is a very general definition, whose principal virtue is to offer a contrast with the first years of the *Académie Montmor*. To provide historical consideration to the radical experimentalism that developed in the period 1661–1666, in the following, I'll first examine its causes; I will then present the transitional period of 1662–1665; I'll conclude by showing that radical experimentalism is behind the project of the *Compagnie des sciences et des arts*, which prefigures the *Académie des sciences*.[62]

In continuity with what has been said above, we shouldn't be surprised that one of the causes of the development of a radical experimentalism was the feeling that the speeches inevitably ended in disputes. In his 1663 speech, Sorbière tends to reconstruct history in this regard by affirming that the *Académie* at first experienced a period of harmony: "there were said in this place excellent things, and there were even carried out several beautiful experiments. And when a topic had been thoroughly considered in two or three sessions, it seemed as if we had exhausted all of its human subtlety."[63] He however soon recognized that dissonance had arrived in the assembly due to "certain spirits who felt themselves above the laws we had established"; because of this, "all others…were as if smothered under torrents of words."[64] In reality, it seems that verbal excess was the normal status of the *Académie Montmor*, and that everyone soon grew tired of it. Huygens evoked the excessive statements of the Aristotelian Antoine de la Poterie, Gassendi's former secretary.[65] Chapelain reports to Nicolas Heinsius (1620–1681) "this bickering among the philosophers and mathematicians of the Montmorian Academy" and attributes it to

---

[61] Denis 1668, 2–3.

[62] McClaughlin 1975. It cannot be contested that this project prefigures more the *Académie des sciences* than do the notes that, probably at the request of Colbert, were written in 1666 by Jean Chapelain and Charles Perrault, both members of the "*Petite Académie*," a small council in charge of proposing initiatives to glorify the King (Perrault 1666). The main goal of Chapelain's note, published in Chapelain 1666, 513 (Collas 1912, 384–388, establishes that Chapelain was the author) was to distinguish "scientists by profession," who are busy only with cabals in the court, and "good faith scientists," who of course were the true scientists. The note from Charles Perrault, who proposed an "*Académie Royale générale*" divided into four sections (Belles-Lettres, History, Philosophy, in the sense of natural philosophy, Mathematics), is very short and the project it promotes was soon abandoned because it faced resistance from already established institutions as the Sorbonne and the *Académie française* (Duhamel 1698, 7–9; Fontenelle 1733, 5–7).

[63] Sorbière 1663, 160–161.

[64] Sorbière 1663, 161.

[65] Huygens, *Journal*, November 9, 1660, in Huygens 1888–1950, XXII, 535; Christiaan Huygens to Lodewijk Huygens, April 26, 1662, in Huygens 1888–1950, IV, 117.

the system of Descartes.[66] Boulliau, who belonged to neither of these two societies, contrasted the urbanity of the literary circle of the Venetian ambassador to the vehemence of the Montmorians:

> From certain persons I have learned that the Venetians are more agreeable, more polite, more urbane, and use complimentary words in discussion. The Montmorians are sharper, and dispute with vehemence, since they quarrel about the pursuit of truth; sometimes they are eager to rail at each other, and jealously deny a truth, since each one, although professing to inquire and investigate, would like to be the sole author of the truth when discovered.[67]

If disputes were an endemic illness of the *Académie Montmor*, it remains to be explained, since so many institutions affected by endemic illnesses endure, why this one became so intolerable that participants came to desire something different. This is because French scientists learned of a different mode of operation. The letters of Oldenburg, first secretary of the Royal Society, and of Huygens, who after his stay in London from March to May 1661 remained a correspondent of Robert Moray, made French scientists aware of the experimental commitment of the Royal Society, which rapidly became a model for them.[68] The correspondence between Chapelain and Huygens is clear on this point.

> Promise that you will teach us about what you will learn of English science. We are told here that the learned people of London have great plans for the advancement of the knowledge of nature. It seems that our Academy is warmed by the emulation they have given it, and that we want to apply ourselves to experiments in favor of all other exercises where only the mind takes part.[69]

Huygens highlighted that the aim of these gentlemen was more to do experiments rather than engage in reasoning, to which Chapelain responds immediately that this "will serve as a strong prompt to the members of the Academy to apply themselves to experiments on which natural sciences are founded, rather than on speculations and conjectures."[70]

We know that the meetings of the *Académie Montmor* officially ended in June 1664, and that Melchisédech Thévenot "received in his home the debris of this assembly."[71] Thévenot (1620–1692), a diplomat in Italy, a traveler, a cartographer and bibliophile, is known for having made a spirit level, written the first treatise in French on swimming, being the King's Librarian as of 1684, and a member of the *Académie des sciences* beginning in 1685.[72] According to the brief autobiography

[66] Chapelain to Heinsius, February 6, 1659, in Chapelain 1880–1883, II, 17.

[67] Boulliau to Heinsius, February 1658, quoted and translated in Brown 1934, 78–79.

[68] On this point, begin with Brown 1934, 119–122.

[69] Chapelain to Huygens, May 30, 1661, in Huygens 1888–1950, III, 273.

[70] Huygens to Chapelain, July 14, and Chapelain to Huygens, July 20, 1661, in Huygens 1888–1950, III, 295 and 299.

[71] Huygens to Moray, June 12, 1664, Huygens 1888–1950, V, 70. Chapelain to Bernier, February 16, 1669, in Chapelain 1880–1883, II, 622.

[72] McClaughlin 1974. The most detailed study of the various activities of Thévenot is currently that of Dew 2009, 81–130.

that he wrote shortly before his death, he supported a "a company of persons known to be very clever," taking advantage of the insight of Frenicle and Steno, who lodged with him at the time, and primarily carried out anatomical and astronomical observations.[73] In his *Discours de la navigation*, he presents this company as if it were the *Académie Montmor* itself: "the assembly formed at the home of Monsieur de Montmor worked at my home the last two years it met."[74] Correspondence from this period confirms this description, and shows in particular that there was no real discontinuity between a part of the *Académie Montmor* in its later years and the *Compagnie des sciences et des arts*. In fact, the scientists who grew tired of the disputes and who by the emulation of the English converted to experimentalism had begun to go their own way with Thévenot even before the dissolution of the *Académie Montmor* in 1664, without the creation of a true academy in an institutional sense.

In 1663, certain sessions of the *Académie Montmor* remained the opportunity to read Cartesian writings, the *Discours du mouvement local* by Géraud de Cordemoy and the *Discours des fièvres* by Rohault, which were published 1 year later, together with Descartes's *World*.[75] The general impression from this correspondence and these journals is that nonetheless from 1662 through 1666, radical experimentalism was present everywhere and in all sorts of manners, while at the same time, if we can say, things became less formalized in the expectation of a new institutional framework. Petit, Auzout and "a bit," Thévenot, felt themselves to be alone as partisans of experiments among the Montmorians, or more generally among French scientists, carried out in the spring of 1662 various astronomical observations.[76] In late 1662 and early 1663, some of the Montmorians began to come together at the

[73] Thévenot 1694, Avertissement, unpaginated.

[74] Thévenot 1681, 8.

[75] It is in this edition that it is said that these two speeches were made at the *Académie Montmor* before being published, without their authors being named.

[76] Petit to Huygens, March 8 and May 5, 1662, in Huygens 1888–1950, IV, 73, 127. From the beginning of his correspondence with Huygens, Petit complained of the way in which, in France, people of quality neglected mechanics, see Petit to Huygens, in Huygens 1888–1950, II, 257. Petit, Auzout and Thévenot are mentioned meeting on Tuesdays in the letter from Petit to Huygens, 17 October 1664, in Huygens 1888–1950, V, 124. The same three would meet Christopher Wren when he came to Paris a few years later (Oldenburg to Boyle, August 24, 1665, in Oldenburg 1965–1973, II, 480). Pierre Petit (1598–1682), born in Montluçon, resided in Paris from 1633 on, wrote objections against the metaphysics of the *Discourse on Method* and against the explanation of refractions in the *Dioptrique*, and communicated Torricelli's experiment to Pascal. *Intendant général des fortifications* from 1649, he was part of the various scientific circles and regretted not being a member of the *Académie des sciences* (see the lettre from Boulliau quoted by Brown 1934, 138). The explanation can perhaps be found in his character; see the cruel portrait made of him in Sorbière to Hobbes, early 1663, in Hobbes 1994, 551–554; and Christiaan Huygens to Lodewijk Huygens, September 28 and November 9, 1662, in Huygens 1888–1950, IV, 241, 256, *passim*. Adrien Auzout (1620–1691), born in Rouen, contributed to Pascal's experiments on the vacuum, worked as an astronomer with Jean Picard at the *Académie des sciences*, of which he was briefly a member (1666–1668) before retiring to Italy and England, apparently for having criticized Charles Perrault's translation of Vitruvius; see Brown 1934, 138, 138–141.

home of the Marquis de Sourdis, "but they go there in smaller numbers, and it is clear that in a short while there will be nobody at all wanting to go there," explained Sorbière, who thought that Sourdis had nothing to give "except his wretched, crude, vulgar hospitality in an unattractive part of his unattractive house," "as if a group of grasping rustic schoolmasters were entertained by an Irish professor [*quemdam professorem*]."[77] In the spring of 1663, Huygens was in Paris, and the issue of the day was to establish "new laws and ordinances" for the *Académie Montmor*; a "general assembly for telescopes" was organized at the same time at the home of Auzout: the power of various telescopes was tested by Auzout, Huygens, Petit, Monconys, and Étienne d'Espagnet, who had the lenses of some of these telescopes made using a new process.[78] In November 1663, after nearly 2 years of unsuccessful attempts, an air pump was finally built for Montmor on the model of Boyle's pump, and was used for experiments, in particular on the phenomena of abnormal suspension shown by Huygens. At nearly the same time, astronomical observations took place in Issy, at the home of Thévenot. A month later, dissections were carried out at the home of Montmor.[79] Shortly before the official dissolution of the *Académie Montmor*, the vacuum machine was paraded to different salons, always seeking to "create a more solid and better regulated establishment for this academy."[80] As for the meetings held at the home of Thévenot, what the Danish scientist and traveler Ole Borch (1626–1690) said seems to indicate that they were devoted to carrying out chemistry experiments on liquors in June–July 1664, then, at least from November 1664 to March 1665, to engaging in anatomical observations under the direction of Nicolas Steno (1638–1686) then Jan Swammerdam (1637–1680).[81] It seems likely that these last observations correspond to the ones found in the three fascicles by Swammerdam that close the *Recueil de divers voyages* by Thévenot, that is to say *Histoire naturelle de l'Ephémère*, *Histoire naturelle du Cancellus ou*

[77] Sorbière to Hobbes, December 23, 1662, in Hobbes 1994, 542. Petit to Huygens, November 8, 1662, in Huygens 1888–1950, IV, 262, mentions also that the *Académie Montmor* was on the verge of moving to the home of Sourdis.

[78] Christiaan Huygens to Lodewijk Huygens, April 6, 1663, in Huygens 1888–1950, IV, 324–325. Christiaan Huygens to [Constantyn Huygens], April 20 and May 4, 1663, in Huygens 1888–1950, IV, 333, 338. Contrary to what the editors of Huygens's *Œuvres complètes* affirm, the d'Espagnet who appears in Huygens correspondence may not be the chemist Jean d'Espagnet (1564–1637?), first Président of the Parlement of Bordeaux: it is more likely his son, Étienne d'Espagnet, counselor at the Parlement of Bordeaux.

[79] Huygens to Moray, November 18 and December 19, 1663, in Huygens 1888–1950, IV, 433, 474. Huygens to Constantyn Huygens, November 30, 1663, in Huygens 1888–1950, IV, 452. Auzout to Christiaan Huygens, December 1663, in Huygens 1888–1950, IV, 481–482. The unreliability of the Montmor pump was noted by Shapin and Schaffer 1985, 265–269.

[80] Huygens to Moray, March 12, 1664, in Huygens 1888–1950, V, 41.

[81] Borch 1983, III, 464; IV, 6–7, 164, 173, 180–181, 186, 274, 283–284. Borch attributes the anatomical observations to Swammerdam, it is Chapelain to Huet, July 31, 1665, in Chapelain 1880–1883, II, 406, who mentions Steno. This anatomical fashion was in no way proper to the meetings of Thévenot: Borch mentions the anatomical preparations that took place at the home of Montmor in February 1665 (when the *Académie* was no longer meeting there) and the ones done by Steno at the home of Bourdelot in May 1665.

Bernard L'Hermitte, and *Le cabinet de Mr Svvammerdam* [sic], *docteur en médecine*. In early 1666, the *Académie Royale des sciences* was founded, and Thévenot retired to Issy, officially to "philosophize and speculate with more liberty in the country," but in fact, we may also conclude, out of spite for not having been among the first members of the *Académie des sciences*.[82]

By the continuist genealogy that I noted above, the *Compagnie des sciences et des arts* is presented as if it were a society along the lines of the *Académie Montmor*. From an institutional point of view, this is incorrect. The rare indications I've been able to collect give instead the impression that the group of scientists advocating radical experimentalism (Thévenot himself, Petit, Auzout, d'Espagnet, and, when they were in Paris, Huygens, Steno and Swammerdam) would gather, at the home of Thévenot and other places, and without a set date.[83] There are no official rules and regulations stating how these meetings were to take place, and I would almost say that such rules couldn't exist. The meetings begin in 1664, that is to say at a time when Louis XIV had already awarded gratifications to a few humanists, including some scientists like Marin Cureau de la Chambre, Johannes Hevelius (1611–1687), and Huygens, and everyone knew that an *Académie des sciences* would be founded. At issue was not about establishing the rules for a private society, but rather the rules of what would become the *Académie des sciences*. From the point of view of the orientation of these meetings, as correspondence and travel journals attest, the scientists participating in them continued and systematized the activities of observation or experimentation that would take center stage from 1662 on.

A final document can show what was the orientation of the *Compagnie des sciences et des arts*, the project which we know thanks to a copy sent to Huygens.[84] The publishers of the *Œuvres complètes* of Huygens hypothesized that this project corresponded to the new rules being sought by the *Académie Montmor* in 1663. The manuscripts reported on by Trevor McClaughlin show that it was more a program from scientists who met at the home of Thévenot.[85] If, as I suppose, the *Compagnie des sciences et des arts* is the product of a scission within the *Académie Montmor* between the more experimentalist of its members and the others, there is no strong contradiction between either position. The very name of the "*Compagnie des sciences*

[82] Chapelain to Steno, March 15, 1666 and May 27, 1667, in Chapelain 1880–1883, II, 447, 514.

[83] Borch also mentions, among those regularly meeting at the home of Thévenot, Vossius (the scholar Isaac Vossius, 1618–1689), Borelli (the chemist and builder of instruments Jacques Borelly (?–1689), later a member of the *Académie des sciences*), Ville Bressé, Bressié or Bressieu (the chemist and engineer Étienne de Villebressieu, who travelled with Descartes at the beginning of the 1630s, and who was the most important source for his first biographer, Pierre Borel, for this period), Frenicle (the mathematician Bernard Frenicle de Bessy (?–1674), who was already living in the home of Thévenot), and Martell (Thomas de Martel (1618–1619–1679–1685?), a bourgeois of Montauban, who was already part of the scientific circles of Paris at the beginning of the 1640s, was a correspondent of Hobbes, then of Oldenburg; the best biography to date is that of Noel Malcolm in his edition of Hobbes' *Correspondence*).

[84] Huygens 1888–1950, IV, 325–329.

[85] McClaughlin 1975, 236; see also Dew 2009, 96. The autobiography of Thévenot found in the Avertissement of Thévenot 1694 affirms that this project was presented to Colbert.

et des arts" is important in this light: bringing together in a company the arts and sciences mean attempting to reform the sciences by putting them through, so to speak, the sieve of the arts. As Thévenot wrote,

> If we have so often call to complain about those who have applied themselves so fruitlessly to these sciences or these studies, the same is not true for those who have cultivated the arts. For they have continually made great progress, and brought them to a very high degree of perfection above that which they were at their beginnings…
>
> Most sciences, as we have them now, and their systems, are just a pure mind game of man, who naturally flees the difficulty of clear reasoning, of finding true precepts, and of drawing the consequences, always ready to admire his work and supporting with a great deal of stubbornness what he advances without foundation.
>
> In the arts, on the contrary, when the worker has poorly reasoned and puts into practice a false reasoning, he is immediately convinced by the lack of success of his work and is corrected by the damages suffered.[86]

The idea that the arts are superior to the sciences because, when confronted with the test of reality, they immediately sanction mistakes, so that, instead of being locked into a dead end, they allow us to see the ways to increase knowledge, is a recurring theme in Bacon.[87] But we also find it in Descartes. Responding to Fromondus who criticized him for creating such a gross (*crassa*) philosophy as mechanics, Descartes defended mechanics with the argument that "as it concerns use and practice, all those who lack the smallest thing are used to being punished with loss of all their expenses."[88] And Thévenot uses Descartes to criticize those who would content themselves with the work of the commentator:

> Those who spoke in good faith of physics or medicine recognized this necessity to carry out experiments and observations to know something about them. Descartes said so everywhere he had the opportunity to talk about it, everyone is now convinced, and that is what the large number of men of letters who today follow his philosophy should primarily occupy themselves, otherwise it would not be of more use for us to have many commentators of Descartes and Gassendi than it has thus far served us to have employed so many centuries commenting the systems of Epicurious, Plato, and Aristotle.[89]

In fact, the project for a *Compagnie des sciences et des arts* speaks only of experiments and techniques, whether for making astronomic observations, learning to understand the human body thanks to chemistry, anatomy, or medicine, inventing new machines and new secrets "both for the manufacture of arts as for curiosity," or facilitating navigation "to increase commerce and to have opportunities to discover the wonders that can be encountered in unknown countries."[90] More important, alongside the traditional functions of the president, "to ensure proper behavior and silence,"

---

[86] Thévenot 1681, 3–6.

[87] See for example Bacon, *Novum organum*, I, 73–74, in Bacon 1996–, XI, 116–119.

[88] Descartes to Plempius for Fromondus, October 3, 1637, AT I 421.

[89] Thévenot 1681, 7.

[90] This point is already highlighted in McClaughlin 1975, 238–242, who notes the common points between the project for the *Compagnie des sciences et des arts* and the practices of the *Académie des sciences*, which is true, but who also suggests that this commonality of doctrine arises from a Gassendist reference, which seems doubtful to me.

and the secretary "to record anything remarkable said or done in the assemblies and to maintain exchanges with the other academies and scientists," which are both found in all academies, they include curators "to keep ready and to execute everything needed for experiments and the machines to design the machines for workers to operate."[91]

Once again we think of Bacon. The allusion is clear when the subject is "making a history of nature as universal as possible, on whose solid foundation one can work to build a physics, and do the same for the history of the arts, and the inventions of men that are in use."[92] If the demand for usefulness is omnipresent (usefulness for individuals and for the general public, a search for commodities for humanity in general and for France in particular) in this project, it does not exclude the fact that one is curious, in particular to learn to distinguish true from false, between what works and what doesn't:

> One will attempt to learn all the well-tested secrets....
>
> One will also test all the important secrets about which one has the description, when there is some likelihood that they could succeed, to approve them and use them if they are good, or to undeceive those who believe in them on their faith in others, without testing them, if they are not true. Lastly one will work to undeceive the world of all the vulgar errors that have passed for so long as truths due to the lack of the necessary experiments to discover their falsehood...
>
> One will also work to learn of all the tricks of tradesmen and merchants and their sophistries with the means to uncover them, which one will publish to prevent the public from being tricked, and to oblige workers to work more faithfully....
>
> If the Compagnie is consulted about any new inventions, about machines, or about major public or private projects, it will deputize those from among it that will be the most conversant in these matters...so that for want of such an examination neither the prince nor the private person will commit themselves to useless expenses and in less than honorable enterprises, as we too often see.[93]

In its proliferation, the project for the *Compagnie des sciences et des arts* constitutes the clearest possible testimony of what was radical experimentalism. More generally, it's time to summarize what has been shown so far about learned societies. The *Académie Montmor* was initially a society where one conferred and made speeches. As of 1662, the degradation of the speeches into arguments and the model of the Royal Society favored, for some of the Montmorians, the development of a radical experimentalism. The *Compagnie des sciences et des arts*, without being a formally instituted academy, brought together in 1664–1665 those who

[91] Huygens 1888–1950, IV, 329.

[92] Huygens 1888–1950, IV, 327. The first reception of Bacon in France was explored in Le Dœuff 1984; on baconianism in mathematics, see Goldstein 2008. The question of the reference to Bacon in late seventeenth century France remains however to be explored, but it may be noted that it was Huygens who in December 1660 lent Thévenot Bacon's *Opuscula varia posthuma, philosophica, civilia, et theologica*, published 2 years prior, and that one finds in his later projects for the assembly of physics in an injunction to "work on natural history more or less following the plan of Verulamius" (Huygens 1888–1950, XXII, 540; VI, 95–96 and XIX, 268).

[93] Huygens 1888–1950, IV, 325–327.

beginning in 1662–1663 distinguished themselves as the members of the *Académie Montmor* partisans of experiments.

How to situate Rohault's *Mercredis* with respect to this evolution, which led to a radical experimentalism? There is no doubt that the *Mercredis* were organized on Wednesdays from 1658 to 1659 and until the death of Rohault in 1671, during which participants could view all sorts of experiments.[94] That means that they began 2 years after the start of the *Académie Montmor*, and that, with the exception of the two first years, were concomitant with the appearance of a radical experimentalism, later institutionalized in the *Académie des sciences*. The dates here are important. The beginning of the *Mercredis* has at times been pushed back to the middle of the 1650s, but I don't think that is supported by the evidence. They are however incontrovertibly attested to in 1659 by the testimony of Clerselier in the preface of volume II of Descartes's *Lettres*.[95] The only document that leads Pierre Clair to "conjecture" that the *Mercredis* began in 1655 is the *Éloge de Pierre-Sylvain Régis* written by Fontenelle.[96] In it, Fontenelle evokes both the disgust theology caused to Régis and the latter's discovery of Cartesian philosophy thanks to Rohault's *Mercredis*: these events are not properly dated, and the only temporal indications given would lead us instead to think that they took place at the beginning of the 1660s. Fontenelle notes that Pierre-Sylvain Régis "had only four or five months left in his stay in Paris" when he discovered Cartesian philosophy, which we take to mean before his departure for Toulouse. And we know from other sources that Régis' lessons in Toulouse began in 1665.[97] If we add to that the fact that Louis Moreri, in his *Grand dictionnaire*, says of Rohault that he "taught ten or twelve years in Paris," but also that the prefaces of two works published in 1657, the preface to volume I of Descartes's *Lettres* and the preface of Savinien Cyrano de Bergerac's *Histoire comique* do not mention the *Mercredis*, it seems to me that we must affirm that Rohault's *Mercredis* began at the very end of the 1650s.[98]

We can also wonder, in a preliminary fashion, about the sources that allow us to reconstitute what these *Mercredis* must have been like, at least insofar as we suppose them to differ from the *System of Natural Philosophy*. The testimony most frequently cited is that of Clerselier in the preface to volume II of Descartes' *Lettres* (1659), and even more in the long preface to Rohault's *Œuvres posthumes* (1682). But such testimony cannot be neutral: if Rohault was the "head of the Cartesian school,"[99]

---

[94] The most exhaustive presentation of Rohault's *Mercredis* remains that of Clair 1978, 42–56.

[95] Clerselier 1659, unpaginated.

[96] Clair 1978, 43.

[97] Fontenelle 1994, 143.

[98] Moreri 1759, 310. Clerselier 1657, unpaginated. Le Bret 1657 (unpaginated) is all the more telling that, regarding Cyrano's illness, he explicitly mentions Rohault, and does not mention the *Mercredis*: "I would do ill to Monsieur Rohault if I didn't add his name to such a glorious list, since this illustrious mathematician who carried out so many beautiful physical proofs...had so great a friendship for Monsieur de Bergerac...that he was the first to discover the true cause of his illness...."

[99] Baillet 1691, II, 442.

Clerselier was the general or the patron. Thus, when he wrote in the 1659 preface that the same people who came to the *Mercredis* to refute the Cartesian doctrine, "after having been convinced by the force of his demonstrations, and fully persuaded by the correctness and the appropriateness that the experiments had with his reasoning, were finally constrained to give each other their hand, and from the open enemies they had been of this doctrine, to declare themselves the partisans and the defenders,"[100] he was not providing for future historians an objective testimony as to what Rohault's *Mercredis* were like. Rather, he was defending Cartesian philosophy against the attacks of his contemporaries. This is all the more true in 1682, in the preface he wrote in the heat of the battle of Cartesianism, when no Cartesian had been named as a member of the *Académie des sciences* and when Cartesianism had been condemned by the religious and political authorities.[101] This preface is a panegyric, like the elogia of academicians that Fontenelle would begin writing a few years later. Clerselier shows that Rohault, dead only some 10 years earlier, was both socially respectable (he was the preceptor of the sons of the Prince de Conti, and should have been the preceptor of the Dauphin; people of quality participated in his *Mercredis*, there were even "thousands" of them), totally orthodox from the point of view of the Catholic religion (those who called him a heretic were envious and malicious scandalmongers), and of course, he was very capable in the sciences, in particular due to the experiments that his "totally mechanical mind" and his "skilled and artful hands" allowed him to carry out, as well as to his ability to stay within the limits of "a peaceful and honest dispute."[102]

We thus should not adopt a superficial reading of testimony like that of Clerselier. We can have doubts, for example, that "thousands of people" truly attended the *Mercredis*. The disputes there were not always "peaceful and honest": the anecdotes reported here and there show that Rohault was regularly in opposition to Adrien Auzout; that he practically came to blows with the gassendist Gilles De Launay (fl. 1656–1677); that "he didn't respond well to difficulties, he got angry."[103] Better, one can reread this testimony in light of the few documents we can find elsewhere. Rohault took the decision to publish his *System of Natural Philosophy* because "his writings were in the hands of an infinite number of persons," and thus that "they had become unrecognizable" and that he was "not sufficiently explained and understood.[104] We also have traces of the physics of Rohault prior to the *System of Natural Philosophy*: his *Fragment de physique* (1662) about which specialists debate whether its authorship should be attributed to Rohault or to Cyrano de Bergerac, his *Phisique nouvelle* (1667), recently edited by Sylvain Matton, and the notes taken by a lawyer, "Monsieur F.," who

[100] Clerselier 1659, unpaginated.

[101] About this battle, see Roux 2012. See also Chap. 2 by Ariew.

[102] Clerselier 1682, unpaginated.

[103] Clair 1978, 46–49.

[104] Clerselier 1682, unpaginated.

attended the *Mercredis* from 1660 to 1669. The latter document is certainly the most interesting, precisely because it consists of notes taken, if not on the spot, surely at the latest a few days after, and without any intention to be published.[105] From the testimony of Clerselier and the notes of Monsieur F., we can draw the following conclusions.

From an institutional point of view, contrary to the meetings of the *Académie Montmor*, the *Mercredis* did not rely on a closed group whose members took turns as speakers and listeners: at the *Mercredis*, Rohault was almost always in front of his listeners or spectators, who, at least officially, were subject to no selection process whatsoever. "At least officially," for a writer reports that on days of lectures, Rohault's first wife, "stood…at the doorstep of her house, and refused entry to any who did not look to be people of quality…; she wanted velvet and let nothing stop her."[106] "Almost always" because it happened that others were in charge of the experiments; Huygens notes, in December 1660: "at Rohault's home, a Spaniard separated silver ore."[107] The *Mercredis* thus differed formally from the encounters that scientific institutions constituted at the time. Whether networks of correspondents or learned societies, these were closed groups; Hobbes reproached, for example, the Royal Society for its "private" character.[108] To use language of our day, they were more like courses given year after year by a professor than seminars bringing together different scientists. And if they did last for more than 10 years, it seems difficult for Rohault to have been able to do anything else than to repeat himself from 1 year to the next, at least for the experiments he was able to do. In fact, Huygens saw experiments on magnetic phenomena in November 1660, which would be viewed by Ole Borch in May 1664.[109]

From the point of view of the orientation of the *Mercredis*, the place given to experiments was as limited as in the beginnings of the *Académie Montmor*. In a conference, by definition one confers, that is to say, one meets to speak together. When in 1682, Clerselier describes the *Mercredis*, he notes that their goal was to "*explain* one after another all the questions of physics," and that to do so, after an improvised speech about an hour long, Rohault would respond at length to the objections made.[110] The aim was, as with Montmor, to have a primarily discursive mode of action, although here without the support of a previously written text. It is not surprising in these conditions that it was as much Rohault's qualities as an experimenter as an orator that were praised by Clerselier himself, as well as by

---

[105] Rohault 1660. McClaughlin 1977, 228n18, attributes this manuscript to René Fédé. Clair 1978, 50–56, provides a few extracts.

[106] Quoted in Clair 1978, 46.

[107] Huygens 1888–1950, XXII, 541.

[108] On the criticism that Hobbes addressed to the Royal Society as a closed private space, see Shapin and Schaffer, 1985, 113–114.

[109] Huygens 1888–1950, XXII, 536; Borch 1983, III, 423.

[110] Clerselier 1682, unpaginated. I highlight.

other contemporaries.[111] It was only as final confirmation that the experiments took place:

> ...to hear him speak about this [the most difficult and curious questions of physics], you would think that he was in concert with nature and that nature took pleasure at revealing her secrets to him....For he directly communicated everything he said about these matters. And so that there would remain no doubt, he added as proof many beautiful experiments that he carried out in front of everyone, and most often he would alert everyone to the effects of the experiments, according to the principles he had previously established, before even getting to the experiment itself.[112]

In the *Mercredis*, the burden of proof was thus not on the experiments: these only took place after everything else, as a sort of complement. Although Clerselier does not hide that Rohault reasoned according to "previously established...principles," he took from the preface to the *System of Natural Philosophy* the affirmation that everyone could anticipate (*prévenir*) the effects of the experiments.

As for the subjects dealt with in the *Mercredis*, all testimony agrees to say that there were great experimental moments with Rohault: there were experiments on phenomena of capillarity (the famous "little tubes" in which water seemed to rise on its own), magnetism (Rohault had a box in which he kept all the objects required for experiments, including a magnet so powerful that it earned the admiration of all), on colors (on certain days an artificial rain would make appear a rainbow against a white cloth).[113] The confrontation of Huygens' travel journal with the notes of Monsieur F. is nonetheless very surprising: while they concern the same period, they do not report the same thing. Huygens reports solely on experiments, while Monsieur F. mentions none, unless in reference to illustrations.

On November 13 and 17, 1660, Huygens saw "experiments with quicksilver" that showed the weight of air and its resistance, then "experiments with magnets." On December 20 and 21, after being invited by a note from Clerselier to go to the home of Rohault, he saw a Spaniard carry out "the separation of silver ore," and then "experiments with tubes and small pipes." When instead we look at the notes

---

[111] See for example Rohault 1660, 1r: "...the conferences were written in a tumult, and at odd hours, he [the person collecting the conferences] was not as able as he would have liked to imitate the correctness and the incomparable precision of the terms of he who had the task of representing the feelings." Foucher 1675, 64–65: "You know that he was intent on reasoning with consequence, and as he perfectly possessed all the subjects he dealt with, he explained them with a great deal of order, and with a certain clarity, accompanied by a natural eloquence that one recognized more in its effect than in the disposition of the terms he would use." Malebranche, *Recherche de la vérité*, Preface to Volume II, in Malebranche 1958–1967, II, 564: "...everyone know with what accuracy and what force this learned man resisted the blows that others wanted to bear to him, and that with two or three words pronounced without heat and without movement, he struck down the imagination of those full of themselves who thought to cover him in embarrassment." Clerselier 1682, unpaginated: "... he summarized so well and in such good order everything objected to him, and responded with such clarity and enlightenment....".

[112] Clerselier 1682, unpaginated. On the meaning of the verb "prévenir," see above, n19. Three pages later, he refers again to the magnet, where the experiments had been anticipated ("prévenues") by the speeches.

[113] Clerselier 1682, unpaginated; Huygens 1888–1950, XXII, 539, 540.

of Monsieur F., we get the impression that the *Mercredis* were a general course on Cartesian philosophy, where they studied knowledge in general in November, arguments in favor for the existence of bodies (it was at this occasion that there was a bit of metaphysics) and the definition of matter as extension and the experiments of Torricelli and Pascal on the vacuum in December, the conservation of momentum in January, then from February to May, the difference, sense by sense between qualities that we feel and the qualities as they are in things. Monsieur F. sometimes mentions questions from the public and the Rohault's responses. For example:

> Someone asked if the matter of the heavens is the same as that of bodies. The answer was yes, since the extension is always extension wherever it is located. Someone asked if a vacuum can exist, and the response was no, since that would mean extension without matter, which is not possible; four walls would be distant from each other by this supposition, there would be no matter, hence no distance, hence they would touch.[114]

Monsieur F. does not explicitly mention experiments, but curiously, he refers to illustrations that are not found in his notes, and which he explicitly says in the case of comets, can be found in the book of Descartes—must we conclude that Rohault distributed the illustrations of the *Principles of Philosophy*? The three last entries in Monsieur F.'s notes are disparate by their dates and topics: on tides (November 1669), on the lodestone (November 17, 1660, which agrees with the affirmations of Huygens' travel journal), on comets (not dated, but an allusion to a comet in the period 1664–1665). One is tempted to say that when a visitor of distinction like Huygens was expected, Rohault prepared to show his most notable experiments, whatever his lesson plan.

In summary, contrary to the first meetings of the *Académie Montmor*, Rohault's *Mercredis* had a vocation for teaching rather than for research, and like them, they at times included moments of experimentation, perhaps less significant that it has been said, both in terms of quality and quantity. What was important was to explain Cartesian philosophy. This form of lectures, which at first was undoubtedly a novelty and an attraction, must soon have been considered passé by scientists, given the radical experimentalism that began in the early 1660s, first in the *Académie Montmor*, in the context of the *Compagnie des sciences et des arts*. This discredit could only have grown after the foundation of the *Académie des sciences*. Rohault was no longer a scientist, he was merely a Cartesian.

We can find a trace of this evolution, I feel, in the attitude of Huygens toward Rohault. During his stay in Paris from 1660 to 1661, Huygens, who had met Rohault at the home of Montmor, came three times to Rohault to see experiments, that is to say much less regularly than he went to the *Académie Montmor*, which Huygens attended nearly every week from November 1660 to February 1661, and one time less than to the *senatulus* of Cartesian ladies who met at the home of Madame de Bonneveaux.[115]

---

[114] Rohault 1660, 12r.

[115] Huygens 1888–1950, XXII, 536, 541; III, 210. One can note in passing that although at times one speaks of the "emancipatory" character of Cartesianism for women, the attitude of Huygens and his correspondents to this *senatulus* should lead one to a more nuanced judgment as to the type of knowledge women could access in this period; and it is significant that there is no sign of the

Until 1665, Rohault, although he was not in direct correspondence with Huygens, is mentioned from time to time in letters to and from Huygens, without any mention of new experiments being done. I however find no trace of him after 1666, including at the publication of the *System of Natural Philosophy* in 1671, or a year later, at the death of Rohault.[116] A factor explaining the progressive disinterest of Huygens with respect to Rohault arises perhaps from Rohault's stubborn insistence that capillary phenomena (the rising of water in thin tubes and the circular shape of the surface of the water) were explained by the weight of air, whereas Huygens had shown that this phenomenon also took place in a vacuum.[117] When Huygens, who had had great difficulties to make his own pump work properly, offered to Rohault to help him build one, he did so only with the explicit condition that Rohault ask for help, for, as he wrote to his brother, "there is always a bit of pedantry in what his does, as you may have noticed."[118] It seems to me that we can see in the lack of interest from Huygens for Rohault the sign that Rohault had fallen away from the experimental community.

As I'll now show, the period 1665–1690 corresponds to the time when the historiographic cliché developed according to which there was an essential contradiction between Cartesianism and experimentalism. This cliché was not totally unfounded at the time it developed, since Cartesians of that period had as their first goal to make known their master's doctrine, and they were as we have shown in the exemplary case of Rohault, much less experimenters and experimentalist than some of their contemporaries. It would however be an error to think that there was a historical necessity here or an ineluctable destiny set from the very first books of Descartes. Cartesianism was never an essence. It's a category that developed historically, though debates, polemics, and controversies, as well as distortions, misunderstandings, and changes in perspective.

3.4 Descartes and the Cartesians: The Constitution of a Cliché

Descartes's philosophy in general, and his natural philosophy in particular, were often attacked. The intensity and the intent of these attacks varied, however. Initially at least, they concerned neither Descartes' experimental abilities or his

presence of women in the most visible learned societies, whether the *Académie Montmor*, the *Académie Bourdelot*, or the *Compagnie des sciences et des arts*.

[116] Huygens 1888–1950, III, 397, 414, 432; IV, 6, 7, 11, 69, 367, 459; V, 29, 41, 101, 105. Aside from correspondence, Rohault's *Traité de physique* is mentioned in the preface of the *Discours de la cause de la pesanteur*; it also appears in certain critical notes on the Cartesian explanation of magnetism (Huygens 1888–1950, XIX, 572).

[117] On Rohault's description and explanation of these phenomena, see Rohault 1681, I, Chap. XXII, Sect. 67–84, 204–214. On Huygens' lack of confidence in Rohault's explanation, see his letters to Moray from December 9, 1663, February 20 and March 12, 1664, resp. in Huygens 1888–1950, IV, 459; V, 29, 41.

[118] Christiaan Huygens to Lodewijk Huygens, January 18, 1662, in Huygens 1888–1950, IV, 11. See as well the letter to the same from January 4, in Huygens 1888–1950, IV, 6–7.

disposition toward experimentalism. Professors who, like Vopiscus Fortunatus Plempius (1601–1671), Libertus Fromondus (1587–1653), or Jean-Baptiste Morin (1583–1656) analyzed what the *Discourse on Method* said about Cartesian physics were not preoccupied with whether this physics used experiments. For them, what was important was to know if, given its principles, it could correctly explain natural things. Astonishingly for us, the Aristotelians thought that Descartes was mistaken, but at least at first, they did not think that his preoccupations were entirely new. Thus Daniel Garber, on considering the reactions of Fromondus and Morin, was able to say about the publication of the *Discourse on Method* and the *Essays* that accompanied it, that this was a revolution that did not happen.[119]

In the following paragraphs, my goal is to sketch out a cartography of the criticisms of Cartesians physics beginning at the last third of the seventeenth century. It can be shown that, alongside the moral critique from faithful Gassendists like Sorbière and Chapelain on the way Descartes communicates the truth, there appears an epistemological critique concerning, if not the absence of experiments in Cartesian physics, at least the secondary status they held. I will begin by presenting the moral critique of the Gassendists in the 1660s, and show that it can be found in Huygens and Leibniz in the 1690s. I will then present the epistemological critique as it appears in Mariotte, a physicist of the *Académie des sciences*. I will finally show something much more surprising, that this epistemological critique is also present in the Jesuit Antoine Rochon.

Let's begin with the criticism from the Gassendists Sorbière and Chapelain, who let it be noted, did not care much for one another.[120] Although they did not really contribute to the advancement of scientific knowledge, they are often considered responsible for the exclusion of the Cartesians from the *Académie des sciences*, as well as important in its experimentalist commitment.[121] At first, that is to say, at the end of the 1630s and the beginning of the 1640s, Descartes was for them an author to be promoted, both for the renewal of philosophy to which he was able to contribute, and for the beauty of his style—which is not without ambivalence, for a beautiful style is also what allows him to pass off uncertain hypotheses as truths.[122] But progressively, and in particular after the metaphysical quarrel between Descartes and Gassendi (1644), the criticism overtook the praise. Although they could have found

---

[119] Garber 1988.

[120] In his letter to Francheville, March 16, 1665, in Chapelain 1880–1883, II, 390, Chapelain accuses Sorbière of having copied Gassendi without understanding him.

[121] On Chapelain's Gassendism, see Collas 1912, 60–64, 151–154, 331–336, 383–388. On Sorbière's Gassendism, see Pintard 1983, 334–348, nuanced however 418–420, 425, 429. On the responsibility of the Gassendists on the fact that the first members of the *Académie des sciences* were not Cartesians, see Taton 1966, 36; on the fact that they would have been *ipso facto* Gassendists and experimentalists, see McClaughlin 1975, 239–240.

[122] See in particular Chapelain to Balzac, May 31 and December 29, 1637, in Chapelain 1880–1883, I, 153, 189; Sorbière to Petit, February 10, 1657, in Sorbière 1660, 691. On the fact that the illustrations and comparisons used by Descartes would calm the crowd, see Sorbière to Mersenne, December 23, 1647, in Mersenne 1932–1988, XV, 585–587. On Gassendi's "too great literature," see Sorbière 1694, 124–126.

in Gassendi arguments against Cartesian physics in terms of experimental practice, whether for the observation of the trajectories of planets, the formation of salts, or the behavior of a weight dropped from the top of a mast, the criticism of Chapelain and Sorbière were primarily on Descartes's dogmatism.[123]

For them, dogmatism designates not only a doctrinal content opposed to skepticism; it also refers to, and mainly to, a moral or political attitude revealed by Descartes' practices in terms of the communication of the truth. In the parallels he establishes between Gassendi and Descartes, Sorbière notes that, while the former does not hesitate to present his thoughts in a familiar fashion with those with whom he conversed, the latter refused all dialog and referred those he spoke with to his writings.[124] According to Sorbière, there was something revealing in the way Descartes behaved when compared with other men: he never treats them as equals, but as their master, the "head of a sect," to whom all intellects will submit, obliging everyone he met to learn his doctrines:

> He did not disapprove in those he wished to instruct what Aristotle asked of a good student, docility and patience to mull over a doctrine in the mind until it was strongly imprinted on the memory.…There is no wonder that after four or five years of assiduous meditation and continuous repetition of certain terms, one does not want to lose the fruit of the pains taken to learn them, and if one convinces oneself in the end, that by the ease that one has acquired in repeating them, they mean more than they are. I am not astonished that Monsieur Descartes, requiring this from his sectators, those who having obeyed him having so shaped their mind to his philosophy, that it seems to me that they hold it closer to heart than he did himself. Even Lullists and Paracelsists end up persuading themselves that the gibberish they have stubbornly learned is founded on good reasons.[125]

The terms "sectators" and "sectarian" recur in the letters of Chapelain in the 1660s to designate the Cartesians, and there is no doubt that, although the first term can simply designate the partisans of a philosophical school, it is used by Chapelain in a negative fashion. Thus, he notes that Descartes was "happy to have sectators who swore on his dogma and believed that which he did not believe himself."[126]

It is not that they were totally uninterested in Cartesian physics. But beyond the fact that they only did so late in the game, their criticism has nothing to do with the

---

[123] On the experimental practice of Gassendi, see Rochot 1964; as for his judgment on Cartesian physics, he responds to Rivet who asked him for a critique (*censura*) of the *Principia philosophiae* that it was superfluous, for such a work should die before its author, see Gassendi 1964, VI, 217.

[124] Sorbière to Petit, November 10 and February 10, 1657, in Sorbière 1660, 679 and 691. Conversely, Clerselier 1667 notes that Roberval refused to put in writing his objections to Descartes.

[125] Sorbière to Petit, November 10, 1657, in Sorbière 1660, 679–680. For other passages where Descartes is described as a "head of a sect," see Sorbière to Saumaise, March 10, 1650 and to Petit, February 10, 1657, in Sorbière 1660, 535, 691. In the correspondence to Mersenne, Descartes is compared to Fludd, see Sorbière to Mersenne, April 15 and December 23, 1647, in Mersenne 1932–1988, XV, 201, 585–587.

[126] Chapelain to Carrel de Sainte-Garde, May 27, 1662, in Chapelain 1880–1883, II, 235–236. The word "sectators" describes Cartesians in Chapelain to Heinsius, February 6, 1659, to Carrel de Sainte-Garde, December 15, 1663 and to Bernier, February 16 and April 26, 1669, in Chapelain 1880–1883, 17, 341, 622, 640.

lack of experiments in the physics of Descartes and the Cartesians. Thus what poses a problem for Sorbière is the possibility, in general physics, of obtaining the three Cartesian elements from the "large indefinite body" that is the extended matter. If it is possible to break it, then nothing can "stop my little bodies of a certain size and shape: rather than allowing indefinite division and imagining all of nature as a large, fluid, permeable body, whose motions can divide it at any point of these dimensions."[127] Likewise, Chapelain believes that the affirmation that a vacuum is impossible leads to contradictions, but does not engage at all in the question of whether the explanations proposed by Descartes are or are not plausible with respect to experiments:

> The quarrel he [Descartes] had with Monsieur Gassendi, my intimate friend, made me want to look again with more attention at his system. I found great brilliance, great novelty, and a happy use of the ancient doctrine of Democritus for the multiplication of worlds and the modern experiments with magnets for the constitution of his machine, as well as beautiful applications of the nature of motion that could only arise from a greatly inventive mind clever at using everything to his ends. But I was stopped first by his *postulata* as to the creation of matter, which is less that of a physicist than a theologian, and of motion, which he made without admitting the least vacuum, which according to his own positions I found impossible, and I consequently saw that everything he deduced for the formation of worldly bodies and for particular generations fell apart, and in falling ruined all his claims. I also judged that what made him exclude the vacuum from his universe was only to have the means of explaining light and the tides, and that without this ambition, by admitting the smallest vacuum in the great plan ["plan" says the French; perhaps one should read "plein," i.e., "fullness"], he could have moved his illusion forward and made it if not true, at least beautiful and plausible enough for it to not be easily shown to be false.[128]

In the last third of the seventeenth century, what I call the "moral" criticism of Descartes and the Cartesians becomes more acute, for example with Huygens and Leibniz, and there is clearly a backlash from the worldly success of Cartesianism and the propaganda of a now well-defined Cartesian party. Huygens, when he reads Baillet's *Vie de Monsieur Descartes*, notes that unlike Galileo, Descartes wanted to establish himself as the "head of a sect" and the "author of a new philosophy." Descartes is at fault not for being mistaken, for all errors can be corrected, but for having presented what he proposed as if it were truths established once and for all, so much so that

> those who believe him and have become his sectators imagine that they possess the knowledge of the causes of everything, as much as it is possible to know them. Thus they often waste time supporting their master's doctrine, and do not study how to penetrate the true reasons for this great number of natural phenomena about which Descartes came out only with illusions.[129]

A year later, in his letters to Gerhard Meier (1646–1703), it is those who claim to be Cartesians that Huygens particularly attacks, for they are mistaken when they

[127] Sorbière to Saumaise, March 10, 1650, in Sorbière 1660, 536.

[128] Chapelain to Carrel de Sainte-Garde, February 16, 1662, in Chapelain 1880–1883, II, 203–204. Carrel de Sainte Garde published in 1663 his *Lettres contre la philosophie de Descartes*.

[129] Huygens 1888–1950, X, 404–405.

think it possible to protect all the theories of this man of great genius (*omnia viri ingeniosissimi dogmata*).[130]

Leibniz's verdict is quite similar. In a letter to Huygens, which also followed the reading of the work of Baillet, Leibniz, who had also noted that "Monsieur Descartes had a strange ambition of becoming the head of a sect," notes that "...the Cartesians are too pre-convinced of their hypotheses. I prefer a Leeuwenhoek who tells me what he sees than a Cartesian who tells me what he thinks," adding, "It is none-theless necessary to join reasoning to observations."[131] When he addresses himself to Paul Pélisson (1624–1693) or Claude Nicaise (1623–1701) in the *Journal des savants*, Leibniz insists that the Cartesians are unfaithful to the genius of Descartes in becoming his sectators and his "paraphrasists":

> I have infinite esteem for Monsieur Descartes, and I recognize his value perhaps better than some who declare themselves Cartesians....But I find that nothing brings more harm to the sciences than the spirit of a sect and than servitude. And indeed, the Cartesians find almost nothing new, and barely progress.[132]
>
> The best response that the Cartesians could make would...to get rid themselves of the spirit of a sect, always contrary to the advancement of science..., of taking on experiments and demonstrations instead of general reasonings that serve only to maintain idleness and cover ignorance. They should try to take a few steps forward and not content themselves with being simple paraphrasists of their master. They should not neglect or despise anatomy, languages, criticism, for want of knowing their importance and value....I would add that I do not know and by what star, whose influence is the enemy of all sorts of secrets, the Cartesians have done almost nothing new, and that almost all the discoveries have been made by people who are not Cartesians. I know only the little pipes of Monsieur Rohault that deserve the name of a discovery by a Cartesian. It seems to me that those who are attached to a single master thus reduce themselves by this sort of slavery, and conceive of almost nothing after him.[133]

But the moral critique of the Cartesian's herd mentality is joined by an epistemo-logical critique on the role of experiments in Cartesian physics. This was in particular the case for the member of the Académie des sciences Edme Mariotte (1620–1684), whose radical experimentalism I have studied elsewhere.[134] Three of the reasons given by him in his *Essai de logique* published anonymously in 1678 for the lack of progress made by physics came from the negligence of experiments by Cartesians. First, notes Mariotte, "several philosophers" stubbornly search for the causes of the principles of experience, when instead it would be better to use these principles as principles, in other words, to use them to deduce several "beautiful consequences."

[130] Huygens to Meier, June 1691, in Huygens 1888–1950, X, 104. See also to Meier, March 26, 1691, in Huygens 1888–1950, X, 54.

[131] Leibniz to Huygens, February 20, 1691, in Huygens 1888–1950, X, 52. The notes of Leibniz on Baillet can be found in Leibniz 1961, IV, 315–324.

[132] Leibniz to Pelisson, March 18, 1692, in Leibniz 1923–, I–7, 292.

[133] *Journal des scavants*, April 13, 1693, 163–164. On the fact that Cartesians did not discover anything, see Leibniz to Gallois, [1677] and to Malebranche, June 22, 1679 in Leibniz 1923–, II–1, 569, 717; to Swelingius, in Leibniz 1961, IV, 329–330.

[134] Roux 2011.

The examples he later gives of cases where this error has been committed leave no room for doubt as to the fact that he's thinking of Cartesians: he mentions the Cartesian type of explanation of elasticity and magnetism.[135] A second reason for the lack of progress in physics can also be attributed to Cartesians: "most philosophers," affirms Mariotte, are infatuated with a few hypotheses that are insufficiently established from an experimental perspective, yet want to use them to explain everything.[136] Although he gives no examples of this fault, from his physics essays we see that one of these poorly established hypotheses was the Cartesians' subtle matter.[137] The case is even clearer for a third cause of the lack of progress in physics, the way in which some claim to explain a natural effect by a single cause, when several causes contribute to producing it. Mariotte follows this diagnostic by a summary of some of the propositions of his *Traité de la percussion* that establish that one cannot explain the communication of motion between bodies solely by the conservation of motion, one of the Cartesian laws of nature, or even the fundamental Cartesian law of nature.[138]

It is known that in their teaching at least, the Jesuits made way for new observations and experiments that were performed in the first half of the century[139]; in these conditions one should not be surprised if, in the polemical texts they wrote against Descartes after the condemnation of his writings, some of them insisted on his inability to take seriously the experimental character of the new science. The *Lettre d'un philosophe à un cartésien de ses amis*, published anonymously in 1672, is here the most telling. It seems to have been published by Antoine Rochon (1637–???), a professor of philosophy in Bordeaux and *prédicateur* in Toulouse who finally quit the Jesuits for the Benedictines in 1685. It is said that another Jesuit scientist, Ignace-Gaston Pardies (1636–1674) polished up Rochon's book.

It's quite interesting that, far from being content to deal solely with matters of faith, Rochon defends a certain way of proceeding in physics. Thus he opposes Cartesians first with the recent experimental discoveries that go against Descartes' physics. Steno and the members of the *Académie des sciences* have shown, "unless one has more deference for the idea of Monsieur Descartes than for the testimony of ones own eyes," that no nerve reaches the pineal gland, that no valve explains the motion of our limbs, and that Mariotte has shown that the optical nerve does not end on the retina.[140] Descartes rested his entire physics on top the laws of motion, but the "author of the *Discours du mouvement local* [Ignace-Gaston Pardies] claims to show that of the seven rules of motion Monsieur Descartes wished to establish, only one is true."[141]

---

[135] Mariotte 1992, 97, 103.

[136] Mariotte 1992, 98.

[137] Mariotte 1717, I, 170–171; II, 341.

[138] Mariotte 1992, 98.

[139] Brockliss 1995a, 454–456; 1995b, 190–194, 199, 209–216.

[140] Rochon 1673, Sect. 47, 120–122.

[141] Rochon 1673, Sect. 48, 122–124.

Rochon then considers the argument of the new philosophers that Aristotelian qualities, virtues and forms do not bring any new knowledge.[142] Against this argument, Rochon insists, in the specific instance of the growing of plants, that Cartesian explanations add nothing to what everybody knows. If the explanation of a given phenomenon is to say that certain corpuscles animated by certain motions produce this phenomenon, it has no informational content.

> Everything you say gives no knowledge of the particular and of what is in fact in a plant. You content yourselves by saying that the pores are arranged *in a certain manner*, that they are *of a certain figure*, that the parts that *conform* to *certain openings* pass through, and that others are stopped. You try to get away with '*a certain*'. But if I ask you what this certain figure is, and what the certain manner and what is this certain juice and these certain parts, you have nothing to say other than that you know no more.[143]

Pointing out that Descartes had challenged the Jesuits to find an issue on which his philosophy would not be more satisfactory than the philosophy of the School, Rochon challenges a whole assembly of Cartesians to explain the simplest thing, the formation of a pumpkin in a single night. His prognosis is, as one might guess, that they will be unable to do better than to parade once again with their "certain figures," "certain shaped pores," "certain motions," and "certain ways."[144]

Rochon notes at this point that the Cartesians claim to anticipate the outcome of experiments (*prévenir l'expérience*):

> It is true that your gentlemen do wonders when they can latch on to an experiment they have done a hundred times to be sure of it. In that case they are happy to show the beauty of their doctrine by anticipating, they say, the experiment [*en prevenant, disent-ils, l'experience*], and by showing that it must follow their principles. This is called divining everything one sees and precisely predicting the past. I never hear of this advantage they give themselves of anticipating the effects of nature without remembering what happened to Cardano.[145]

If Cartesians are to be compared to Cardano, it's because Cardano would defend his rules in astrology by drawing up the horoscopes of the dead. When he tried however to extend his computations to future events concerning the living, what actually happened obliged him to resume his computations retroactively to adapt them to the actual events. Thus, if it is true that "in the experiments they have already done," "everything is in marvelous agreement with nature," to be truly convincing, Cartesians would need to be able to say "what would happen if one carried out a certain experiment that I suggest to them, and that they probably had never done."[146]

Thus it is in the period 1670–1690 that was constituted the affirmation that Cartesianism is essentially in contradiction with experimentalism, an affirmation

---

[142] Rochon 1673, Sect. 50, 128–129.

[143] Rochon 1673, Sects. 59–60, 140–144, and 142 for the quotation.

[144] Rochon 1673, Sect. 84, 194–196.

[145] Rochon 1673, Sect. 85, 197–198. "Prévenir l'expérience" is what Rohault claimed to do with his third sort of experiment, see above.

[146] Rochon 1673, Sect. 85, 202.

destined to become a cliché in the eighteenth century.[147] Where there's smoke there's fire applies to the Cartesians when faced with the first scientific institutions focused on experimentation and observation, but this certainly does not apply to Descartes himself, unless insofar as he had become in the hands of the Cartesians the object of a paraphrase and a Scholastic teaching. This explains in part the contradictory verdicts given to the function of experiments in the Cartesian sciences.

The fate of an expression can illustrate this. We know that Descartes presented the *World* as a fable, and that in the letter-preface that he wrote in 1647 for the *Principles of Philosophy* he recommended to his readers that they read the book "first in whole as if a novel," that is to say, as he says himself, all at once, not interrupting ones reading, and suspending the question of truth.[148] Readers critical of Descartes quickly adopted this expression to ridicule him. Thus we know the witticism of Pascal as reported by Antoine Menjot: "the late Monsieur Pascal called Cartesian philosophy the novel of nature, more or less the same as the story of Don Quixote."[149] In the years that concern us, we can find this idea everywhere, in Chapelain, Sorbière, Huygens, Leibniz, or Charles Perrault.

3.5 Conclusions

In the great silence of their libraries and classrooms, today's historians no longer see the conflicts that marked the history of science, or at least, they cannot perceive what was so clear and distinct about them. Even when they develop an irenic vision of this history, they can be tempted to project on certain moments of the past lines of conflict that did not yet exist, or that on the contrary, had in the meantime disappeared.

The doctrines of empiricism and experimentalism, even if they could be defined in absolute terms, as I did at the beginning of this chapter, must be the object of a historical modulation when they are used to characterize a philosopher with respect to the philosophers of his time. Indeed, this kind of characterization is necessarily relative, for it refers to positions that existed at that time. The main idea of this chapter is in particular that the kind of experimentalist commitment there was at the time of Descartes was no longer the case in the 1660s: the very existence of the *Académie des sciences* and the scientific movement that had prepared its foundation introduced significant modifications in the field of natural philosophy. Rohault and Descartes had the same epistemological positions, but they are modulated differently

[147] Voltaire to Maupertuis, October 1, 1738, in Voltaire 1968–1977, V, 307–308: "if one had to get into this other and no less frivolous question, which one nonetheless agitates, of knowing who was the greater physicist, Descartes or Newton, it would be enough to consider that Descartes almost never carried out experiments…. If one wanted to discuss the physics of Descartes, what could one perceive there other than hypotheses?"

[148] AT XI 31, 48.

[149] See the references given in Roux 2006, n40.

when compared to the field formed by the philosophical positions of the 1630s or that of the 1660s.

It would nonetheless be a mistake to draw from this chapter ammunition to strengthen the historiographic prejudice according to which France was blinded by a rationalism so blinkered that it could claim to be universally applicable, while England benefited from the lights of full-fledged experimentalism. I believe on the contrary, as I indicated in the second part of this chapter, that the epistemological styles of France of the first *Académie des sciences* and of England of the first Royal Society are identical, at least with regard to the use of experiments. Many things have masked this similarity: the philosophical confusion between empiricism and experimentalism, or the confusion between carrying out experiments and having a doctrine of experimentation, the difficulty of distinguishing 300 years later between the social and intellectual issues of the 1630s and the 1660s, and finally, the existence of founding national myths, with their pantheons of great men. It was not however my aim in this chapter to explore all these issues.

References

Azouvi, François. 2002. *Descartes et la France. Histoire d'une passion nationale*. Paris: Fayard.

Bacon, Francis. 1996. *The Oxford Francis Bacon*, 15 vols, ed. G. Rees and L. Jardine. Oxford: Clarendon Press.

Baillet, Adrien. 1691. *La vie de M. Descartes*, 2 vols. Paris: Daniel Horthemels.

Béguin, Katia. 1999. *Les Princes de Condé. Rebelles, courtisans et mécènes dans la France du Grand Siècle*. Seyssel: Champ Vallon.

Borch, Ole. 1983. *The Journal of the Danish polyhistor Ole Borch, ed. with introduction and indices by H.D. Schepelern*, 4 vols. Copenhagen/London: Reitzels Forlag, Brill.

Bougerel, Joseph. 1737. *Vie de Pierre Gassendi*. Paris: Jacques Vincent.

Boyle, Robert. 2001. *The correspondence of Robert Boyle*, ed. Michael Hunter, Antonio Clericuzio, and Lawrence M. Principe. London: Pickering & Chatto.

Brockliss, Laurence W.B. 1995a. Descartes, Gassendi, and the Reception of the Mechanical Philosophy at the French Collèges de Plein Exercice, 1640–1730. *Perspectives on Science* 3(4): 450–479.

Brockliss, Laurence W.B. 1995b. Pierre Gautruche et l'enseignement de la philosophie de la nature dans les collèges jésuites français vers 1650. In *Les Jésuites à la Renaissance*, ed. Luce Giard, 187–219. Paris: Presses Universitaires de France.

Brown, Harcourt. 1934. *Scientific organizations in seventeenth century France*. New York: Russel and Russel.

Cassini, Giovanni Domenico. 1693. De l'origine et du progrès de l'Astronomie, et de son usage dans la Geographie et la Navigation. In *Recueil d'observations faits en plusieurs voyages par ordre de sa Majesté... par Messieurs de l'Académie Royale des sciences*. Paris: Imprimerie Royale.

Chapelain, Jean. 1662. Liste de quelques gens de lettres vivans en 1662. Composée par ordre de M. Colbert, par M. Chapelain. In *Memoires de littérature et d'histoire. Tome second*, ed. P. Desmolets. Paris: Noyons, 1749.

Chapelain, Jean. 1666. Lettre de M. Chapelain à M. Colbert dans lesquelles il expose son sentiment sur les divers projets de création et d'encouragement que méditait le ministre en faveur des arts et des sciences. In *Lettres, instructions et mémoires de Colbert*, vol. V, ed. Pierre Clément, 513–514, 8 vols. Paris: Perrin, 1861–1882.

Chapelain, Jean. 1880–1883. *Lettres de Jean Chapelain*, ed. Philippe Tamizey de Larroque, 2 vols. Paris: Imprimerie Nationale.

Clair, Pierre. 1978. *Jacques Rohault (1618–1672). Bio-bibliographie. Avec l'édition critique des Entretiens sur la philosophie*. Paris: CNRS.

Clarke, Desmond M. 1982. *Descartes' philosophy of science*. Manchester: Manchester University Press.

Clarke, Desmond M. 1989. *Occult powers and hypotheses. Cartesian natural philosophy under Louis XIV*. Oxford: Oxford University Press.

Clerselier, Claude. 1657. Préface. In *Descartes, Lettres de Monsieur Descartes*. Tome Premier, ed. Claude Clerselier. Paris: Charles Angot.

Clerselier, Claude. 1659. Préface. In *Lettres de Monsieur Descartes. Tome second*, ed. Claude Clerselier. Paris: Charles Angot.

Clerselier, Claude. 1667. Préface. In *Lettres de Monsieur Descartes. Tome troisième et dernier*, ed. Claude Clerselier. Paris: Charles Angot.

Clerselier, Claude. 1682. Préface. In *Œuvres posthumes de Rohault*, ed. Claude Clerselier. Paris: Guillaume Desprez.

Collas, Georges. 1912. *Jean Chapelain. 1595–1574. Étude historique et littéraire d'après des documents inédits*. Paris: Perrin.

Dear, Peter. 1995. *Discipline and experience. The mathematical way in the scientific revolution*. Chicago: Chicago University Press.

Denis, Jean-Baptiste. 1668. *Lettre écrite à Monsieur Sorbière... par Jean Denis, touchant l'origine de la transfusion du sang et la manière de la pratiquer sur les hommes*. Paris: Jean Cusson.

Des Chene, Dennis. 2001. *Spirits and clocks. Machine and organism in Descartes*. Ithaca: Cornell University Press.

Descartes, René. 1667. *Lettres de Monsieur Descartes. Tome troisième et dernier*, ed. Claude Clerselier. Paris: Charles Angot.

Dew, Nicholas. 2009. *Orientalism in Louis XIV's France*. Oxford: Oxford University Press.

Duhamel, Jean-Baptiste. 1698. *Regiae scientiarum Academiae historia*. Paris: Jean-Baptiste Delespine.

Fontenelle, Bernard Le Bovier de. 1733. *Histoire de l'Académie royale des Sciences. Tome I. Depuis son établissement jusqu'à 1686*. Paris: Gabriel Martin, Jean-Baptiste Coignard et Hyppolite-Louis Guérin.

Fontenelle, Bernard Le Bovier de. 1994. *Œuvres complètes. Tome VI*. Paris: Fayard.

Foucher, Simon. 1675. *Critique de la recherche de la vérité. Où l'on examine en même temps une partie des principes de Mr Descartes. Lettre par un académicien*. Paris: Martin Coustelier.

Garber, Daniel. 1988. Descartes, the Aristotelians, and the revolution that did not happen in 1637. *The Monist* 71(4): 471–486.

Garber, Daniel. 2001a. Descartes and experiment in the *Discourse* and *Essays*. In *Descartes embodied. Reading Cartesian philosophy through Cartesian science*, 85–110. Cambridge: Cambridge University Press.

Garber, Daniel. 2001b. Experiment, community, and the constitution of nature in the seventeenth century. In *Descartes embodied. Reading Cartesian philosophy through Cartesian science*, 296–328. Cambridge: Cambridge University Press.

Gassendi, Pierre. 1658. *Opera Omnia*, Lyon, 6 vols. Repr. Stuttgart: Friedrich Fromman Verlag, 1964.

Gassendi, Pierre. 1964. *Opera Omnia*, 6 vols. Lyon 1658. Repr. Stuttgart: Friedrich Fromman Verlag.

Goldstein, Catherine. 2008. Écrire l'expérience mathématique au XVIIe siècle; la méthode selon Bernard Frenicle de Bessy. In *Réduire en art. La technologie de la Renaissance aux Lumières*, ed. Pascal Dubourg-Glattigny and Hélène Vérin, 213–224. Paris: Éditions de la Maison des Sciences de l'Homme.

Hahn, Roger. 1971. *The Anatomy of a Scientific Institution. The Paris Academy of Sciences, 1666–1803*. Berkeley/Los Angeles/London: University of California Press.

Hirschfield, John Milton. 1981. *The Académie royale des sciences. 1666–1683*. New York: Arno Press.

Hobbes, Thomas. 1994. *Thomas Hobbes. The correspondence: 1660–1679*, The Clarendon edition of the works of Thomas Hobbes, vol. 2. Oxford: Oxford University Press.

Huygens, Christiaan. 1888–1950. *Œuvres complètes de Christian Huygens*, 22 vols, ed. Société Hollandaise des Sciences. La Haye: Martinus Nijhoff.

Koyré, Alexandre. 1953. An experiment in measurement. *Proceedings of the American Philosophical Society* 97(2): 222–237.

Kuhn, Thomas. 1976. Tradition mathématique et tradition expérimentale dans l'évolution des sciences physiques. *Journal of Interdisciplinary History* 7: 1–31.

Le Bret, Henri. 1657. Préface. In *Histoire comique par Monsieur de Cyrano Bergerac. Contenant les Estats & Empires de la Lune*, ed. Henri Le Bret. Paris: Charles de Sercy.

Le Dœuff, Michèle. 1984. Bacon chez les Grands au siècle de Louis XIII. In *Francis Bacon. Terminologia e fortuna nel XVII secolo*, ed. Marta Fattori, 155–178. Roma: Ateneo.

Leibniz, Gottfried Wilhelm. 1923–. Leibniz. Sämtliche Schriften und Briefe, Preussische (then Deutsche) Akademie der Wissenschaften ed. Darmstadt, then Leipzig, then Berlin.

Leibniz, Gottfried Wilhelm. 1961. *Die philosophischen Schriften von G. W. Leibniz* (1875–1890), 7 vols. Repr., ed. C.I. Gerhardt. Hildesheim/New York: G. Olms.

Malebranche, Nicolas. 1958–1967. *Œuvres de Malebranche*, 20 vols., ed. André Robinet. Paris: Vrin.

Mariotte, Edme. 1717. *Œuvres de Mr Mariotte, de l'Académie Royale des Sciences, divisées en deux tomes, Comprenant tous les Traitez de cet Auteur, tant ceux qui avoient déjà paru séparément, que ceux qui n'avoient pas encore été publiez.* Leiden: P. van der Aa.

Mariotte, Edme. 1992. *Essai de logique*, ed. Alan Gabbey and Guy Picolet. Paris: Fayard.

Mazauric, Simone. 2007. *Fontenelle et l'invention de l'histoire des sciences à l'aube des Lumières*. Paris: Fayard.

McClaughlin, Trevor. 1974. Une lettre de Melchisédec Thévenot. *Revue d'histoire des sciences* 27(2): 123–126.

McClaughlin, Trevor. 1975. Sur les rapports entre la Compagnie de Thévenot et l'Académie royale des Sciences. *Revue d'histoire des sciences* 28(3): 235–242.

McClaughlin, Trevor. 1977. Le concept de science chez Jacques Rohault. *Revue d'histoire des sciences* 30(3): 225–240.

McClaughlin, Trevor. 1996. Was there an empirical movement in mid-seventeenth century France? Experiments in Jacques Rohault's *Traité de physique*. *Revue d'histoire des sciences* 49(4): 459–481.

McClaughlin, Trevor. 2000. Descartes, experiments, and a first generation Cartesian, Jacques Rohault. In *Descartes' natural philosophy*, ed. Stephan Gaukroger, John Schuster, and John Sutton, 330–345. London/New York: Routledge.

Mersenne, Marin. 1932–1988. *Correspondance du P. Marin Mersenne, religieux minime, Paul Tannery, Cornélis de Waard and Armand Beaulieu eds.*, 17 vols. Paris: Éditions du CNRS.

Mesnard, Jean. 1963. Pascal à l'Académie Le Pailleur. *Revue d'histoire des sciences* 16(1): 1–10.

Milhaud, Gaston. 1921. *Descartes savant*. Paris: Alcan.

Monconys, Balthazar de. 1665–1666. *Journal des voyages de Monsieur de Monconys*, 3 vols. Lyon: Horace Boissat & George Remeus.

Moreau, Denis. 1999. *Deux cartésiens. La polémique entre Antoine Arnauld et Nicolas Malebranche*. Paris: Vrin.

Moreri, Louis. 1759. *Le Grand dictionnaire historique ou Mélange curieux de l'histoire sacrée et profane. Tome Neuvième*. Paris: Les Libraires associés.

Mouy, Paul. 1934. *Le développement de la physique cartésienne, 1646–1712*. Paris: Vrin.

Oldenburg, Henry. 1965–1973. *The correspondence of Henry Oldenburg*, 9 vols., ed. and trans. Alfred Rupert Hall et Marie Boas Hall. Madison, Wisconsin-London: University of Wisconsin Press.

Pascal, Blaise. 1663. *Traitez de l'quilibre des liqueurs et de la pesanteur de la masse de l'air*. Paris: Guillaume Desprez.

Perrault, Charles. 1666. Note de Charles Perrault à Colbert pour l'établissement d'une académie générale. In *Lettres, instructions et mémoires de Colbert*, vol. V, ed. Pierre Clément, 512–513, 8 vols. Paris: Perrin, 1861–1882.

Pintard, René. 1951. Autour de Pascal: l'Académie Bourdelot et la querelle du vide. In *Mélanges d'histoires littéraires offerts à Daniel Mornet*, 73–81. Paris: Nizet.

Pintard, René. 1983. *Le libertinage érudit dans la première moitié du xviie siècle*, 2nd ed. Paris/Genève: Slatkine.
Rochon, Antoine. 1673. *Lettre d'un philosophe à un cartésien de ses amis*. Paris: Estienne Michallet.
Rochot, Bernard. 1964. Gassendi et l'expérience. In *Mélanges Alexandre Koyré*, vol. II, 411–422. Paris: Hermann.
Rohault, Jacques. 1660. Conférences, recueillies par M. F. avocat qui y a, dit-il, ajouté du sien. Mss 2225, Bibliothèque Sainte-Geneviève, Paris.
Rohault, Jacques. 1681. *Traité de physique*, 2 vols. Lyon: Claude Galbit.
Rohault, Jacques. 1987. *System of natural philosophy, illustrated with Dr. Samuel Clarke's notes, taken mostly out of Sir Isaac Newton's philosophy*. New York/London: Garland Publishing.
Rohault, Jacques. 2009. *Physique nouvelle (1667)*, ed. Sylvain Matton. Paris/Milan: Séha, Archè.
Roux, Sophie. 1998. Le scepticisme et les hypothèses de la physique. *Revue de synthèse* 119(2–3): 211–255.
Roux, Sophie. 2006. La philosophie naturelle à l'époque de Le Nôtre. Remarques sur la philosophie mécanique et sur le cartésianisme. In *Fragments d'un paysage culturel, André Le Nôtre, Institutions, arts, sciences et techniques*, ed. Georges Fahrat, 98–111. Sceaux: Musée de l'Île-de-France.
Roux, Sophie. 2011. *L'Essai de logique de Mariotte. Archéologie des idées d'un savant ordinaire*. Paris: Classiques Garnier.
Roux, Sophie. 2012. A French Partition of the Empire of Natural Philosophy (1670–1690). In *The mechanisation of natural philosophy*, ed. Daniel Garber and Sophie Roux, 55–98. New York/Dordrecht/Boston/London: Kluwer Academic Publishers.
Roux, Sophie. 2013. Pour une conception polémique du cartésianisme. Ignace-Gaston Pardies et Antoine Dilly dans la querelle de l'âme des bêtes. In *Qu'est-ce qu'être cartésien?* ed. Delphine Kolesnik, 315–337. Lyon: Presses de l'ENS de Lyon.
Savérien, Alexandre. 1783. *Histoire des philosophes modernes. Tome VI. Histoire des physiciens*. Paris: Bleut and Guillaume.
Schmaltz, Tad. 2002. *Radical Cartesianism: the French reception of Descartes*. Cambridge: Cambridge University Press.
Shapin, Steven, and Simon Schaffer. 1985. *Leviathan and the air-pump: Hobbes, Boyle, and the experimental life*. Princeton: Princeton University Press.
Sorbière, Samuel. 1660. *Lettres et Discours sur diverses matières curieuses*. Paris: François Clousier.
Sorbière, Samuel. 1694. *Sorberiana sive Excerpta ex ore Samëlis Sorbiere*. Toulouse: Colomeis.
Sorbière, Samuel. 1663. Discours prononcé le 3 d'avril 1663. À l'ouverture de l'Académie de physiciens qui s'assemblent tous les mardis chez Monsieur de Montmor. In Guillaume Bigourdan, Les premières réunions savantes de Paris au XVIIe siècle, *Comptes rendus hebdomadaires des séances de l'Académie des sciences*, 164, séances du 22 et du 29 janvier 1917:159–162 and 216–218.
Sorbière, Samuel, and Du Prat, Abraham. 1657. Règlement de l'assemblée de physiciens, qui se fit à Paris, chez Monsieur de Montmor l'an 1657. In Sorbière à Hobbes, 1er fév. 1658, in Sorbière 1660: 633–634. To be found also in Huygens 1888–1950, IV, 514–515.
Taton, René. 1966. *Les origines de l'Académie royale des sciences, Conférence du Palais de la Découverte, 15 mai 1965*. Paris: Palais de la Découverte.
Thévenot, Melchisédech. 1681. Discours sur l'art de la navigation. In *Recueil de voyages de Mr Thévenot dédiés au Roi*. Paris: Estienne Michallet.
Thévenot, Melchisédech. 1694. *Bibliotheca Thevotiana sive Catalogus impressorum et manuscriptorum Liborum Bibliothecae viri Clarissimi D. Melchisedecis Thevenot*. Paris: Florentin et Pierre Delaulne.
Voltaire, François-Marie Arouet. 1968–1977. *Les Œuvres complètes de Voltaire*. Geneva/Banbury/Oxford: Institut et Musée Voltaire & Voltaire Foundation.
Zittel, Claus. 2009. *Theatrum philosophicum. Descartes und die Rolle ästhetischer Formen in der Wissenschaft*. Berlin: Akademie Verlag.

Chapter 4
Dutch Cartesian Empiricism and the Advent of Newtonianism

Wiep van Bunge

Abstract At least since Voltaire, the perception of Cartesianism has often suffered from comparison to Newtonianism. In particular, Descartes' 'Rationalism' has been regarded as basically flawed on account of its incompatibility with Newton's approach to natural philosophy, which was to dominate much of eighteenth-century thought. In this paper it is argued that, on the contrary, both Descartes and some of his most tenacious Dutch admirers did not eschew Empiricism at all, but were actually instrumental in the early dissemination of Newtonianism on the Continent, and at Leiden University in particular.

4.1 Voltaire Versus Descartes

In his fourteenth *lettre philosophique*, entitled "Sur Descartes et Newton," written in 1728 and first published in 1733 in English and in 1734 in French, Voltaire launched a deliberate effort to destroy the reputation of Descartes as a natural philosopher. Voltaire had been present at Newton's burial on March 28, 1727 in Westminster Abbey, and the occasion had made a lasting impact on the budding *philosophe*.[1] Put succinctly, Voltaire was convinced Newton was right and Descartes had been wrong. Descartes' only genuine achievement, Voltaire pointed out, had been his geometry: the way in which he used algebraic equations in order to calculate curves had been particularly useful in his *Dioptrique*, but all his other works "abound with errors," as Voltaire mused:

> Geometry was a guide that he himself had created in a way and which would have been of sure guidance in his physics; yet in the end he abandoned this guide and surrendered

[1] Or so experts have traditionally inferred from Voltaire 1964, esp. 71, but there is no clear evidence that he actually attended the funeral of Newton, whom he incidentally never met either. See Barber 1979. For more on Voltaire and Newton in general see Rousseau 1976.

W. van Bunge (✉)
Erasmus University, Rotterdam, The Netherlands
e-mail: vanbunge@fwb.eur.nl

M. Dobre and T. Nyden (eds.), *Cartesian Empiricisms*, Studies in History and Philosophy of Science 31, DOI 10.1007/978-94-007-7690-6_4,
© Springer Science+Business Media Dordrecht 2013

himself to the spirit of system. His philosophy then was merely an ingenious novel and at most the ignorant found in it a semblance of truth.[2]

Apart from his obvious and manifold mistakes in physics, Voltaire continued, referring to Descartes' theory of "la création des vérités éternelles," his metaphysics resulted in the absurd conclusion "that two and two make four only because God has willed it so."[3] Still, Descartes must be credited with terminating Scholasticism, for thus he laid the foundations on which Newton was able to build.

Voltaire's assessment of Descartes would become hugely influential during the eighteenth century as well as among twentieth-century scholars. In particular the pejorative way in which Voltaire referred to the Cartesian "esprit de système" would be reiterated by many *philosophes*, including D'Alembert, who in his *Discours préliminaire* (1751) to the *Encyclopédie* would turn the distinction between the "esprit systématique" and the "esprit de système" into one of the key insights of eighteenth-century science. In his wake, Ernst Cassirer set out to reconstruct "The Mind of the Enlightenment" on this very opposition.[4]

Voltaire himself would return repeatedly to Descartes, for instance in the *Dictionnaire philosophique* (1764), in which he stressed once more the insufficient empirical content of Cartesian natural philosophy: "True physics, then, consists in successfully identifying all effects. We shall know the first causes when we have become gods." Clearly, Descartes was not divine: "To us has been given the power to calculate, to weigh, to measure, to observe: that is natural philosophy; nearly all the rest is a chimera."[5] In fact, Voltaire claimed, everything that was new in Cartesianism was flawed, and even the earlier praise lavished on his countryman on account of his destruction of Peripateticism made way for the more sobering conclusion that Descartes halted the intellectual progress of mankind for half a century: "thus he merely replaced Aristotle's chaos with another chaos."[6] In his last major work, the massive *Questions sur l'Encyclopédie* of 1770, Voltaire even claimed that the chaos Descartes had called his "natural philosophy" had been of no use whatsoever to Newton who had hardly taken the trouble to read Descartes'

[2] Voltaire 1964, 75: "La Géometrie était un guide que lui-même avait en quelque façon formé, et l'aurait conduit sûrement dans sa Physique; cependant il abandonna à la fin ce guide et se livra à l'esprit de système. Alors sa Philosophie ne fut qu'un roman ingénieux, et tout au plus vraisemblable pour les ignorants."

[3] Voltaire 1964, 76: "que deux et deux ne font quatre que parce que Dieu l'a voulu ainsi."

[4] Cassirer 1951, Chap. 1.

[5] Voltaire 1826, 90: "La véritable physique consiste donc à bien déterminer tous les effets. Nous connaîtrons les causes premières quand nous serons des dieux....Il nous est donné de calculer, de peser, de mesurer, d'observer: voilà la philosophie naturelle; presque tout le reste est chimère."

[6] Voltaire 1826, 95: "Il ne substitua donc qu'un chaos au chaos d'Aristote."

work. Again, Voltaire pointed to Descartes' negligence in failing to acknowledge the data supplied by experience.[7]

This is hardly the occasion to analyse the precise nature of Voltaire's assessment of Newtonian physics and its empirical contents, which he probably failed to grasp, nor should I like to comment on Newton's attitude towards Descartes, an issue which has continued to vex Newton scholars to the present day.[8] This much is clear, that Voltaire's comments on Newton's lack of interest in Cartesian physics and mathematics are quite simply untrue.[9] Even Alexandre Koyré, who could hardly be accused of holding any Empiricist bias, felt Newton's *Principia* should not primarily be read as a critique of Descartes' philosophy but rather as an attempt to correct the latter's scientific theories and hypotheses.[10] Yet it remains to be seen what the kind of textual analysis supplied by such comparative scholarship actually provides us with. Of course, we need to ascertain what Descartes' and Newton's *Principia* are actually saying, and surely a careful comparison of the role played by experience in both texts could yield important insights, but from a historical perspective we could also try to catch a glimpse of the part played by experience in the philosophical *movement* Descartes' work initiated: how about the experiential dimension of seventeenth-century "Cartesianism"? A special reason for highlighting the case of *Dutch* Cartesianism lies in the fact that after Cartesianism had made its first real impact at the universities of Utrecht and Leiden, the latter was also the first university, in 1717, to appoint a Newtonian professor of natural philosophy, turning Leiden into a crucial institution in the early dissemination of Newtonianism. In fact, as will only be too familiar, in 1734, shortly after the publication of the *Lettres philosophiques*, Voltaire moved to Leiden in order to study with Willem Jacob 's Gravesande and the famous medical professor Herman Boerhaave (1668–1738).

4.2 Dutch Cartesianism and Newtonianism

Much has been written recently on the history of Dutch Cartesianism, and in particular Theo Verbeek has established that its relevance has traditionally been underestimated due to Descartes' own misgivings concerning his prolonged stay in the Netherlands: Descartes' decision to leave the Dutch Republic in the summer of 1649 after having lived and worked there for more than two decades appears to

[7] [Voltaire] 1771, 218–225. In a passage added to the 1739 edition of the (fifteenth of the) *Lettres philosophiques* Voltaire specifically mentions Bacon as the man of whom Descartes should have taken heed. Voltaire 1964, 232.

[8] Von Borzeszkowski and Wahsner 2000.

[9] Herivel 1965; McGuire 1995; Guiccardini 2009, 59–136 and 293–308; Ducheyne 2012, esp. 253–263 and 269–278.

[10] Koyré 1968, 115. In addition Koyré stressed the religious nature of Newton's opposition to Descartes: 127. See more recently I. Bernard Cohen's highly detailed "Guide to Newton's *Principia*" in Newton 1999; Janiak 2008; Janiak and Schliesser 2011.

have sprung from his failure to recognize the reality as well as the magnitude of his own achievements. Or, to put it differently, he failed to comprehend the way in which Dutch "toleration" actually worked, for more often than not the solemnity with which official decrees were issued did not at all correspond with real measures taken. Despite the decrees issued against the proliferation of Cartesianism, first in 1642 at Utrecht, and then in 1647 at Leiden university, we now know that annoying as such declarations may have been to the pride of the very proud René Descartes, they did not stop the subsequent appointment, especially in Leiden, but also in Franeker, Groningen and even in Utrecht, of several dozens of professors in philosophy, medicine and even theology, who all harboured considerable sympathy for the *philosophia nova* propounded by Descartes.[11] Indeed, the fact that in the Dutch Republic Descartes' views continued to provoke violent polemics well into the early eighteenth century only confirms the reality of "Dutch Cartesianism" as a powerful school of thought, which gradually grew into something of a political faction. In the Dutch Republic religious and theological concerns were always bound to make themselves felt beyond the faculties of theology, but for a while the academic practice of philosophy also inspired a series of quarrels. In addition to the (meta)physical and methodological issues Descartes' philosophy gave rise to, it provoked fierce debates on the autonomy of philosophy within the academic curriculum.

By the time Willem Jacob 's Gravesande (1688–1742) was appointed at Leiden as professor of mathematics and astronomy, this issue had been settled in favour of the "Cartesians"—a rather loose group of philosophers, mathematicians, physicists, physicians and theologians with an outspoken sympathy for the cause of an autonomous practice of philosophy and the natural sciences along the lines set out by Descartes. In 1717 's Gravesande was a lawyer and essentially a well connected journalist, who 2 years previously had met Isaac Newton in London, and apparently he had passed the test, for he was appointed following a very favourable report Newton wrote on his behalf to Dutch diplomats. 'S Gravesande's inaugural lecture was the first explicitly Newtonian academic lecture of its kind.[12] It should be added, however, that, as early as 1715, his Leiden colleague Herman Boerhaave, a former student of Burchard de Volder (1643–1709), one of the last major Leiden Cartesians, had led the way by lecturing in favour of Newton's method.[13] In particular's Gravesande's two-part *Physices elementa* (1720–1721) and his *Philosophiae Newtonianae Institutiones* (1723) as well as his *Introductio ad philosophiam* (1736) soon became very popular across Europe, earning him the respect of his colleagues both in Britain and on the Continent.[14] The *Physices elementa* were translated into English within a year of its publication.

[11] Thijssen-Schoute 1989; Dibon 1990; Verbeek 1992; Van Bunge 2001; Vermij 2002. For a detailed analysis of some of the early critiques, see Van Ruler 1995.

[12] 's Gravesande 1717.

[13] Boerhaave 1715. On Boerhaave, see most recently Knoeff 2002; Kooijmans 2012.

[14] 's Gravesande 1720–1721, 1723, 1736.

Although over the past few years much has been written on the early history of Dutch Newtonianism, the literature still shows considerable lacunae: it is odd, to say the least, that to this day the only available monograph on 's Gravesande is an Italian dissertation written over 40 years ago by Giambattista Gori.[15] This is the more startling in view of its contemporary, European impact. Contrary to what is implied in Voltaire's insistence that Newton basically destroyed Descartes and his scientific heritage, there is actually quite some evidence to suggest that by the early eighteenth century the main attraction of Newtonianism resided in its ability to serve as an antidote to the ungodly philosophy of Spinoza, which was widely held to be atheist, materialist and fatalist.[16] More specifically, the first Dutch "Newtonians," including such "amateurs" as the devout Mennonites Adriaan Verwer (ca. 1655–1717) and Lambert ten Kate (1674–1731), who were mainly known as linguists, explicitly sought to counter the "geometrical" pretensions of Spinozism. Another member of this Amsterdam circle of friends was Bernard Nieuwentijt (1654–1717), who as a student at Leiden has "succumbed" to Spinozism, but who would make up for the sins of his youth by publishing two majors assaults on Spinoza's atheism.[17]

Newton's reputation as a mathematician turned him into an ideal ally in the battle against those who still admired Spinoza's *Ethica ordine geometrico demonstrata.* Both Boerhaave and 's Gravesande presented Newton as proof of a very reassuring fact: mathematicians do not have to be atheists. On the contrary, *real* mathematicians enable us to discern God's providential reign over His creation.[18] So the first Dutch Newtonians joined the last Dutch Cartesians in a *common* effort to stem the tide of the radical Enlightenment, for the "Spinozists" were widely held to destroy the careful separation between philosophy and theology, engineered by Descartes and his supporters. Moreover, the future popularity of the experimental philosophy propounded by Boerhaave and 's Gravesande seems to have been well prepared by several generations of natural philosophers and physicians who were still largely sympathetic with the breakthrough Cartesian philosophy had realised by the middle of the previous century. Perhaps we have failed to recognise the contribution made by Cartesianism to the growing interest in "experimental philosophy" because much of the finest recent literature on late Dutch Cartesianism was produced by historians of science. More often than not they share an anti-Empiricist interpretation of Descartes' philosophical project, which they see confirmed in what they present as the inevitable failure of Cartesian natural philosophers to accommodate the experiential turn characteristic of the dying decades of the century.[19] Yet it is obvious, first, that Dutch natural philosophers such as Burchard de Volder and Christiaan Huygens

[15] Brunet 1926; Gori 1972; Ruestow 1973; de Pater 1979, 1994; Schuurman 2004, Chap. 8; Van der Wall 2004.

[16] Israel 2001, Chap. 27 and 2006, Chap. 8; Vermij 2003; Jorink 2009. See also Ruestow 1973, Chap. 7.

[17] Vermij 1991.

[18] Jorink 2009.

[19] See Chap. 1 by Dobre and Nyden.

who had been raised on a steady diet of Cartesianism were among the first Dutch readers of Newton's *Principia*, and two of the few scholars in Europe *able* to do so, and second, that Gori already forcefully argued that De Volder was actually crucial to the future rise of Newtonianism at Leiden, recording for instance 's Gravesande's use of the equipment first put in place by De Volder.[20] De Volder, incidentally, also wholeheartedly joined the attack on the rise of "atheism" by publishing his own *Disputationes philosophicae omnes contra Atheos* (1685).

4.3 Burchard de Volder

According to Gerhard Wiesenfeldt, however, the breakthrough of Newtonianism at Leiden filled a *Leerer Raum*, "an open space" or a void, left by the demise of Cartesianism, which he dates to as early as 1675 on the evidence of the inauguration, on January 26, of that year, of a *Theatrum physicum*. In his view 1675 heralds "the repudiation of a dogmatic natural philosophy and an empirical method of philosophical teaching."[21] The "dogmatic" philosophy in question is, of course, Descartes' and the operative word in Wiesenfeldt's account is "and." The man who installed the Leiden *Theatrum physicum*, however, was Boerhaave's professor, Burchard de Volder, whose affiliation to Dutch Cartesianism Wiesenfeldt consistently tries to downplay, for instance by pointing to the fact that the first university De Volder attended as a student was Utrecht instead of Leiden.[22] In Utrecht, Descartes' philosophy was less popular than it had become in Leiden. In Utrecht, however, De Volder was a pupil of the philosopher and mathematician Johannes de Bruyn (1620–1675), whose dedication to Cartesianism is beyond dispute, as Wiesenfeldt has to admit. What is more, after having taken his Utrecht doctorate in Philosophy in 1660, De Volder moved to Leiden in order to study medicine with the Cartesian professor De Le Boë Sylvius (1614–1672).[23]

Having completed his studies in 1664 he established a medical practice in his native city of Amsterdam, where he made the acquaintance of Johannes Hudde (1628–1704), mayor, former collaborator of Franciscus van Schooten (1615–1660), friend of Spinoza, and one of the most powerful allies of the *philosophia nova* in

[20] See Chap. 10 by Nyden; Gori 1972, 20–42. See, more recently, also Feingold 2004, 69: "The person to put Holland on the Newtonian map, so to speak, was Burchard de Volder." Wim Klever has tried to turn De Volder into a covert supporter of Spinoza: Klever 1989. This has been refuted by Lodge 2005.

[21] Wiesenfeldt 2002, 1: "die Abkehr von einer dogmatischen Naturphilosophie und eine empirische Ausrichtung der philosophischen Lehre."

[22] Wiesenfeldt 2002, 56.

[23] See, on De Bruyn and De Le Boë Sylvius, as well as on most of the Dutch authors mentioned in this paper: Van Bunge et al. 2003.

the Dutch Republic. It was Hudde who in 1670 suggested to the *curatorium* of Leiden university to appoint De Volder as professor of Philosophy, which it did that same year. As a consequence, the Cartesian faction at Leiden was only boosted further, as became clear in June 1674 when De Volder joined his elder colleagues Abraham Heidanus (1597–1678) and Christopher Wittichius (1625–1687) in a meeting with Caspar Fagel, the pensionary of Holland, trying to convince this successor of Johan de Witt that Cartesianism did not pose a threat either to the political constitution of the Republic or to the Reformed creed.[24] Only a few weeks later, in August, De Volder made a journey to England where he made new friends among the members of the Royal Society. This journey prompted De Volder to ask the Leiden curators to designate a location where the experimental approach to natural philosophy as it was practiced in London could be emulated.[25] As Wiesenfeldt suggests, De Volder was not only concerned to liven up his lectures, but also to steer the practice of natural philosophy into quieter waters.[26] In January 1675 the Leiden *Theatrum physicum* duly opened its doors, and in Wiesenfeldt's perspective this event heralded "the end" of Cartesianism in Leiden.[27]

To all intents and purposes, De Volder himself would have been baffled by such an account, for as Wiesenfeldt himself points out, between 1680 and 1700 he dedicated at least 32 disputations to the explicit defence of Cartesianism.[28] Until the 1690s, De Volder again and again rose to the defence of Cartesianism, for instance in reply to Pierre Daniel Huet's (1630–1721) *Censura Philosophiae Cartesianae* of 1689, arguing that while some of Descartes' theories might well have been mistaken, the general principles of his philosophy remained true.[29] By this time his favourite textbook had become Jacques Rohault's *Traité de physique* (1671).[30] In his 1709 *Eloge de feu*, written years after De Volder's retirement, Jean Le Clerc (1657–1736), who held little sympathy for Cartesianism, wrote that its defence by De Volder was the finest he had ever read.[31] As a journalist, Le Clerc was actually one of the first propagators of Newton's natural philosophy in the Dutch Republic, as well as a Lockean Empiricist.[32] Le Clerc had come to know De Volder personally, and by the

[24] Le Clerc 1709, 356–359.

[25] Wiesenfeldt 2002, 61: the Royal Society itself was on summer break, so De Volder will not have been present at one of its sessions, but he appears to have met both Robert Boyle and Robert Hooke. In Cambridge he visited Newton.

[26] Wiesenfeldt 2002, 62.

[27] Wiesenfeldt 2002, 89.

[28] Wiesenfeldt 2002, 90.

[29] They were collected in De Volder 1695. De Volder was unhappy with this book since he claimed the disputations involved were not meant to be published and were merely part of an academic exercise. They did not necessarily reflect his personal views. See Wiesenfeldt 2002, Chap. 7.

[30] Le Clerc 1709, 398. See Chap. 9 by Dobre.

[31] Le Clerc 1709, 383.

[32] Vermij 2002, 350–352; Schuurman 2004, Chap. 5.

end of his life, or so, Le Clerc wrote, he had indeed lost much of his former confidence in Descartes' views:

> toward the end of his life, and even some years before, he had recognized the weakness of *Cartesianism*, apparently as much through his own reflection as with the help of some able Englishmen who have established other principles. I have heard him make fun, more than once, of a good deal of Descartes' *Meditations*, even though he had long been expounding them.[33]

It remains to be seen whether this account can be trusted, since Le Clerc was hardly an impartial spectator, and to the extent that it can, it would seem to convey at best the misgivings of a retired professor, dating from several decades after his founding of the *Theatrum physicum*. In 1664 De Volder had argued confidently that the way in which mathematicians deduced clear and distinct ideas should serve as an example to the applied science of medicine.[34] In 1682 he still expressed similar sentiments regarding the quest for the laws of motion in natural philosophy.[35] De Volder's mature assessment of the roles to be played by reason and experience in the sciences was delivered in his *Oratio de rationis viribus et usu in scientiis* (1698), in which he reached the conclusion that while reason ruled supreme both in mathematics and in metaphysics, natural philosophy and medicine had to be built on experience.[36]

De Volder's intellectual *Werdegang* is probably best understood as a gradual recognition that natural philosophy in general and medicine in particular were in need of an experimental basis yet to be laid. This insight, however, originated in a perfectly "Cartesian" context: there is nothing to suggest that De Volder's enthusiasm for the experimental philosophy grew out of growing opposition against Descartes' views. As it happened, some of his experiments done during the 1670s and 1680s, such as the arguably most famous ones in which he used an air pump, built for him by Samuel van Musschenbroek, yielded results contradicting Descartes' physics.[37] *As a consequence*, De Volder eventually changed his mind on the usefulness to natural philosophy and medicine of the deductive, conceptual procedure to be followed in metaphysics and mathematics. Thus, near the end of his career De Volder could perhaps indeed be called a "disenchanted" Cartesian, but in his important correspondence with Leibniz, which took place from 1698 to 1706, in which the latter tirelessly tried to win the famous professor over for his own "monadic" notion of substance, he continued to write in defence of Descartes' metaphysics.[38]

[33] Le Clerc 1709, 398: "sur la fin de ses jours, et même quelques années auparavant, il avoit reconnu le foible du *Cartesianisme*; autant apparement, par sa propre méditation, que par le secours des habiles Anglois, qui ont établi d'autres principes. Je l'ai ouï se moquer, plus d'une fois, d'une bonne partie des *Méditations* de Descartes, quoi qu'il les eût expliquées pendant long-tems."

[34] De Volder 1664; Wiesenfeldt 2002, 225.

[35] De Volder 1682; Wiesenfeldt 2002, 225–226.

[36] De Volder 1698; Ruestow 1973, 106–112; Wiesenfeldt 2002, 227–230.

[37] Wiesenfeldt 2002, 108–132.

[38] Lodge 2004. A critical edition of this correspondence is forthcoming.

Henri Krop has even argued that the picture painted by Le Clerc and further explored by Ruestow and Wiesenfeldt of De Volder as an elderly philosopher who had come to regret his former "Rationalism" fails to do justice not only to De Volder's continuing loyalty to Cartesianism, but also to the Empiricist strand in his thinking: Krop shows it can be traced as far back as De Volder's medical thesis of 1664, in which he had already recommended to take heed of what experience tells us about the workings of our body.[39] While it is true that De Volder's mature statements on the relationship between metaphysics and the sciences differ crucially both from Descartes' *Discours* and the *Meditations*, in view of the recent reassessment of Descartes' own scientific career there is every reason to consider at least the possibility that what Krop dubbed De Volder's "empirical Cartesianism" was not at all unconnected to Descartes' personal development.

4.4 Cartesian "Rationalism"

One of the obvious problems of using terms such as "Cartesianism" is of course the a-historical suggestion implied that it refers to a "closed" and more or less coherent set of propositions concerning the world at large, while it is clear that the many seventeenth-century authors who were generally identified as being "Cartesians" actually held views which can often hardly be attuned to any single work of Descartes. What is more, Descartes' own thought actually developed considerably, and it is far from clear in what sense Descartes' deductive "Rationalism" was "inconsistent" with his "Empiricism," for as Daniel Garber has argued, even the early, "Rationalist" Descartes who wrote the *Discours* felt that experiment had a crucial role to play in the deductive procedure of science. In particular Descartes' treatment of the rainbow in the *Meteorology* reveals a scientist at work, whose deductive reasoning on the causes of optical phenomena is *guided* by a familiarity with light and its refraction, acquired by observation.[40] Having completed his work on method and metaphysics in the early 1640s, Descartes' interests turned to medicine and subsequently to a physiological account of the passionate life of man. The extent to which Descartes' work as a "scientist" had something to do with his much earlier work on method, is a moot point, to say the least, for there is little in his work following the *Discours de la Méthode* which actually refers to his own method, which he first formulated by the end of 1619. Garber has pointed to two separate factors which made this method redundant to Descartes' own intellectual progress. First, Descartes' general development from a solver of individual problems to a "system builder," as is evident from the *Principia* in particular, and second, the gradual recognition that the concepts of intuition and deduction crucial

[39] Krop 2003, 187–189. Taking his cue from the history of applied mathematics, in the same volume, Vanpaemel 2003 also stresses the continuity between Dutch Cartesianism and Newtonianism.

[40] Garber 2001, Chaps. 6 and 5.

to Descartes' initial conception of method only raise further questions regarding the validity of the mental operations involved.[41]

As a matter of fact, several of the earliest Dutch supporters of Cartesianism already defy the traditional Rationalist-Empiricist dichotomy, including the Utrecht physician Henricus Regius (1598–1679), who argued against the notion of innate ideas, and the Leiden philosopher Adriaan Heereboord (1614–1659), whose pleas in favour of *libertas disputandi* explicitly included Bacon.[42] So De Volder's appreciation of the part to be played by experience in science was hardly exceptional among Descartes' Dutch admirers.[43] As Harold Cook recently put it, to Dutch Cartesian physicians such as Regius, but also Florentius Schuyl and Francis de le Boë Sylvius (De Volder's professor of medicine at Leiden):

> what Descartes' writings promised was not so much a new metaphysics—much less one that said the best theories came from deductive reasoning—but a demonstration that the physical investigation of natural bodies on which they had long been engaged was indeed the path to a true understanding of nature.[44]

In Cook's reconstruction of Descartes' development, the youthful "dreams" of a single Rational method disciplining scientific knowledge as such made way for a more practical and empirical outlook. By the late 1620s the study of anatomy and animal physiology had become Descartes' main ambition. His publication of the *Meditations* in 1641 could easily be presented as a diversion from his medical studies, but the medical overtones of the sixth meditation in particular suggest the contrary in its new insistence on experience informing us about the union of body and mind and the confidence it oozes about our ability to know the external world. His subsequent work on the passions only further vindicates not only the practical, that is, moral aims of the mature Descartes but more in particular his conviction that no progress could be made in this domain without taking recourse to our senses: "Descartes had begun doing philosophy with a proof of the existence of God, but during his many years in the Dutch Republic he became something of an Empiricist."[45]

When in 1645 Descartes was accused by Huygens of leaving too little room for experiential observation, he claimed to be confused by this criticism, and there are no indications of any insincerity on Descartes' part.[46] As early as the *Discours*, he appears to recognise the necessity to supplement the a priori deduction of the general laws of nature with an a posteriori account of the behaviour of individual phenomena.[47] The "rift," as Paul Schuurman has called it, running right through

---

[41] Garber 2001, Chap. 2 and Garber 1992, Chap. 2. See also, for instance, Schuster 1993.

[42] For Regius, see Chap. 7 by Bellis.

[43] See, for instance, Verbeek 1993, 2002.

[44] Cook 2010, 26.

[45] Cook 2007, 259.

[46] AT IV 224–225, CSMK 358–359.

[47] Clarke 1982, 17; Schuurman 2004, 26–33.

what is supposed to be Descartes' unitary "mathematical" method, is obvious in the *Principia*, for the first two parts of his "natural philosophy," in which Descartes deduces from the *Cogito* both the existence and the nature of the material world as well as its most fundamental laws, is followed by the third and fourth parts in which he uses essentially hypothetical accounts, based on sensory experience, in order to explain individual physical phenomena.

When Descartes scholars first started to explore the nature and relevance of Descartes' Empiricism, many of their colleagues remained sceptical, but more recently both Garber and Cook have been able to point to expert predecessors: in 1945, Jean Laporte, on the final pages of his magisterial *Le rationalisme de Descartes* boldly concluded that as far as he was concerned Descartes was an *Empiricist*:

> To adjust oneself, in all things, to *what one sees*; to describe it *as it is observed*, to whatever category it may belong, without mixing in anything that comes from our own sensibility: that is the Cartesian attitude as it manifests itself in the theory of method as well as in the *Cogito* and the development that precedes it. That is the *empirical* attitude, in the primary and authentic meaning of the word: only they dislike to admit this who are always led by their inveterate associations to confound *empirical* and *sensualistic*. So that, if we must characterize the philosophy of Descartes by a label, the one that would best fit would be—all paradox apart—that of empiricism, radical and complete *empiricism*.[48]

This lead was further developed by Desmond Clarke, who has pointed to the ambiguities involved in delineating the semantic domain covered by Descartes' usage of "experience," "observation," and related concepts. As it turns out, Descartes' distrust of empirical evidence is fairly limited, for it appears to be mainly directed towards the many errors made in setting up experiments.[49] Next, Clarke addressed the nature of Descartes' conception of "deduction" and the Cartesian distinction between metaphysics and physics, after which he finally analysed his account of the requirements to be met by a properly scientific explanation. According to Clarke, for instance, Descartes' understanding of an "a priori" explanation should not be understood in any Kantian sense since it does not in any way exclude empirical evidence. In fact, Descartes' epistolary exchanges with several fellow "scientists" strongly suggest he had causal explanations in mind, which were to be used in the context of a hypothetico-deductive methodology.[50]

Thus, now some 30 years ago, a thorough and sophisticated reading of Descartes became available, allowing for an assessment of his philosophy of science that

---

[48] Laporte 1988, 477: "Se plier en toutes choses, à *ce qu'on voit*; l'enregistrer *comme on le voit*, à quelque ordre qu'il appartienne, sans rien y mêler de sa sensibilité propre: voilà l'attitude cartésienne, telle qu'elle se manifeste dans la théorie de méthode, comme aussi dans le *Cogito* et dans les démarches qui en precedent. C'est l'attitude *empirique*, au sens premier et authentique du mot: ceux-là seuls répugnent à en convenir, que des associations invétérées conduisent toujours à confondre *empirique* et *sensualiste*. En sorte que, si nous voulons à toute force caractériser la philosophie de Descartes par un nom, le nom qui lui siérait le mieux serait, tout paradoxe à part, celui d'empirisme—*empirisme* radical et intégral."

[49] Clarke 1982, Chap. 2.

[50] Clarke 1982, Chap. 5.

precluded any easy characterisation of Cartesianism as an essentially Rationalist philosophy.[51] Looking for explanations of why Descartes and seventeenth-century Dutch Cartesians came to be associated with "Rationalism" in the way that they did, the first name which springs to mind is, of course, Kant's, whose late eighteenth-century proposal to turn Philosophy into a "critical" endeavour was underpinned with a "History of Pure Reason," in which the histories of "Rationalism" and "Empiricism" had reached a deadlock.[52] But closer at home Dutch Cartesians were facing a remarkably vigorous Aristotelian tradition whose main protagonists presented themselves as "Empiricists," fighting off the Rationalist scepticism of the *novatores*. A particularly interesting example is supplied by the metaphysician Gerard de Vries (1648–1705), the last major opponent of Cartesianism at Utrecht, who accused Descartes of ignoring the philosophical relevance of sensory experience and whose rejection of innate ideas showed a remarkable resemblance to the opening chapters of Locke's *Essay concerning Human Understanding*.[53]

Questionable as it may be to read Descartes as the author of an anti-Empiricist methodology and turn Dutch Cartesianism into an instance of early modern Rationalism waiting to be superseded by Newtonianism, Paul Schuurman was no doubt right to direct attention to the absence in Dutch intellectual life of a genuinely Empiricist *epistemology*, prior to Locke.[54] Perhaps a final remark is in place in this context, for the idea that early modern philosophy somehow received its main impetus by the opposition between "Rationalists" and "Empiricists" has already been criticised for a long time. Few experts have been as lethal in their commentary as Knud Haakonssen, a major authority on the early modern history of natural law, who has argued forcefully that before Kant epistemology was not at all as important as modern commentators make it out to be. According to Haakonssen much of our fascination with seventeenth and eighteenth-century theories of knowledge suffers from the view—the rise of which he attributes to both Kant and Reid—according to which the theory of knowledge is the proper subject of philosophy as such: the very concept of "early modern philosophy," Haakonssen has argued, is essentially part of what he calls an "epistemological paradigm," which fails to capture the contemporary self-assessment of early modern philosophy.[55]

Ethics, politics and aesthetics have suffered most obviously from this, and more generally, Haakonssen feels, the narrowing of the moral dimension of pre-Kantian moral thought to its ability to *justify* moral propositions completely ignored the widely shared conviction that philosophy was to contribute to the shaping of the self, that it should first and foremost be *lived*. Most importantly, however, Haakonssen takes the epistemological paradigm to task over its "individualism and mentalism—the assumption that knowledge has to be accounted for in terms of the activity (or passivity)

[51] See also Buchdahl 1963.

[52] See most recently Garber and Longuenesse 2008.

[53] Schuurman 2007. See also Ruestow 1973, 78–87; Wiesenfeldt 2002, 82–89, 162–184.

[54] Schuurman 2004, 68–69.

[55] Haakonssen 2004, 103–104.

of the individual person's mind,"[56] which obscures the fundamental debates on historical testimony as well as the non-mentalist aspects of much of seventeenth- and eighteenth-century philosophy of language. It remains to be seen, meanwhile, how Haakonssen's perspective on the rise of modern philosophy is able to account for the crucial importance attached, not only by Descartes and Newton, to contemporary *natural* philosophy, which from an Aristotelian perspective equally constituted the heart of Philosophy. As it happens, Dutch Cartesians, including most notably Spinoza, had much to say on moral philosophy, language, and (biblical) testimony. But it is to be doubted whether their efforts will impress critics such as Haakonssen, for from a strictly philosophical perspective this much seems clear, that their commitment to "individualism and mentalism" remained, indeed, perfectly intact.[57]

References

Aalderink, Mark. 2010. *Philosophy, scientific knowledge, and concept formation in Geulincx and Descartes*. Utrecht University: Publications of the Department of Philosophy.

Barber, William H. 1979. Voltaire et Newton. In *Voltaire and the English*, ed. Mason Haydn, 193–202. Oxford: Voltaire Foundation (*SVEC* 176).

Boerhaave, Herman. 1715. *Sermo academicus de comparendo certo in physicis*. Leiden: Petrus van der Aa.

Brunet, Pierre. 1926. *Les physiciens hollandais et la méthode expérimentale en France au XVIIIe siècle*. Paris: Albert Blanchard.

Buchdahl, Gerd. 1963. The relevance of Descartes' philosophy of science for modern philosophy of science. *The British Journal for the History of Science* 1: 227–249.

Cassirer, Ernst. 1951. *The philosophy of the enlightenment*. Princeton: Princeton University Press.

Clarke, Desmond M. 1982. *Descartes' philosophy of science*. Manchester: Manchester University Press.

Cook, Harold J. 2007. *Matters of exchange. Commerce, medicine, and science in the Dutch golden age*. New Haven: Yale University Press.

Cook, Harold J. 2010. Victories for Empiricism, failures for theory: Medicine and science in the seventeenth century. In *The body as object and instrument of knowledge. Embodied Empiricism in early modern science*, ed. Charles T. Wolfe and Ofer Gal, 9–32. Dordrecht: Springer.

de Pater, C. 1979. *Een newtoniaans natuuronderzoeker*. Utrecht: Proefschrift Rijskuniversiteit Utrecht.

de Pater, C. 1994. Willem Jacob 's Gravesande (1688–1742) and Newton's *Regulae Philosophandi*, 1742. *Lias* 21: 257–294.

de Volder, Burchard. 1664. *Disputatio medica inauguralis de natura*. Leiden: Severinus Matthias.

de Volder, Burchard. 1682. *Oratio de conjugendis philosophicis et mathematicis diciplinis*. Leiden: Voorn.

de Volder, Burchard. 1685. *Disputationes philosophicae omnes contra Atheos*. Middelburg: Johannes Lateranus.

de Volder, Burchard. 1695. *Exercitationes academicae quibus Renati Cartesii philosophia defenditur adversus Petri Danielis Huetii Censuram philosophiae Cartesianae*. Amsterdam: Arnoldus van Ravestein.

[56] Haakonssen 2004, 114.

[57] See for a very interesting recent account, Aalderink 2010.

de Volder, Burchard. 1698. *Oratio de rationis viribus et usu in scientiis*. Leiden: Frederik Haring.
Descartes, René. 1991. *Philosophical writings of Descartes*, vol. III, Trans. John Cottingham, Dugald Murdoch, Robert Stoothoff, and Anthony Kenny. Cambridge: Cambridge University Press
Descartes, René. 1996. *Oeuvres de Descartes*, 2nd ed. C. Adam and P. Tannery. Paris: Vrin
Dibon, Paul. 1990. *Regards sur la Hollande du siècle d'Or*. Naples: Vivarium.
Ducheyne, Steffen. 2012. *"The main business of natural philosophy." Isaac Newton's natural-philosophical methodology*. Dordrecht: Springer.
Feingold, Mordechai. 2004. *The Newtonian moment. Isaac Newton and the making of modern culture*. Oxford: Oxford University Press.
Garber, Daniel. 1992. *Descartes' metaphysical physics*. Chicago: Chicago University Press.
Garber, Daniel. 2001. *Descartes embodied. Reading Cartesian philosophy through Cartesian science*. Cambridge: Cambridge University Press.
Garber, Daniel, and Béatrice Longuenesse (eds.). 2008. *Kant and the early moderns*. Princeton: Princeton University Press.
Gori, Giambattista. 1972. *La fondazione dell'esperienza in 's Gravesande*. Florence: Olschki.
Gravesande, Willem Jacob 's. 1717. *Oratio inauguralis de Matheseos in omnibus scientiis, praecipue in Physicis, Usu, nec non de Astronomiae perfectione ex Physica haurienda*. Leiden: Samuel Luchtmans
Gravesande, Willem Jacob 's. 1720–1721. *Physices elementa mathematica, experimentis confirmata, sive Introductio ad Philosophiam Newtonianam*, 2 vols. Leiden: P. and B. J. van der Aa.
Gravesande, Willem Jacob 's. 1723. *Philosophiae Newtonianae Institiones in usos academicos*. Leiden: Petrus van der Aa.
Gravesande, Willem Jacob 's. 1736. *Introductio ad philosophiam, metaphysicam et logicam continens*. Leiden: J. and H. Verbeek.
Guiccardini, Niccolò. 2009. *Isaac Newton on mathematical certainty and method*. Cambridge, MA: Harvard University Press.
Haakonssen, Knud. 2004. The idea of early modern philosophy. In *Teaching new histories of philosophy*, ed. J.B. Schneewind, 99–121. Princeton: University Center for Human Values.
Herivel, John. 1965. *The background to Newton's Principia. A study of Newton's dynamical researches in the years 1664–1684*. Oxford: Oxford University Press.
Israel, Jonathan I. 2001. *Radical enlightenment. Philosophy and the making of modernity, 1650–1750*. Oxford: Oxford University Press.
Israel, Jonathan I. 2006. *Enlightenment contested. Philosophy, modernity, and the emancipation of man, 1670–1752*. Oxford: University Press.
Janiak, Andrew. 2008. *Newton as philosopher*. Cambridge: Cambridge University Press.
Janiak, Andrew, and Eric Schliesser (eds.). 2011. *Interpreting Newton. Critical essays*. Cambridge: Cambridge University Press.
Jorink, Eric. 2009. "Honouring Sir Isaac, or, exorcising the ghost of Spinoza". Some remarks on the success of Newton in the Dutch Republic. In *Future perspectives on Newton scholarship and the Newtonian legacy in eighteenth-century science and philosophy*, ed. Steffen Ducheyne, 23–34. Brussels: Koninklijke Vlaamse Akademie van België voor Wetenschappen en Kunsten.
Klever, Wim. 1989. A Crypto-Spinozist on a Leiden Cathedra. *Lias* 15: 191–241.
Knoeff, Rina. 2002. *Herman Boerhaave (1668–1738). Calvinist chemist and physician*. Amsterdam: Koninklijke Nederlandse Akademie van Wetenschappen.
Kooijmans, Luuc. 2012. *Het orakel. De man die de geneeskunde opnieuw uitvond: Herman Boerhaave (1669–1738)*. Amsterdam: Balans.
Koyré, Alexandre. 1968. *Études newtoniennes*. Paris: Gallimard.
Krop, Henri. 2003. Medicine and philosophy in Leiden around 1700: Continuity or rupture? In *The early enlightenment in the Dutch Republic, 1650–1700*, ed. Wiep van Bunge, 173–197. Leiden: Brill.

Laporte, Jean. 1988. *Le rationalisme de Descartes*. Paris: Presses universitaires de France. First published in 1945.

Le Clerc, Jean. 1709. Eloge de feu de Mr. De Volder, Professeur en Philosophie et aux Mathematiques dans l'Academie de Leide. *Bibliotheque choisie* 18: 346–401.

Lodge, Paul. 2004. Leibniz's close encounter with Cartesianism in the correspondence with De Volder. In *Leibniz and his correspondents*, ed. Paul Lodge, 162–192. Cambridge: Cambridge University Press.

Lodge, Paul. 2005. Burchard de Volder: Crypto-Spinozist or Disenchanted Cartesian? In *Receptions of Descartes. Cartesianism and anti-Cartesianism in early modern Europe*, ed. Tad M. Schmalz, 128–146. London: Routledge.

McGuire, J.E. 1995. *Tradition and innovation. Newton's metaphysics of nature*. Dordrecht: Kluwer.

Newton, Isaac. 1999. *The Principia. Mathematical principles of natural philosophy*, ed. and trans. I. Bernard Cohen and Anne Whitman. Berkeley: University of California Press.

Rohault, Jacques. 1671. *Traité de physique*. Paris: Veuve de Charles Sevreux.

Rousseau, André Michel (ed.). 1976. *Voltaire en Angleterre*, 3 vols. Oxford: Voltaire Foundation (*SVEC* 145–147).

Ruestow, Edward G. 1973. *Physics at seventeenth and eighteenth-century Leiden. Philosophy and the new science in the University*. The Hague: Martinus Nijhoff.

Schuster, John A. 1993. Whatever should we do with Cartesian method? – Reclaiming Descartes for the history of science. In *Essays on the philosophy and science of René Descartes*, ed. Stephen Voss, 195–223. Oxford: Oxford University Press.

Schuurman, Paul. 2004. *Ideas, mental faculties and method. The logic of Descartes and Locke and its reception in the Dutch Republic, 1630–1750*. Leiden: Brill.

Schuurman, Paul. 2007. Continuity and change in the Empiricism of John Locke and Gerard de Vries (1648–1705). *History of European Ideas* 33: 292–304.

Thijssen-Schoute, C.L. 1989. *Nederlands cartesianisme*. Utrecht: HES. First published in 1954.

van Bunge, Wiep. 2001. *From Stevin to Spinoza. An essay on philosophy in the seventeenth-century Dutch Republic*. Brill: Leiden.

van Bunge, Wiep, Henri Krop, Bart Leeuwenburgh, Han van Ruler, Paul Schuurman and Michiel Wielema (eds.). 2003. *The dictionary of seventeenth and eighteenth-century Dutch Philosophers*, 2 vols. Bristol: Thoemmes.

van der Wall, Ernestine G.E. 2004. Newtonianism and religion in the Netherlands. *Studies in History and Philosophy of Science* 35: 493–514.

van Ruler, J.A. 1995. *The crisis of causality. Voetius and Descartes on God, nature, and change*. Leiden: Brill.

Vanpaemel, Geert. 2003. The culture of mathematics in the early Dutch enlightenment. In *The early enlightenment in the Dutch Republic, 1650–1700*, ed. Wiep van Bunge, 197–211. Leiden: Brill.

Verbeek, Theo. 1992. *Descartes and the Dutch. Early reactions to Cartesian philosophy, 1637–1650*. Carbondale: Southern Illinois University Press.

Verbeek, Theo (ed.). 1993. *Descartes et Regius. Autour de l'Explication de l'Esprit humain*. Amsterdam: Rodopi.

Verbeek, Theo. 2002. Dutch Cartesian philosophy. In *A companion to early modern philosophy*, ed. Steven Nadler, 167–182. Oxford: Blackwell.

Vermij, Rienk. 1991. *Secularisering en natuurwetenschap in de zeventiende en achttiende eeuw*. Amsterdam: Rodopi.

Vermij, Rienk. 2002. *The Calvinist Copernicans. The reception of the new astronomy in the Dutch Republic, 1575–1750*. Amsterdam: Koninklijke Nederlandse Akademie van Wetenschappen.

Vermij, Rienk. 2003. The formation of the Newtonian natural philosophy. The case of the Amsterdam mathematical Amateurs. *The British Journal for the History of Science* 26: 183–200.

[Voltaire]. 1771. *Questions sur l'Encyclopédie, par des amateurs*, vol. III. [Paris: s.n] (1st ed. 1770).

Voltaire. 1826. *Dictionnaire philosophique*, vol. 53, *Oeuvres completes de Voltaire*. Paris: s.n.

Voltaire. 1964. *Lettres philosophiques*, ed. Raymond Naves. Paris: Garnier.

von Borzeszkowski, Horst-Heino, and Renate Wahsner. 2000. Voltaire's Newtonianism. A bridge from English Empiricism to Cartesian rationalism and its implications for the concept of mechanics in German idealism. *Preprints Max Planck-Institut für Wissenschaftsgeschichte* 147: 5–27.

Wiesenfeldt, Gerhard. 2002. *Leerer Raum in Minervas Haus. Experimentelle Naturlehre an der Universität Leiden, 1675–1715*. Amsterdam: Koninklijke Nederlandse Akademie van Wetenschappen.

Chapter 5
Heat, Action, Perception: Models of Living Beings in German Medical Cartesianism

Justin E.H. Smith

Abstract In this chapter I characterize the reception and development of Cartesian medical philosophy in Germany, through a reading of a number of its principal exponents. I begin by briefly showing why, for Descartes himself, medicine was central to the project of philosophy as a whole, and on the basis of this I show that German Cartesian medical philosophy remained fundamentally true to the basic concerns of Descartes himself. I show, nonetheless, following the groundbreaking work of Trevisani, that there were widely divergent views held on specific Cartesian doctrines within the Duisburg school. I analyze, in particular, the theory of composite bodies held by Johannes Clauberg and a number of his contemporaries and immediate successors. I go on, finally, to analyze G. W. Leibniz's eventual theory of corporeal substance and organic body, suggesting that German medical Cartesianism constitutes an important middle term for understanding the development of Leibniz's radically modified mechanist view of living bodies. Heat, action, and perception, characterize, respectively, Descartes', Clauberg's, and Leibniz's respective views of what is distinctive in living bodies, and by charting the shift from one to the next we are able to clearly grasp what was at stake in the mechanist engagement with the problem of life.

5.1 Introduction: The Idea of Medical Philosophy

Where an early modern philosopher places medicine in the hierarchy of human endeavors is no trivial matter. It is, in fact, an important indication of the scope and orientation of his or her entire philosophical project. Descartes, to take one prominent example, had thought of medicine as the art of maintaining health, and understood

J.E.H. Smith (✉)
Université Paris Diderot - Paris 7, Paris, France
e-mail: justin.smith@concordia.ca

M. Dobre and T. Nyden (eds.), *Cartesian Empiricisms*, Studies in History and Philosophy of Science 31, DOI 10.1007/978-94-007-7690-6_5,
© Springer Science+Business Media Dordrecht 2013

health in turn as "undoubtedly the chief good and the foundation of all the other goods in this life."[1] Or, as he puts it in a letter to William Cavendish of October, 1645: "From the beginning, the principal aim of my studies has been the conservation of health."[2] While some recent scholarship has begun to take these claims seriously, for the most part for the past several centuries scholars have chosen to selectively edit out Descartes' own characterization of his philosophical project. They have been reluctant to see that Descartes' philosophy was, to use Vincent Aucante's apt phrase, fundamentally a "medical philosophy."[3]

This is not to say that Descartes' work was itself medicine, or that it took up medicine as the focus of a detached philosophical inquiry, in the way we might today engage in the "philosophy of medicine." Rather, Descartes' was a medical philosophy to the extent that he saw medicine as integral to the project of philosophy, insofar as (i) it was the key to health and longevity, and thus to the realization of the good life;[4] and (ii) it was conceived as including rules of diet, hygiene, and bodily comportment, and to this extent was nothing less than the corporeal flip-side, so to speak, of ethics. Medical philosophy takes care of the body and care of the soul to be two aspects of a broader eudaimonistic project. This two-sided project is, in essence, a *mise-en-pratique* of a basic philosophical commitment about the relation between mind and body. That is, sights, thoughts, and experiences should be avoided, in so far as they are capable of provoking harmful passions; passions, in turn, are the result of the possibility of real influence of the mind on the body.

This practical orientation is in no way at odds with the deeper theoretical concerns in Descartes with which we are more familiar; in fact, by turning our attention to what might be called the "medical eudaimonism" of early modern philosophers—that is, their belief that studying the means toward health and longevity is a requisite of the good life—we are able to gain new light on some of the deepest theoretical questions with which they were concerned. In the context of the Cartesian school, in particular, we see that the fundamental philosophical problem of the mind-body connection is never very far from the concern with medical questions. By paying attention to these medical questions we may come to more fully understand what was at stake in the solutions to the mind-body problem that were put forth by different thinkers.

In this article, I will not attempt to sketch out the broad outlines of the impact of Cartesianism on medicine in Germany in the late seventeenth century, for this would be an impossibly large undertaking for such a short investigation. Instead, I will attempt to show the fruitfulness, for our understanding the early legacy of Descartes, and indeed for understanding Descartes himself, of considering the French philosopher as first and foremost a medical philosopher. If we take seriously Descartes' claim to

---

[1] AT VI 61–62, CSM I 142–143.

[2] AT IV 329, CSMK 173–174: "La conservation de la santé a esté de tout temps le principal but de mes études."

[3] See Chap. 4 by van Bunge. See also Aucante 2006; Lindeboom 1978; Manning 2013.

[4] See Grmek 1968.

Newcastle, and if we proceed to consider his immediate successors as more or less faithful to this priority, then we are in a position to understand a good deal more about the scope and aims of the Cartesian philosophical project than when we take this project to be first and foremost one of ontology, theology, and other endeavors supposedly more fundamental than medicine. But there's more: by paying due attention to the medical philosophy, and by approaching Cartesian philosophy *as* medical philosophy, we are also able to unravel some persistent questions about ontology. In particular, we may learn a great deal about the history of the concepts of material or bodily substance (which are not necessarily the same thing), and of the problematic connection of these with thinking substance.

Here I will begin with a treatment of the early German legacy of Descartes as a medical philosopher by discussing the importance of Cartesian medical philosophy in the work of Johannes Clauberg (1622–1665). I will show in particular that for Clauberg there is a clear distinction between matter on the one hand and living bodies on the other, and that these latter are marked off ontologically from the things of the inanimate world, in view of their high degree of internal organization and their capacity for functioning as integral wholes. This unity and activity bring it about that for Clauberg individual corporeal beings are themselves substances, rather than these bodies simply belonging to one undifferentiated, universal *res extensa* (as sometimes appears to be the case for Descartes). And this substantial character of organized bodies has significant implications for Clauberg's understanding of the "conjunction," to use his language, of mind and body. I will proceed to briefly chart the development of the idea of mind-body conjunction in some relatively minor German Cartesian medical philosophers who followed Clauberg, and I will conclude with an account of the theory of organic unity in Leibniz, arguing that Clauberg's work may be seen as an important intermediary between Descartes' and Leibniz's respective accounts, and indeed that all three are alike to the extent that their respective philosophies of corporeal substance are clear examples of what I have been calling early modern medical philosophy.

5.2 Johannes Clauberg: "Cartesianam Philosophiam naturae Germanicae nationis maxime convenire"

Johannes Clauberg was the principal figure in the early introduction of Cartesianism in Germany. Having completed his studies at Groningen in The Netherlands, his education and subsequent career and publication record all unfold as much in the Netherlands as in northwest Germany. Clauberg is perhaps best known for contributing to the widespread acceptance of the concept of "ontology" as a distinct field of philosophy in his *Elements of Philosophy, or Ontosophia* of 1647,[5] composed

[5] There is an earlier occurrence of the term 'ontology' in a marginal note to the entry 'Abstractio' of Rudolf Goclenius' 1613 *Lexicon philosophicum.*

while still in Holland and published in Groningen. Clauberg was a disciple of Descartes in Germany, but rather than adhering dogmatically to the Cartesian program, he in fact helped to expand it, most notably by developing a system of Cartesian logic. This undertaking had much to do with the late-Scholastic context in which Clauberg was working, and with the enduring foundational character of logic for those working within the Aristotelian tradition.

Clauberg's first extended defense of Cartesianism was published a year after his arrival in Germany in 1651, the *Defense of Descartes against Jacob Revius* (Clauberg 1652). This work was followed by several subsequent treatments of various aspects of Cartesian philosophy, many of which were intended as pedagogical *abregés* of the French philosopher's work, such as the 1658 *Paraphrase of René Descartes' Meditations on First Philosophy*. At least one work, the *Difference between Descartes and the Philosophy that Is Otherwise Common in the Schools*,[6] highlights the differences between Scholasticism and Cartesianism, yet Clauberg is widely seen as having synthesized the two traditions rather than being overly concerned to highlight their points of divergence from one another.[7] Thus Francesco Trevisani notes, in allusion to the classic study by Josef Bohatec, that "[t]he zeal with which the perfect compatibility between Cartesian physics and Aristotelian physics was shown, and the privileged role that ontology has in the foundation of the metaphysics of thought [*conoscenza*], are able to induce us to believe that Clauberg was not a *pure Cartesian*, but rather a *Scholastic Cartesian*."[8]

Clauberg enhanced the Cartesian program with a logic, but in another respect, perhaps, he remained very true to the central concerns of Descartes himself: to the extent that he recognized the centrality of medicine to the advancement of philosophy. Put differently, Clauberg understood that Descartes' philosophy was a medical philosophy. It may thus be that Clauberg's greatest contribution was to propagate a version of Cartesianism in Germany, after his arrival at the University of Duisburg in 1651, that emphasized the medical and physiological side of this new tradition. Certainly, the emphasis on this side of Descartes may, like Clauberg's logic, also be seen as a necessary adaptation to a distinctly German context, this time not to late Scholasticism, but rather to Paracelsianism, iatrochemistry, and to the scattered and difficult-to-classify traditions of practical natural philosophy in Germany. According to Clauberg himself, the German approach to knowledge is eminently practical, and in this respect Cartesianism is a particularly well-adapted import. "Cartesian philosophy," he writes, "is perfectly suited to the peoples of a Germanic nature, since the true German is not in the habit of using many words, nor does he love

[6] Originally published in German in 1658 under the title *Unterschied zwischen der cartesianischen und der sonst in den Schulen gebräuchlichen Philosophie*.

[7] For other examples of synthesis, see Chap. 2 by Ariew and Chap. 10 by Nyden.

[8] Trevisani 1992, 97. Trevisani expands a number of the central themes of this work in a subsequent German edition. See Trevisani 2012. See also Bohatec 1966; Trevisani 1982; Coqui 2009; Hurson 2009; Mehl 2001.

inane exclamations or loquacity, but rather offers from the things themselves more than he promises in words."[9]

It is not simply that Clauberg continues the practical-medical side of Descartes' program, while supplementing it with the logic it had previously lacked. Rather, Clauberg is interested in underlining the fundamental unity of the different domains of the philosophical project. In medicine, as perhaps the most practical of disciplines, there is a particular concern to avoid the *praejudicia autoritatis*, the unwarranted reliance on the claims of ancient authorities. "Whatever is good in medicine does not come from those things that were frequently repeated by physicians in the Schools…but rather from experiment and observation."[10] A few decades later, Christian Thomasius (1655–1728) would make this point in a more general way when he wrote that "Descartes is to be praised for having been the first in these recent times to subject prejudice to observation."[11] With at least a faint *soupçon* of self-contradiction, Thomasius even maintains that it is God himself who sent Descartes to eradicate superstition and blind reverence for authority: "Divine providence has made use of Descartes," he writes, "in order to awaken learned men from the deep sleep of superstition and of the prejudice of authority."[12]

To a great extent, for Clauberg, it is this sort of blind reliance on authority that perpetuates the single greatest error of both physics and medicine, and that is the confusion of the respective domains of mind and body. The "particular cause," he writes,

> of the infinite confusions and of the darkest shadows in physics and in medicine was, up until now, that they do not consider the properties of the mind alone through the mind alone taken separately, but because they always adhere to the senses and confuse the mind with the body.[13]

In this respect, for Clauberg, the advancement of medicine and physics goes right together with the promotion of Cartesian dualism. To advance in these disciplines requires a rejection of ancient authority in favor of observation, together with an adherence to the doctrine of clear and distinct ideas. These commitments together reveal a world of bodies operating according to its laws, and a world of minds according to its laws.

---

[9] Clauberg 1680, XIII lxvii; cited in Trevisani 1992, 41: "…Cartesianam Philosophiam naturae Germanicae nationis maxime convenire siquidem genuinus Germanus multis verbis non solet uti, nec amat inanem jactantiam aut loquacitatem, sed potius plura rebus ipsis praestat, quam verbis promittit."

[10] Trevisani 1992, 95: "Medicina quodcunque boni habuit non ex illis, quae in Scholis Physicis frequentabantur…; sed potius ab experientia & observatione."

[11] Thomasius 1719, 8, 24, 26, 115: "Cartesius ist deshalben billich zu loben, dass er in diesen letzten Zeiten der erste gewesen, der auf die Beobachtung…Vorurtheil gedrungen hat."

[12] Thomasius 1699, Vorrede, Sect. 16, 26 "[D]ie göttliche Vorsehung sich Cartesii bedienet, die Gelehrten aus dem tiefen Schlaff des Aberglaubens und des praejudicii autoritatis zu erwecken."

[13] Cited in Trevisani 1992, 108: "Et praecipua causa infinitarum confusionum & densissimarum tenebrarum in Physica & Medicina fuit hactenus, quod neque solius mentis proprietates sola mente seorsim considerant, sed quia perpetuo adhaerent sensibus & mentem cum corpore confundunt."

5.3 Matter and Body from a Medical Point of View

But can we just leave it at that? Or is there a more complex story to be told about the living body, and the way it differs from the rest of the world of *res extensa*? It is here that we may be well served by looking at Clauberg as a member of the tradition of medical philosophy. As is the case with Descartes, Clauberg's concerns as a medical philosopher are typically given short shrift. For example, in a 1999 volume intended to correct the general neglect of Clauberg's philosophy, there is virtually no mention, other than in the bio-bibliographical sketch at the end of the volume, of Clauberg's engagement with the problems of physiology, anatomy, and medicine.[14] Much of our familiarity with the medical concerns of Clauberg and his contemporaries is due to Francesco Trevisani's *Descartes in Germania: La ricezione del cartesianesimo nella Facoltà filosofica e medica di Duisburg (1652–1703)*, a magisterial study, already cited above, of the Cartesian school of philosophy and medicine at the University of Duisburg in the second half of the seventeenth century. The present investigation is deeply indebted to Trevisani's important work.

In her insightful contribution in the 1999 volume, Christia Mercer notices that Clauberg has a clear response to the unresolved problem in Descartes as to whether a body is an individual substantial thing, or rather whether *res extensa* is only one general, non-individuable substance. Mercer argues that Clauberg gives a clear response, in favor of the view that "the individual corporeal thing is a substance."[15] She draws principally on the "Disputationes" which form a part of the *Physica quibus rerum corprearum* of 1664, in which a distinction is made between matter or *res extensa* on the one hand, and body on the other. Mercer sees this distinction in Clauberg as being articulated in terms of the traditional Scholastic distinction between primary and secondary matter. As Clauberg writes:

> [W]e must distinguish between prime Matter, what is extended simpliciter and considered universally, what is pure substance, depending on God alone; and secondary matter, namely, what is extended in this way or that, what is provided with this or that form, what is placed in a certain class of things.[16]

Mercer explains that for Clauberg "corporeal properties always exist in extension and moreover each individual body is a particular arrangement of matter or extension." It follows that each individual corporeal thing will have corporeal properties inhering in it. This implies, in turn, "that an individual corporeal thing will function as a subject and hence will be a substance."[17]

Mercer is certainly correct here. However, I would like to argue that the full scope and significance of the theory of corporeal substance only becomes apparent when we turn our attention to the philosopher's engagement with the life sciences,

---

[14] Verbeek 1999.

[15] Mercer 1999, 151.

[16] Cited in Mercer 1999, 151.

[17] Mercer 1999, 151.

in particular his effort to provide a model of the living being. This is something Clauberg does in extensive detail in the *Theoria corporum viventium*, also published as part of his *Physica, quibus rerum corporearum* in 1664.

Why does Clauberg set himself the task of developing a theory of living bodies? Until very recently, insufficient attention was paid in the scholarship to the ontological distinctions early modern philosophers themselves sought to make between animate and inanimate entities.[18] One significant example of this inattention is in the treatment of Leibniz's purported nominalism. As I have argued extensively elsewhere,[19] the German philosopher's views on the ontological status of species are entirely contingent on whether the species in question is mathematical, physical, or, as we would put it today, "biological," and in the last case there most certainly are real, non-conventional kinds of individual living beings which are members as a result of their shared descent. Similarly, in the case of Clauberg, different ontological considerations need to be brought to bear for living beings than for non-living ones. Mercer notes in her treatment of Clauberg that the world does not only consist in bare *res extensa*, but also in "amoebas, baboons, whirlpools, and xylophones,"[20] yet for a philosopher such as Clauberg the final two items are not at all like the first two. The xylophone is an artifact, like a clock. As to the whirlpools, some recent metaphysicians have attempted to assimilate living beings to meteorological entities—and in this respect hydrological entities such as whirlpools are not significantly different—to the extent that both are cases of, in the words of Peter van Inwagen, a "homeodynamic storm of simples, a self maintaining, well individuated, jealous event."[21] Here, the final word is the most significant for getting at a crucial ontological difference that appears, at least in this description, to evade van Inwagen's attention: a whirlpool may remain a mere event, but a baboon needs somehow to be accounted for as an *entity*, and for Descartes, Clauberg, and Leibniz alike this account will presumably have to involve more than just the way the parts or the simples involved happen to be working together for a time. The full account is going to have to involve some engagement with the problem of life, as something over and above organization, and, while much effort is going to be spent by all of these philosophers on accounting for the life of the body without having to take recourse to its connection to a soul (whether rational or merely animating), nonetheless a living being is always the sort of thing for which inherence of or independence from a soul is always a problem in need of treatment.

Clauberg takes up a clear position on the question of the boundary between the animate and the inanimate. He believes that no animating principle needs to be invoked in order to explain mineralogenesis and other cases of spontaneous

---

[18] Much of the recent scholarship on the conceptualization of living beings in early modern philosophy has been focused on the work of Leibniz. See in particular Duchesneau 1998, 2010; Smith 2011.

[19] See Smith 2011, Chap. 7.

[20] Mercer 1999, 149.

[21] Van Inwagen 1995, 121.

organization in nature. He writes: "On the earth we find some bodies that are lacking in life, such as all fossils, stones, metals; and some that are alive, such as plants and animals."[22] Clauberg identifies the bodies of living beings as "organic," and he begins to elaborate the difference between the organic and the inorganic in terms of relative degrees of complexity or partedness: "Hence an organic body is ascribed only to living beings, that is, a body endowed with various organs, of which sort are, in plants, the root, the stem, the branch, the shoot, etc.; in animals the head, the mouth, the stomach, etc." But he quickly goes on to add an important aesthetic distinction: "the beauty of living bodies is greater," he writes, "…in all events such variety is not perceived by sense in other bodies."[23] The idea that organism—as an abstract noun comparable to "mechanism," rather than as a count noun comparable to "machine"—consists in organization below the level of perceptibility, and that this degree of detail is a mark of beauty in nature and a reflection of divine wisdom, would be a leitmotif of much later German natural philosophy. In particular, Leibniz would come to describe organism as "a more exquisite mechanism,"[24] characterized by the fact that there is no lower level of organization in the structure of the living body.

The organic unity of a living body is something that is analogous to the proper functioning of a machine. "And we judge the body of any living human or animal," Clauberg writes, "to differ from a dead body, as much as a clock that is correctly set and that has in itself a corporeal principle of motion…differs from the same clock when it is broken, in which the principle of motion has ceased to operate." He continues by saying that the difference, however, lies in the degree of complexity of the organization:

> In order that this comparison should appear less paradoxical, one must consider how various motions in automata that are made by human industry can be destroyed by the work of certain little wheels and other instruments, whose number is very small, if they are compared with the nearly infinite multitude of bones, muscles, nerves, arteries, veins and other parts, which are repeated in the body of any animal….[25]

This difference of complexity, moreover, is not just one of degree, but is actually a mark of a significant ontological divide between the living body and the machine—the former is something that can only have come into existence through divine creation: "Thus the engine of the animal body is to be investigated as an automaton made by the hands of God, which is infinitely better ordered, and have more

---

[22] Clauberg 1664, Prolegomena, 283: "In terra inveniuntur corpora alia vita expertia, ut fere omnia fossilia, lapides, metalla; alia viva, ut plantae & animalia."

[23] Clauberg 1664, Prolegomena, 283: "Hinc solis Viventibus adscribitur corpus organicum, hoc est, variis organis praeditum, cujusmodi sunt in planta radix, caulis, ramus, surcus &c. in Animali caput, os, venter &c. 8. Major itaque corporum viventium pulchritudo est…Saltem in aliis corporibus tanta varietas sensu non percipitur."

[24] See e.g., Leibniz, writing in Stahl 1720, 9: "[O]mnis organismus revera sit mechanismus, sed exquisitior" ("All organism is in truth mechanism, but more exquisite").

[25] Clauberg 1664, 317–318.

admirable motions in them than any that can be made by human art..."[26] The term that Leibniz would later use for this infinitely better ordered automaton is "divine machine," but we should make no mistake: the rudiments of Leibniz's elaborate model are already there in Clauberg, and indeed in Descartes. What Leibniz will offer over and above his predecessors is a detailed account of the nature of infinity, and one that is integrated with his other philosophical concerns (e.g., the famous "labyrinth of the continuum"), while Descartes and Clauberg allowed the concept to function in a vague and somewhat poetic way to mean something like "immense."

We will return to the Leibnizian model of the organic body soon. It is sufficient, for now, to have established that, for Clauberg, the nature of the organic body itself sets it apart from mere universal *res extensa*, and this quite apart from any consideration of the living body's connection with a soul or immaterial animating principle. Of course, a dead body has all the same parts a living body does, at least before it begins to decay. And yet there is nothing organic, in the technical early-modern sense, about a corpse. For Clauberg, there is a straightforward way of understanding the difference: namely, life. "I call *life*," Clauberg writes, "that which cannot be understood without action."[27] And here we see a significant deviation from the far more austere characterization Descartes gives of life as mere heat: "I do not deny life to animals," Descartes writes, "since I regard it as consisting simply in the heat of the heart; and I do not even deny sensation, insofar as it depends on a bodily organ."[28] For Clauberg, by contrast, not only is life action, it is precisely in the life of a being that we may discover the nature of the mind-body *conjunctio*: "The conjunction of the mind [*animi*] and the body should consist in certain acts...because it should constitute the life *of the whole*, which is attributed to the composite Man, distinct from the life of the component *parts*."[29] It is life, then, that is the difference between winking and blinking, or between me raising my arm and my arm going up. And such action is possible because the human body functions together as a whole, with the parts integrated together organically.

Now one obvious problem for Clauberg is to carve out a position for himself that does not go too sharply against Descartes' animal-machine doctrine, that is, the French philosopher's belief that non-human animals do not partake at all of any soul-like, immaterial principle. Clauberg echoes Descartes' view that "*all corporeal life consists in heat and moisture*," and he clarifies that "by the name of 'heat'

[26] Clauberg 1664, 317–318: "Corporis igitur animalis machinamentum inspiciendum est tanquam automatum quoddam manibus Dei factum, quod infinities melius sit ordinatum, motusque in se admirabiliores habeat, quam ulla quae arte humana fabricari possint."

[27] Clauberg 1664, X 5; cited in Trevisani 1992, 112: "*Vitam dico*, quae sine actu intelligi nulla potest."

[28] AT V 278–279, CSMK 366–367.

[29] Clauberg 1664, X 5; cited in Trevisani 1992, 112: "Conjunctio animi & corporis in actibus quibusdam...debuerit consistere, quia debuit constituere vitam *totius*, quae Homini composito tribuitur, a vita *partium* componentium distincta."

moreover you will understand that ardent fire without light."[30] But how are we to understand the claim that life is heat and moisture, in relation to the claim that life is action? And is action only characteristic of human life, or of life in general? The answer appears to be that, for Clauberg, many vital processes, including all of the processes we observe in non-human animals, can be accounted for without any invocation of a conjunction with the mind, insofar as many of these processes happen without the mind so much as noticing:

> Thus there are many motions in us that do not depend in any way on the mind, such as the beating of the heart, the coction of foods, nutrition, convulsions, respiration…and likewise walking, singing, and like things in those who are awake, when these are done without the mind being made aware…[31]

And for Clauberg, this means in turn that apparent action in non-human animals can be explained in entirely mechanical terms: "And since we experience these things in ourselves, we should not wonder so much if the light reflected from the body of the wolf into the eyes of the sheep should have the power to excite in it the motion of fleeing."[32]

So we have two different definitions of "life," one identifying it as heat and moisture, the other as action. In the former case, this is something in which animals can plainly share, while in the latter case it would seem that human beings alone can be said to be alive. Yet recall that Clauberg distinguishes between the "life of the whole" of a human being, on the one hand, and "the life of the component parts" on the other. If he is unwilling to accept that a sheep is alive in the fullest sense, presumably he would also be unwilling to identify, say, a human kidney as a living being. But at the same time in this passage Clauberg clearly wishes to describe the component part as living not in view of its heat, but in view of the action it contributes to the action of the whole. So whether Clauberg is willing to accept this or not, willy-nilly he is inclining toward a view on which action would include organic function that contributes to the functioning of the integrated whole of any living being, even where that action would not have to be explained through conjunction with a mind. And this is a sort of life that, in addition to the *feu sans lumière*, must be attributed to organically integrated animal bodies and human bodies alike.

The capacity for action is conceptually inseparable from the capacity for passion, and Clauberg takes the unwilled character of many of our passions as an indication that they must be caused by something other than the mind, namely, bodies. He characterizes these as nothing other than a variety of perception: "In general I call a passion any sort of *perception* that we have, since often it happens that our

---

[30] Clauberg 1664, 284: "…omnis *vita corporea in calido & humido consistere* vulgo dicitur… *Calidi* autem nomine intelliges Ignem illum ardentem & non lucentem."

[31] Clauberg 1664, 316: "Ita plurimi dantur in nobis motus, qui nullo pacto a mente pendent, ut pulsus cordis, ciborum coctio, nutritio, convulsiones, respiratio…, atque etiam in vigilantibus ambulatio, cantio & similai, cum fiunt animo non advertente…"

[32] Clauberg 1664, 316: "Et quoniam haec in nobis ipsis experimur, non est quod tantopere miremur, si lumen e lupi corpore in ovis oculos reflexum eandem vim habeat ad motum fugae in ipsa excitandum."

mind [*animus*] does not make it such as it is, and we always receive it from things that are represented in awareness."[33]

Now as already mentioned, we see in Clauberg's account of organic structure the rudiments of Leibniz's eventual model of organism as an infinitely nested structure. Leibniz would in turn offer a solution to the problem we have identified in Clauberg—of both wishing to ascribe life to component parts of a living body, but also hesitating on the grounds that these parts can function without any need for conjunction with the mind, while life can only consist in that conjunction—by positing *petites perceptions* throughout all matter. These are perceptions that at the same time function as the metaphysical grounding of the organization that elevates matter to the level of body. As Mercer has already shown, and as becomes clearer through an investigation of Clauberg's theory of living beings, the Duisburg philosopher was plainly committed to an ontological distinction between matter and body. There is some evidence that Descartes had been interested in making such a distinction, yet here Clauberg is plainly pulling away from his French predecessor. The direction in which he pulls is towards a robust theory of corporeal substance, of the sort that Leibniz would exhaustively develop, and in the case of both of these German philosophers we might suppose that it was in no small measure the legacy of Aristotelian philosophy that was still so vital in Germany that led them towards a modified Cartesianism and a greater concern for the salient ways in which living beings come forth as different from mere matter. Before moving on to a consideration of the further development of the doctrine of corporeal substance in Leibniz, let us first go on to consider some of Clauberg's immediate successors at Duisburg, with an eye in particular to their understanding of the problem of living bodies.

5.4 Clauberg's Successors

Francesco Trevisani has done a splendid job of following out in great detail all of the complex ramifications of Cartesianism in the Duisburg school, in particular by paying close attention to the lesser known but nonetheless important physicians at Duisburg, and also by cataloguing and studying the medical dissertations written by students under their supervision. Consistently, what we find over the course of several decades, in an otherwise very diverse group of people, is a firm commitment to the spirit of Cartesianism, mixed with an eclectic interest in local traditions, both academic and popular. Theodor Craanen (1620–1690) is a fine example of this tendency. Craanen began his studies of medicine in Utrecht in 1651, and moved to Duisburg in 1656, where he became a doctor of medicine in 1657. He was the author of numerous medical works, including the *On the Suffocation of the Uterus* (1651) the *Inaugural Medical Disputation on Vertigo* (1672), and the *Medical Disputation on the Flow of the Menstrual Blood* (1676). Craanen is principally

[33] Clauberg 1664, 652.

noteworthy for his *Physico-Medical Treatise on Man*,[34] in which he argues that the movement of the heart is entirely dependent on the blood rather than on spirits. Thus the cause of cardiac motion, as Trevisani puts it, is hydropneumatic and thermodynamic, rather than neuromuscular.[35] Craanen disputes, however, the idea that the heart is the source of an innate heat, placing this view among the "lovely things invented by the ancients, nay more, the chimeras, the explanation of which is so difficult that they could never extricate themselves therefrom."[36] Finally, he is particularly adamant in his opposition to a role for any vital spirits in the body, and in particular in the circulation of the blood. These commitments are all, from Craanen's point of view, central to his identity as a Cartesian, and one detects a fairly rigid, dogmatic commitment to them, rather than a commitment based on internally compelling arguments.

Other adherents of the Cartesian school will echo Craanen's dogmatism, even as they drift from the core commitments of Descartes himself. Craanen is followed by Tobias Andreae (1604–1676), who is active as a professor of medicine at Duisburg from 1662 to 1669. Andreae is noteworthy for the position he defends on the connection between mind and body, which approaches a sort of pre-established harmony. He writes: "Thus indeed Nature formed us in such a way that to like thoughts there correspond like bodily motions, and to like motions of the spirits, like thoughts."[37] However, the parallelism between mind and body does not so much arise from the fact that these are running on two separate, perfectly syncopated tracks, but rather that mind and body together constitute a single "substantial psychophysical unity."[38] In this respect, it is not Leibniz, but rather Spinoza, with his doctrine of psychophysical parallelism, who is most suitably called to mind for comparison.

The absence of efficient causation between mind and body does not prevent Andreae from appealing to a more familiar Aristotelian causal scheme when explaining particular natural processes. In particular, he adopts a theory of generation that explicitly reverts to an Aristotelian account of the "contagious" influence of the semen upon the menstrual blood:

> …all of the female blood is impregnated by the seminal spirit of the male, which is disposed to producing its colliquamentum…The more perfect animals, therefore, excited by orgasm, emit the seminal spirit, which is so subtle that (as Harvey observes), granted that the visible semen does not enter the innermost reaches of the uterus, nevertheless it pervades.[39]

---

[34] See Craanen 1689.

[35] Trevisani 1992, 68–69.

[36] Craanen 1689, 148; cited in Trevisani 1992, 68–69.

[37] Andreae 1659, Sect. 20; cited in Trevisani 1992, 118: "Ita enim Natura nos formavit, ut similibus cogitationibus similes motus corporei, similibusque spirituum motibus similes cogitationes respondeant."

[38] Trevisani 1992, 118.

[39] Andreae 1669, Sect. 20; cited in Trevisani 1992, 137: "…totus sanguis foemineus per *spiritum seminalim* maris imprægnatus, ad ejusdem colliquamentum proucendum aptus reddatur…Oestro igitur venereo animalia perfectiora irritata, spiritum seminalem tam subtilem emittunt, ut licet (observante Harveo) visibile semen penetralia uteri non intret, hic tamen pervadens spiritus."

Andreae supposes, very much in contrast with Craanen, that the blood contains an active principle that is, among other things, responsible for the proper development of the fetus. The active principle of the blood "emits something spirituous from itself" that has something in common with "the spirit of the world or with subtle matter," and that is "a sort of architect of [the fetus's] nutrition and growth."[40]

Plainly, plastic natures, internal architects, and the world soul have no place in Descartes' account of the formation of organized beings, nor in Clauberg's or Craanen's account, for that matter. Andreae provides a clear example of the flexibility of the thinkers responsible for the early incorporation of Descartes into German medical philosophy, and also the strong tendency towards syncretism, with deeply rooted traditions, such as Aristotelian generation theory, continuing to thrive, and without any acknowledged contradiction with the new philosophy.

Two of the handful of medical dissertations completed at Duisburg under Andreae's supervision are of particular interest here: Heinrich Dulcken's *Philosophico-Medical Disputation Man the Microcosm* (1665), and Friedrich Gottfried Barbeck's (1644–1703) *Inaugural Medical Dissertation on the Royal Disease* (1669).

Dulcken was heavily influenced by Andreae, and as is the case with his mentor, in Dulcken's work we see a high degree of syncretism. He was a committed Cartesian, yet was also attached, as the title of his dissertation announces, to the central Renaissance trope of the microcosm and macrocosm. Dulcken adapts this trope to a broadly Cartesian context by describing the two *kosmoi* as analogous automata that—and this is the key point—are "regulated by the same natural laws and subjected to the same evolutionary processes."[41] Thus Dulcken writes:

> The human body arises from the same principles as the corporeal world, of which it is a part; what is more, we see extended matter, and in it quantity, figure, location, motion and rest, in both instances. We observe the same laws of nature in the human body that are in the Universe.[42]

Strikingly, Dulcken extends the microcosm-macrocosm analogy to the political realm as well, and chooses a surprising analog in the human body for the political sovereign:

> ...we can by no means lay aside the most elegant analogy of man with any given political kingdom...Without doubt, our soul resides as a king in the capital city of the brain; it has its throne in the pineal gland, and its close counsellors in the perceptions of the internal and external senses.[43]

---

[40] Andreae 1669, Sect. 17; cited in Trevisani 1992, 138: "...ad nutritionem concurrit, praecipue ut principium activum, quatenus spirituosum quid ex se emittit, quod cum spiritu mundi seu materia subtili...hujus nutritionis & augmentationis quasi architectus est."

[41] Trevisani 1992, 133.

[42] Dulcken 1665, Sect. 4: "*Corpus hominis* ex iisdem principiis cum mundo corporeo, cujus partis est, constat; enimvero materiam extensam, inque ea quantitatem, figuram, situm motum & quietem utrobique videmus. Eaedem leges naturae in corpore humano, quae in Universo observantur."

[43] Dulcken 1665, Sect. 19: "...elegantissimam hominis & regni alicujus *Politici* analogiam omittere haudquam possumus...Nimirum: Anima nostra *Regis* instar in cerebro *metropoli* residet; thronum habet glandulam pinealem, *consiliarios intimos* perceptiones internorum & externorum sensuum."

Dulcken plainly means this as a mere comparison, rather than the more robust analogy between the human being and the world as a whole, which he appears to believe involves a real, objective likeness. Nonetheless, it is significant that he is able to receive Descartes' problematic doctrine of the *sedes animae*, not only without acknowledging its problematic character (namely, that if body and soul have nothing in common, then there ought to be no point at which they, so to speak, meet up), but even elevating the problematic part to a much more exalted role than Descartes had, perhaps hesitantly, ascribed to it. In this Dulcken once again underlines another important feature of German medical Cartesianism: while syncretistic and flexible in the way it mixed Cartesianism with other indigenous traditions, it was also in another sense dogmatic, to the extent that elements of the Cartesian program were adopted uncritically and transformed into mere signifiers of affiliation.

Friedrich Gottfried Barbeck, a one-time student of Franciscus Sylvius in Leiden,[44] provides another interesting variation on the reception of Descartes in Germany. He rejects Andreae's central idea that to similar thoughts there must correspond similar motions of the body. For Barbeck, "it is of the greatest difficulty, in what way the human mind has its sensations [*sensus*] from the motion of the body, as it does not perceive motion as motion, clearly and distinctly, but rather by arbitrary signs constituted by God."[45] The answer for Barbeck is to argue that there is a sort of *tertium quid*, an "interpreter," between the two.[46] In his *Philosophical Disputation on Human Nature* of 1684, Barbeck argues that:

> Corporeal animal life consists in the heat that is produced in the heart by means of a hot effervescence. The central role that the heart has in the functions of the organism is confirmed by the fact that, mediately or immediately, all the bodily humors flow into it, and also by the fact that it is a muscle and, as such, is its own internal cause of motion.[47]

For Barbeck, however, the production of heat in the animal body is not ultimately explicable in mechanical terms: "In fact it does not suffice to know that the constitution of all bodies can be explained mechanically; it will be necessary rather to state with clarity, in the particular knowledge of things, what the mechanical disposition is."[48] Here, by "mechanical disposition," Barbeck appears to mean something like "the ends to which the machine is disposed, as a result of its structure or conformation." As we will see, the idea that machines have such dispositions will reappear in Leibniz as an explicit theory of mechanical teleology, and will be an important aspect of Leibniz's modified mechanism, as against the standard view of Cartesian mechanism,

---

[44] On Sylvius, see Chap. 10 by Nyden.

[45] Barbeck 1684, Sect. 13; Trevisani 1992, 161: "...maximae difficultatis est, quomodo mens humana ex corporis motu sensus suos habeat, cum ne quidem motum ut motum, clare & distincte, sed sub alio sensu tanquam per signa arbitraria a Deo constituta, percipiat."

[46] Barbeck 1687, Sect. 13; cited in Trevisani 1992, 161: "...creatorem inter mentem ac corpus tanquam mediatorem ac *interpretem* existere ac significare."

[47] Barbeck 1684; cited in Trevisani 1992, 197.

[48] Barbeck 1684; cited in Trevisani 1992, 197.

according to which it does not admit any end-governedness at all in the execution of bodily processes.[49]

This all-too rapid survey of some of the lesser-known figures in the Duisburg school of Cartesian medicine reveals, in fact, some of the very important background to the much better known mature philosophy of Leibniz. We see, in particular, some important pre-Leibnizian expressions of a desire to appropriate the revolutionary work of Descartes, but to do so in a way that does not dispense with the important insights of the Aristotelian tradition too hastily. German philosophers, from Clauberg through Leibniz, value the gains of the revolution, yet are much more conservative, and concerned to protect the treasury of philosophy's history, than their revolutionary leader. In particular, they are all, without exception, interested in preserving a fundamental distinction between living beings and things. In Clauberg, this distinction is sometimes expressed in terms, which would be echoed in Leibniz, of prime matter and secondary matter. In Clauberg and his successors alike, there will be a persistent preoccupation with spelling out precisely what is unique about entities endowed with the property of life. Some will drift very far from Descartes—such as Andreae, who maintains that living bodies have an "internal architect" or immaterial formative principle—while all, in a syncretistic fashion that clearly anticipates Leibniz, believing of themselves that their various deviations do not at all constitute grounds for expelling them from the fold of the Cartesians. Let us now turn briefly to a consideration of how Leibniz's mature commitments position him as a philosopher who continues and develops these aspects of early German Cartesianism.

5.5 Leibniz

There can be no room in this rapid *survol* to provide any adequate treatment of Leibniz; it will be enough to briefly spell out some of the evidence in favor of thinking of his mature corporeal-substance metaphysics as a further development of certain tendencies already on display in Duisburg medical Cartesianism. We have already identified a number of important features of the Cartesianism of Clauberg and his successors. First, like Descartes himself, they adhere to a distinctly medical philosophy, according to which the study of medicine, and philosophical reflection on the results of this study, constitutes an integral part of philosophy itself. Second, the early German Cartesians have a highly syncretistic approach to philosophy, adopting elements of the new philosophy while not hesitating to preserve older traditions where these are deemed valuable. Third, there is a strong concern—much stronger than any Descartes himself was willing to express—to mark off living beings from mere things, and living bodies from mere matter; and the study of medicine is an important aspect of the way the German Cartesians responded to this concern.

[49] Barbeck 1686, Sect. 5; cited in Trevisani 1992, 230–231.

Fourth, there is a tendency to back-track from Descartes' original, austere insistence that the machines of animal bodies could not possibly be understood in terms of their ends, with the result that German Cartesianism opens up the possibility of a theory of mechanical teleology. All of these are also very significant elements of Leibniz's thought.

As to medical philosophy, as we have already seen Descartes himself thought of medicine as the art of maintaining health, and understood health in turn as "undoubtedly the chief good and the foundation of all the other goods in this life."[50] Leibniz will tend to characterize medicine's role in more moderate terms, holding that health is merely the *second* most important thing in human life, after virtue, and that therefore medicine is the second most important human art, after philosophical theology. Thus Leibniz writes to Bouvet in 1697:

Medicine is the most necessary of the natural sciences. For just as theology is the summit of the knowledge of things pertaining to spirit, and just as it contains both good morals and good politics, one can say that medicine as well is the summit and as it were the principal fruit of our knowledge of bodies, to the extent that they are related to our own. But all of physical science and medicine itself have as their final goal the glory of God and the supreme happiness of men, for in conserving them, medicine gives them the means to work for the glory of God.[51]

This point of difference between the two philosophers is certainly not the only one, and may indeed be a reflection of deeper philosophical differences about the limits of mechanical explanation. Here, Leibniz is certainly continuing the spirit of the earlier German Cartesians in adopting a qualified approach to the body of commitments associated with Cartesianism. At least since the important work of Christia Mercer, Leibniz has been widely and rightly understood to be a "conciliatory eclectic."[52] As Mercer was certainly aware, this sort of eclecticism was widespread among Leibniz's countrymen.

Nowhere is Leibniz's conciliatory spirit clearer than in his famous attempt to synthesize mechanism and teleology. This synthesis is typically seen as having emerged over the course of the 1690s in Leibniz's mature theory of pre-established harmony, which entails, among many other very significant consequences, that the series of states of bodies unfold as a sort of "kingdom within a kingdom," the greater, encompassing kingdom being nothing other than God's end-driven kingdom of grace. What is however less well-known is that prior to the development of this mature theory, Leibniz was working out models of mechanical teleology that bear a more direct affinity to the sort we have already seen in Barbeck. In a cluster of recently translated texts on medical and physiological topics from the early 1680s, Leibniz develops a notion of mechanical ends according to which "any machine…is best described in terms of its final cause, so that in the description of its parts it is therefore apparent in what way each of them is coordinated with the

[50] AT VI 61–62, CSM I 142–143.

[51] Leibniz 1718, 115.

[52] See Mercer 2001.

others for the intended use."[53] Here we have what might be called a "heuristic" or "epistemic" teleology, where the ends of machines are identified as an analytic tool, in the absence of more exact knowledge of how their parts work together mechanically to bring about effects.

It is in the analysis of the mutual functioning or conspiracy of the organic parts of a living being's body that Leibniz develops his most sophisticated and in some respects novel views about organic structure and teleology. Recall, now, that for Clauberg an animal body is distinguished from a mere machine to the extent that it is "infinitely better ordered," and has "more admirable motions…than any that can be made by human art."[54] As already mentioned, Clauberg does not seem to have a very precise understanding of the concept of infinity as deployed here; it seems to mean nothing much more than "immense." Leibniz, by contrast, would develop a very technical and rigorous conception of infinity, and would make it the centerpiece of very many different aspects of his philosophy, not least his theory of the organic structure of living bodies. For Leibniz, to cite a well-known and very representative passage from the *Monadology* of 1714, an organic body is precisely matter that has been "arranged by divine wisdom" so that it is "essentially organized throughout and thus that there are machines in the least parts of the machine to infinity."[55] It is central to this conception of organic structure that there be literally no lower limit to the structural complexity of the animal body, no final stage of decomposition at which the analysis of the structure would come to an end. This commitment, moreover, is grounded in Leibniz's deepest metaphysical principles concerning the ultimate basis of reality in immaterial nodes of perceiving substance, or monads, and concerning the derivation of all of corporeal reality from these basic elements. Leibniz has effectively taken an old familiar trope of the analysis of animal bodies—that they are very complex or "infinite"—and both embedded it within a highly sophisticated and rigorous account of infinity, and moreover grounded it in his most basic metaphysical principles. It is unlikely however that he would have done this if the old, familiar trope had not been developed by Clauberg and others in the way it was.

In the end, Leibniz will identify life neither with bodily heat (as Descartes did), nor with bodily action (as Clauberg had), but instead with perception alone, which is not a bodily function at all, but rather the exclusive activity of immaterial monads. This account was, however, long in developing. In a little known text of the early 1680s, entitled "On Writing the New Elements of Medicine."[56] Leibniz hopes to account for "life" in terms of the "economy" of the animal body alone. Decades later, in contrast, he will come to decouple the faculty of perception from the body, and will root it in the perceptions of the simple immaterial substance. As Leibniz writes in his polemic against G. E. Stahl of 1709–1710, the action of the

[53] See Smith 2011, Appendix 3.

[54] Clauberg 1664, 317–318.

[55] Leibniz 1849–1860, VI, 543–544.

[56] See Smith 2011, Appendix 2.

animal body can be exhaustively accounted for in terms of its "vegetative structure alone," and thus Stahl is making a great mistake in "call[ing] 'life' what others call 'vegetation'."[57]

5.6 Conclusion

The development from Descartes through Clauberg to Leibniz of the concept of life, which moves from heat through action to perception, maps in intriguing if inexact ways onto Leibniz's own motion in the course of his career from mechanism through dynamics or the science of self-active, force-endowed bodies in motion, to, finally, a metaphysics of immaterial perceivers, or monads, which serve to ground all of corporeal reality. However much more might be said about the parallels suggested here, one thing is already clear: a great deal can be learned about the basic philosophical commitments of Descartes and Leibniz through close attention to their shared concern to understand the nature and structure of the bodies of living beings.

References

Andreae, Tobias. 1659. *Disputatio philosophica inauguralis explicans naturam & phaenomena cometarum*, University of Duisburg, Duisburg.
Andreae, Tobias. 1669. *Brevis explicatio corporis humani prout anima vegetativa pollet* (respondent: Petrus Krebs), University of Duisburg, Duisburg.
Aucante, Vincent. 2006. *La philosophie médicale de Descartes*. Paris: Presses Universitaires de France.
Barbeck, Friedrich Gottfried. 1669. *Dissertatio medica inauguralis de morbo regio, seu ictero*, University of Duisburg, Duisburg.
Barbeck, Friedrich Gottfried. 1684. *Disputatio philosophica de natura hominis, ac mentis imperio in corpus et corporis actu in mentem* (respondent: Johannes Antonius Stock), University of Duisburg, Duisburg.
Barbeck, Friedrich Gottfried. 1686. *Disputatio medica prima de sanguinis corporis humani purificatione* (respondent: Johannes Bernhardus Becher), University of Duisburg, Duisburg.
Barbeck, Friedrich Gottfried. 1687. *Disputatio de spectris, lemuribus ac ostentis ex medicinae & philosophiae principiis deducta* (respondent: Albertus Witten), University of Duisburg, Duisburg.
Bohatec, Josef. 1966. *Die cartesianische Scholastik*. Hildesheim: Olms.
Clauberg, Johannes. 1647. *Elementa philosophiae, sive Ontosophia*. Groningen: Johannes Nicolai.
Clauberg, Johannes. 1664. *Physica quibus rerum corporearum vis & natura; Mentis ad Corpus relatae proprietates; denique Corporis ac Mentis arcta & admirabilis in Homine conjunctio explicantur*. Amsterdam: Danielem Elzevirium.
Clauberg, Johannes. 1652. *Defensio cartesiana adversus Iacobum Revium*. Amsterdam.
Clauberg, Johannes. 1658. *Paraphrasis in R. Descartes meditationes de prima philosophia*, University of Duisburg, Duisburg.

[57] Leibniz, in Stahl 1720, 11.

Clauberg, Johannes. 1680. *Differentia inter Cartesianam et alias in Scholis usitatam Philosophiam*. Groningen.

Coqui, Guillaume. 2009. Le cartésianisme de Clauberg. In *Descartes und Deutschland: Descartes et l'Allemagne*, ed. Robert Theis et al., 29–44. Hildesheim: Olms.

Craanen, Theodor. 1651. *De suffocatione uteri*, University of Duisburg, Duisburg.

Craanen, Theodor. 1672. *Disputatio medica inauguralis de vertigine*, University of Leiden, Leiden.

Craanen, Theodor. 1676. *Disputatio medica de fluxu sanguinis menstrui*, University of Leiden, Leiden.

Craanen, Theodor. 1689. *Opera omnia*, vol. 1. Tractatus physico medicus de homine. Antwerp.

Descartes, René. 1970. *Oeuvres de Descartes*, ed. Charles Adam and Paul Tannery. Paris: J. Vrin.

Descartes, René. 1985–1991. *The Philosophical Writings of Descartes*, 3 vols., ed. John Cottingham, Robert Stoothoff, and Dugald Murdoch. Cambridge: Cambridge University Press.

Duchesneau, François. 1998. *Les modèles du vivant de Descartes à Leibniz*. Paris: Vrin.

Duchesneau, François. 2010. *L'organisme et le vivant chez Leibniz*. Paris: Vrin.

Dulcken, Henrius. 1665. *Disputatio philosophico-medica de homine microcosmo... examini subjicit Henrius Dulcken, Dusseldorpiensis*, University of Duisburg, Duisburg.

Grmek, Mirko Dražen. 1968. Les idées de Descartes sur le prolongement de la vie. *Revue d'histoire des sciences et de leurs applications* 21(4): 285–302.

Hurson, Didier. 2009. Rezeptionsgeschichte der philosophie Descartes' im Deutschland des 17. Jahrhnderts. In *Descartes und Deutschland: Descartes et l'Allemagne*, ed. Robert Theis et al., 17–27. Hildesheim: Olms.

Leibniz, G.W. 1718. *Otium hanoveranum, sive Miscellanea ex ore & schedis illustris viri, piae memoriae Godofr. Guilielm. Leibnitii*, ed. Joachim Friedrich Feller. Leipzig: Johann Christian Martin.

Leibniz, G.W. 1849–1860. *Die philosophischen Schriften von G. W. Leibniz*, 7 vols., ed. C.I. Gerhardt. Berlin: Weidmannsche Buchhandlung.

Lindeboom, Gerrit. 1978. *Descartes and medicine*. Amsterdam: Rodopi.

Manning, Gideon. 2013. Descartes' healthy machines and the human exception. In *The mechanization of natural philosophy*, ed. Sophie Roux and Daniel Garber, 237–262. Leiden: Brill.

Mehl, Édouard. 2001. *Descartes en Allemagne 1619–1620: le contexte allemand de l'élaboration de la science cartésienne*. Strasbourg: Presses Universitaires de Strasbourg.

Mercer, Christia. 1999. Johann Clauberg, corporeal substance, and the German response. In *Johannes Clauberg (1622–1665) and Cartesian philosophy in the 17th century*, ed. Theo Verbeek, 147–159. Dordrecht: Kluwer.

Mercer, Christia. 2001. *Leibniz's metaphysics: Its origins and development*. Cambridge: Cambridge University Press.

Smith, Justin E.H. 2011. *Divine machines: Leibniz and the sciences of life*. Princeton: Princeton University Press.

Stahl, G.E. 1720. *Negotium otiosum, seu Skiamachia*. Halle: Impensis Orphanotrophei.

Thomasius, Christian. 1699. *Versuch vom Wesen des Geistes oder Grund-Lehre*. Halle: Christoph Salfelden.

Thomasius, Christian. 1719. *Einleitung zu der Vernunfft-Lehre*, 5th ed. (First edition 1691). Halle: Christoph Salfelden.

Trevisani, Francesco. 1992. *Descartes in Germania: La ricezione del cartesianesimo nella Facoltà filosofica e medica di Duisburg (1652–1703)*. Milan: Franco Angeli.

Trevisani, Francesco. 1982. Medizin und Cartesianismus in Deutschland am ende des 17. und zu beginn des 18. jahrhunderts. In *Heilberufe und Kranke im 17. und 18. Jahrhundert*, ed. W. Eckart and Johanna Geyer-Kordesch. Münster: F. Angeli.

Trevisani, Francesco. 2012. *Descartes in Deutschland: Die Rezeption des Cartesianismus in den Hochschulen Nordwestdeutschlands*. Münster: Lit Verlag.

Van Inwagen, Peter. 1995. *Material beings*. Ithaca: Cornell University Press.

Verbeek, Theo (ed.). 1999. *Johannes Clauberg (1622–1665) and Cartesian philosophy in the seventeenth century*. Dordrecht: Kluwer.

Chapter 6
Could a Practicing Chemical Philosopher Be a Cartesian?

Bernard Joly

Abstract When Descartes touches upon objects and operations of chemistry in the fourth part of *Principia philosophiae* (1644), he destroys any possibility of chemistry becoming a specific science. He reduces all chemical operations to matters of size, shape and motion of particles. In the frame of Cartesian natural philosophy, chemistry vanishes into mechanics. In this chapter, I would like to examine how, under these conditions, some philosophers or chemists who were sometimes regarded as Cartesian thinkers introduce chemical discourse in their natural philosophy. Some of them, such as Boyle, said that the mechanical structure underlies all chemical operations. Therefore, chemistry can only exist as empirical knowledge. Others however, such as Lémery (father and son) proposes new mechanical explanations specific to chemistry, in a way which is opposite to the theories of *Principia philosophiae*. Chemistry, which cannot be developed without laboratory work, leads to an unusual empirical Cartesianism which I suggest we examine in the light of the debate between some French chemists at the *Académie royale des sciences* at the end of the seventeenth century and the beginning of the eighteenth century.

Descartes did not like chemists, whom he also called alchemists. Like most of his contemporaries, he considered the two words to be synonymous. He often jested at alchemists and their little secrets, their vain quest after the philosopher's stone. But he levelled at them more fundamental criticisms, concerning both the nature of their knowledge, and the shortcomings of their theorizing. This led him, in the fourth part of his *Principia philosophiae*, to operate a grand scale reduction of the workings of chemistry to the principles of his mechanistic conception of matter,

B. Joly (✉)
Université Lille 3, UMR 8163 "Savoirs, textes, langage," Lille, France
e-mail: bernard.joly@univ-lille3.fr

M. Dobre and T. Nyden (eds.), *Cartesian Empiricisms*, Studies in History and Philosophy of Science 31, DOI 10.1007/978-94-007-7690-6_6,
© Springer Science+Business Media Dordrecht 2013

such as had been expounded in previous sections of his work. Under such conditions, one could have imagined that chemistry, in the eyes of Cartesian philosophers, could have no future. Fully integrated into Cartesian physics, it seemed doomed to extinction as a specific discipline.[1] This proved in no way to be true. It will be examined here why and how chemists were in a position to claim that they belonged to Cartesianism in the late seventeenth and early eighteenth centuries, a claim that induced them to develop a conception of Cartesianism which granted an important place to laboratory experimentation.[2] But eventually it will have to be asked whether such a standpoint made these Cartesian chemists genuine empiricists.

6.1 The Mechanistic Reduction of the Objects of Chemistry

Descartes' attitude towards chemistry may appear paradoxical, since he criticized it while indulging in its practice. On the one hand, he seemed to reject it. Everyone remembers the famous passage from *Discours de la méthode*, in which he dismissed alchemy, astrology and magic as belonging to the "false science," claiming that henceforward it would no longer be possible to be abused by "the promises of an alchemist or the predictions of an astrologer, the tricks of a magician or the frauds and boasts of those who profess to know more than they do."[3] In the same period, he wrote to Marin Mersenne (1588–1648): "I am of a mind with you to laugh at the imaginings of this chemist of whom you write, and believe that such chimeras do not deserve to engage for one single moment the thoughts of an honest man."[4] As a result of the spread of Paracelsian ideas, the doctrines of alchemists were very influential at the beginning of the seventeenth century. They could be associated with the vast movement of criticism of Aristotelian ideas and his philosophy of nature. Descartes had met alchemists, first during his journey to Germany, then at the beginning of the 1620s in Paris, before he turned away from their doctrines.[5] His dismissal of a subject to which he had been attracted in his youth, at a period when he was trying to get away from the patterns of Scholasticism, seems to have

---

[1] One can notice that, in the tree of philosophy presented in the *Lettre-Préface* of 1647, mechanics, but also medicine, which Descartes often associated with chemistry, keeps a distinct place from physics, which is not the case for chemistry.

[2] Our enquiry is limited to the evocation of some figures among those who dealt with chemistry in a would-be spirit of Cartesianism and has no vocation to cover the whole field of the interactions between Cartesianism and chemistry (or alchemy) in the seventeenth and eighteenth centuries throughout Europe. A whole book would be needed for such a study.

[3] See *Discours de la méthode* I. AT VI 9, CSM I 115.

[4] Letter to Mersenne, March 1637. AT I 351: "Je me moque avec vous des imaginations de ce chymiste dont vous m'écrivez, et croie que semblables chimères ne méritent pas d'occuper un seul moment les pensées d'un honnête homme."

[5] See Gouhier 1958; Maillard 1998; Mehl 2001.

become a lifelong standpoint in his doctrine. Thus, in 1649, he criticized in parallel terms "those who boast of possessing secrets, in chemistry or judicial astrology."[6]

At the same time as he criticized it, Descartes showed an interest in the chemistry of his time.[7] On April 15, 1630, he announced to Mersenne that he now studied "chemistry and anatomy simultaneous; everyday I learn something that I cannot find in any book."[8] Two years later he explained to him that he undertook "various experiments to discover the essential differences between oils, ardent spirits, ordinary waters and acidic liquids, salts, etc."[9] These experiments served as a preparatory work for the writing of *Le Monde*, a book which was to remain unpublished during his lifetime. However, he gave a summary of this work in part 5 of the *Discours de la méthode*, referring to chemical developments which left no trace in the manuscript found after his death, and which gave us partial access to his work. It is also in the *Meteorology*, the second scientific essay introduced by *Discours de la méthode*, that we may find the result of his chemical research. Descartes describes water "carrying towards the top of an alembic the small particles of these oils which Alchemists were in a practice of extracting from dry plants,"[10] or "these spirits or 'eaux de vie' which always rise first from the distilled bodies."[11] It is mainly in the third discourse of *Meteorology*, devoted to salt, that Descartes refers to the works of chemists, notably when he mentions "this extremely strong and sour water which can dissolve gold, and which is named salt spirit or oil by Alchemists."[12]

Later on, Descartes frequently broached chemical topics with correspondents asking for his advice. In this respect, Mersenne, Constantin Huygens (1596–1687) and William Cavendish (1593–1676) were his main interlocutors. He let Mersenne know about his various experimental attempts, and the latter communicated to him the letters of Christophe de Villiers (1585–1650), a physician from Sens who was taking an interest in alchemy and whose theories about salt were criticized by Descartes. Chemical questions crop up several times in his correspondence with Huygens, to whom he gave his advice in August 1638 about the dissolution of metals by *aqua fortis* which do not attack the other mineral substances, or on "quick silver [which] resolves gold, tin, and lead, though it can hardly fix itself on the other metals, and even less on the bodies which are not metallic."[13] At that point,

[6] Letter to Chanut, March 31, 1649. AT V 327, CSMK 370.

[7] I have developed this point in Joly 2011, Chap. II: "Descartes et les 'chymistes' de son temps."

[8] AT I 137, CSMK 21.

[9] Letter to Mersenne, April 5, 1632. AT I 243, CSMK 37.

[10] AT VI 241: "emporte vers le haut d'un alambic les petites parties de ces huiles que les Alchimistes ont coutume de tirer des plantes sèches."

[11] AT VI 247: "ces esprits ou eaux de vie, qui s'élèvent toujours les premières des corps qu'on distille."

[12] AT VI 263–264: "cette eau extrêmement aigre et forte, qui peut soudre l'or, et que les Alchimistes nomment l'esprit ou l'huile de sel."

[13] Letter to Huygens, August 1638. AT II 351: "le vif argent [qui] résout l'or, l'étain et le plomb, bien qu'il ne se puisse presque pas attacher aux autres métaux, et encore moins aux corps qui ne sont point métalliques."

Huygens considered Descartes a genuine expert on chemical questions, for in 1645 he asked Descartes to send him a little treatise on chemical operations similar to the treatise on mechanics that he sent him a few years earlier, entitled *Traité des engins*.[14] Descartes declined the invitation, saying that everything he knew about the question had been expressed in the fourth part of *Principia philosophiae*.[15] This did not prevent him from taking up the subject again in 1646 in a letter to Cavendish in which chemical developments occupy several pages. After renewing his criticism of "chemists" who "use words in an uncommon sense only in order to make it seem that they know what in fact they do not know,"[16] he expounds his views on the principles of chemists, on the generation of stones, and on the nature of quicksilver.

It would be wrong to believe that Descartes had set up an opposition between alchemy, whose illusory nature he denounced, and chemistry, whose practices he accepted. On the contrary, he used each of the two terms interchangeably to refer to the theories he condemned or to the practices of which he gave a new interpretation: in 1649, he calls "chymistry" the theory and practice of those whom he called "alchimists" in 1637. Such a situation should not surprise us, since the historians of ancient chemistry have shown in recent years that the distinction between chemistry and alchemy did not really make sense in the seventeenth century.[17] More than that, Descartes' diffidence concerning "chymistry" or alchemy does not mean that he was opposed to the transmutation of metals. Like many of his contemporaries, he thought that the question of the philosopher's stone, like that of squaring of the circle, must be "searched by human understanding."[18] It is not the operation in itself that seems unreasonable to him, but the doctrine one uses to justify its possibility.

We should thus wonder about the reasons for the Cartesian paradox about chemistry, which must be distinguished from the ambivalence around that question which was rather widespread in his time[19]: why did Descartes strenuously reject a knowledge whose operations he approved of? The various reasons for Descartes' diffidence of alchemy have been documented elsewhere.[20] They could be gathered around two main lines here: the rejection of the founding principles of the alchemists' doctrine, but also a defiance of laboratory work, whose practical interest was recognized, but which could be of no use as a means of justifying theory. Thus, in Descartes' eyes, the chemical doctrine can claim neither doctrinal soundness, nor experimental basis. Two historical supports of alchemy, the book and the alembic, are thus put to question.

---

[14] Letter from Huygens to Descartes, July 7, 1645; AT IV 244 and 779. The *Traité des engins* had been sent by Descartes to Huygens on October 5, 1637; AT I 435–447.

[15] Letter to Huygens, August 4, 1645; AT IV 260 and 780. More about this later.

[16] Letter to the Marquess of Newcastle, November 23, 1646; AT IV 569, CSMK 302.

[17] See Halleux 1979; Principe and Newman 1998; Principe and Newman 2001; Joly 2007a.

[18] *Notae in programma*; AT VIII 353: "la recherche du raisonnement humain."

[19] Contrary to Francis Bacon or Mersenne, Descartes had no wish to reform alchemy, or to sort out the various doctrinal standpoints. What he wanted to do was deny all chemical practices the possibility of any theoretical basis. On Bacon's attitude, see Joly 2003; on Mersenne's, see Joly 2001b.

[20] Joly 2011, Chap. IV: "Le refus cartésien d'une science chimique."

Descartes rejects the theories of matter on which chemists founded their work. Challenging the fact that what they called their principles (Mercury, Sulfur, and Salt) really had the status of principles, he disrupts one essential aspect of the theory of matter which was at the heart of the alchemical doctrine. As he briefly, but very clearly states in the *Lettre-préface* of the French edition of the *Principes de la philosophie*, the Mercury-Sulphur-Salt triad no more deserves the status of principle than void and atoms, or hot, cold, dry and wet, and "all other similar things which some people have proposed as their first principles."[21] Neither Democritean atomism, nor the Aristotelian theory of sensitive qualities, nor the *tria prima* of Paracelsians are compatible with the Cartesian theory of matter, such as expounded in the second part of the *Principia philosophiae*. He expatiates on the subject in several passages of his correspondence. Already in 1631, even though he conceded some sort of interest in the alchemical research of his friend Villebressieu, he insisted that "there is only one material substance which receives from an external agent its action or its ability to move from one place to another."[22] Far from being founding principles, the elements and principles of chemists are nothing but the result of these movements of matter. He comes back to this argument 10 years later, when he explains to Mersenne the reasons for his disagreement with the theses of de Villers, the physician from Sens. Expatiating upon the whole of the alchemists' doctrines, he writes:

> These Principles are nothing less than a wrong fantasy, deriving from the fact that, in their distillations, they extract waters which are all the most slippery and flexible parts of the bodies from which they extract them, and refer them to Mercury. They also extract from them oils, whose parts are shaped like branches, loose and flexible enough to detach themselves, and these they refer to Sulphur; and they refer to Salt the loosest parts of what remains, which can mix and merge, so to say, with water; last, the grossest parts which remain are their *Caput mortuum*, or *Terra damnata*, which they only account for as a useless thing. As a matter-of-fact, I cannot conceive of these parts as indivisible or different from one another, but through the diversity of their figures.[23]

This way of justifying the existence of principles by invoking experiences of distillation supposed to extract as such, could be found at that time in numerous "chymical" texts, for instance those by Joseph Du Chesne (1544–1609) or Etienne De Clave.[24] What Descartes objected to was not so much the operation itself, and the results it entailed, as the way it generally used to be interpreted. Far from being

---

[21] AT IXb 8, CSM I 183.

[22] Letter to Villebressieu, Summer 1631; AT I 216, CSMK 33.

[23] Letter to Mersenne, July 30, 1640; AT III 130–131: "Ces Principes ne sont rien qu'une fausse imagination, fondée sur ce qu'en leurs distillations, ils tirent des eaux qui sont toutes les parties plus glissantes et pliantes des corps dont ils les tirent, et qu'ils les rapportent au Mercure. Ils en tirent aussi des huiles, dont les parties sont en forme de branches, assez déliées et pliantes pour pouvoir être séparées, et ils les rapportent au Soufre; et ils rapportent au Sel les parties les plus déliées de ce qui reste, qui se peuvent mêler et comme incorporer avec l'eau; puis enfin les parties plus grossières, qui demeurent, sont leur *Caput mortuum*, ou *Terra damnata*, qu'ils ne comptent que comme une chose inutile. Au reste, je ne conçois point ces parties indivisibles ni autrement différentes entre elles, que par la diversité de leurs figures."

[24] See Joly 2002.

ontologically distinct principles that institute an irreducible plurality in the heart of matter, Mercury, Sulfur, and Salt are nothing but various parts of one similar matter and differ through nothing but their respective configuration. Thus Descartes' concern here is not with substituting the "wrong fantasy" of alchemists with the "right" one, but rather with contrasting two methods with each other. Whereas alchemists aimed, through an inductive approach, to justify their would-be principles by resorting to experience, Descartes, on the contrary, meant to justify the empirical properties of chemical substances by resorting to principles, which he did not imagine but conceived, that is to say, deduced from their metaphysical foundations.

The same idea is repeated in his letter to Cavendish dated November 23, 1646:

> In my view, the chemists' salt, sulphur and mercury are no more different from each other than the four elements of the philosophers, and not much more different from each other than water is from ice, foam and snow. I think that all these bodies are made of the same matter, and that the only thing which makes a difference between them is that the tiny parts of this matter which constitute some of them do not have the same shape or arrangement as the parts which constitute the others.[25]

This doctrine was central to the 88 articles (45–123) devoted to the objects and operations of chemistry in the fourth part of *Principia philosophiae*.[26] Resuming the topic of the identification of the would-be principles of chemists with the "three kinds of bodies" whose formation he accounted for by mechanical arguments, Descartes carries out a process of reduction of chemistry to mechanics, which ought to result in its disappearance as a scientific discipline.

Yet, we are under restraint of reaching such a conclusion straight away, because of Descartes' constant interest in chemical operations, and particularly in distillation. As has just been shown, the latter keeps the status of a privileged experience. If it is no longer destined to exhibit vain principles, it nevertheless remains an essential practice, which fully brings into relief the existence of clearly distinct categories of chemical bodies whose natural formation mechanical interpretation accounts for by operations of hammering, tearing apart or rubbing against one another. Descartes does not deny the importance of experimental practices, but he keeps repeating his concern with a knowledge which rests on experience and laboratory work. Several times he complains of his lack of the full technical competence for carrying out chemical experiments under good conditions, and expresses his wish of working in company with a chemist able to carry out the various laboratory experiments. He says so to Mersenne in 1642, referring to his former friend Villebressieu:

> I have a good opinion of him for that he has the skills to put into practice what he could be ordered in that respect [the "little secrets of chemistry"], and I hold him to be good-natured enough. He offered to come and stay here with me, which I cannot accept at this moment, because I do not wish to stop for any experiment till my Philosophy is published.[27]

[25] AT IV 570, CSMK 302.

[26] AT VIII 231–269 for the Latin edition, AT IXb 225–266 for the French edition.

[27] Letter to Mersenne, December 7, 1642; AT III 598: "Ce que j'estime en lui est qu'il a des mains pour mettre en pratique ce qu'on lui pourrait prescrire en cela [les 'petits secrets de chymie'], et que je le crois d'assez bon naturel. Il m'offre de venir ici, ce que je ne voudrais pas maintenant, à cause que je ne veux point arrêter à faire aucunes expériences, que ma Philosophie ne soit imprimée."

In so far as the fourth part of the *Principia philosophiae* rests partly on the evocation of the experiences made by chemists about fire or distillation, such words could be surprising. At any rate they are an indication that, at that time, Descartes deemed he had sufficient knowledge of chemistry to be able to launch into his mechanistic interpretation of the operations and objects of chemistry. No doubt Descartes does not say everything: it is established that, at that time, he frequented Cornelis van Hoghelande (1590–1646), who used to practice chemical medicine. If one may believe the testimony of Daniel Georg Morhof (1639–1691), this Dutch chemist had "made a lot of experimentation in company with Descartes."[28] However, in later days, Descartes kept invoking his lack of experience when invited by friends to proceed with his chemical experimenting. He said so to Huygens in 1645 to motivate his refusal to write out the "little treatise of chymistry" that the latter had asked from him:

> As I have already given out the little knowledge I have of this subject in the fourth part of my principles, when dealing with the nature of minerals and fire and with all the diverse effects to which almost the whole of Chemistry can be related, it is impossible for me to write anything more without incurring the risk of running into error, for lack of making experiments necessary to acquire the particular knowledge of each thing; and as I have no practical opportunity to make them, I henceforward decide to put an end to this study, and to all similar studies, which could only be carried on to my satisfaction with the help of other people; for I still have enough researches to make, for which I need no other resources than those of my own mind, to be able to live pleasantly enough for the rest of my days.[29]

No doubt Descartes was no longer the same since he had announced his laboratory experimenting in 1630. He had realized that his capacities as an experimenter were limited, and wished to devote his time to more rewarding activities such as would not impinge upon his solitude as a philosopher. Yet this lack of experience did not prevent him from writing the fourth part of *Principia philosophiae*, which confirms the essentially deductive nature of his reasoning in that work. A point which is further confirmed, in the following year, by his answer to Cavendish, to whom he acknowledged that he had not yet made "all the necessary experiments" to establish the nature of quicksilver. But he thought he knew enough to account for all its properties, whether it be its fluidity, its faculty to shape into tiny balls on a table, or not to stick to the hands (alchemists would say that it is a kind of water that does not wet), or to mix easily with lead or gold.[30] More than ever, it seemed that Descartes thought it enough to account for the chemical properties of observable

---

[28] Morhof 1673; testimony mentioned in Matton 1998, 112: "…multa cum Cartesio operatus est."

[29] Letter to Huygens, August 4, 1645; AT IV 260–261 and 780–778: "Ayant déjà écrit tout le peu que je savais touchant cette matière, en la quatrième partie de mes principes, lorsque j'y ai traité de la nature des minéraux et de celle du feu, et de tous les divers effets auxquels se peut quasi rapporter toute la Chymie, il ne m'est pas possible d'en rien écrire davantage, sans me mettre en hasard de me méprendre, à cause que je n'ai point fait les expériences qui m'auraient été nécessaires pour venir à la connaissance particulière de chaque chose; et n'ayant point la commodité de les faire, je renonce dorénavant à cette étude et à tous les autres semblables, touchant lesquels je ne pourrais entièrement me satisfaire sans l'aide d'autrui; car il en reste encore assez d'autres, auxquels je n'ai besoin que de moi seul, pour occuper agréablement le reste de ma vie."

[30] Letter to the Marquiss of Newcastle, November 23 1646; AT IV 572, CSMK 302.

bodies by means of his mechanical physics, without confronting himself with new hypotheses that could be confirmed by experience. Thus one may be struck by the absence of any heuristic dimension in his discourse about chemistry, which reduces chemistry to a number of empirical observations with no status as scientific knowledge, since the principles which account for the observed phenomena remain exterior to it: as a matter of fact, they belong to the mechanistic system and the geometrical properties of matter, which have been expounded in the general part of his physics.

6.2 What Is a Cartesian Chemistry?

The Cartesian critique of chemistry could but find an echo at that time. It meets halfway the strong reserves that had been expressed for centuries toward a knowledge whose legitimacy was constantly put to question. Introduced into Europe in the twelfth century, at a time when the works of Arabian scientists were being translated, alchemy could never manage to find its place in the network of medieval sciences. As it could not separate theory from practice, it needed the presence of a laboratory, which would have been out of place in universities dedicated to intellectual pursuits. Related to *Artes Mechanicae,* it could not boast the status of a speculative science.[31] Such a point of view, shared by a wide proportion of intellectuals until the seventeenth century, was of course strongly reproved by alchemists' intent to show that alchemy was a genuine form of natural philosophy. Thus, from the *Summa Perfectionis* by pseudo-Geber in the thirteenth century, to the *Abrégé des secrets chimiques* (1636) by Pierre Jean Fabre (1588–1658) or the *Traité de la chymie* (1661) by Nicaise Le Febvre (1610–1669), a series of works were published and their aim was to argue that alchemy was rooted in a theory of matter, which would have conferred it the rank of a philosophy.

Alchemy, which had been able to resist the onslaught of Aristotelianism and Galenism, was not more threatened by the growth of mechanistic theories. One could even say that those who referred to Cartesian ideas and wished to develop a science based on the principles of mechanics, were obliged to take the resistance of chemistry into account, so that what took place eventually was as much an adaptation of Cartesianism to the claims of chemistry, as a submission of the latter to the constraints of mechanism. It is true that the first Cartesians remained faithful to their master's spirit. As was shown, respectively, by Luc Peterschmitt and Mihnea Dobre, there is no denying that authors like Géraud de Cordemoy (1626–1684), Jacques Du Roure (fl. 1653–1683), Robert Desgabets (1610–1678), or Jacques Rohault (1618–1672) were dealing with the objects of chemistry, yet they no way considered the latter as a science.[32]

In his *Traité de physique* Jacques Rohault, admittedly, granted much more importance to laboratory work than Descartes did in the fourth part of his

[31] See Mandosio 1990–1991, 1998.

[32] Peterschmitt 2007; Dobre 2011.

Principes de la philosophie. The latter, as we have seen, had confessed, in a letter to Huygens, that he had given up experimenting to gain access to the "individual knowledge of each thing," thus acknowledging the fact that the singular knowledge of the properties of the diverse objects of chemistry had been of no use to him to operate the mechanical reduction of chemistry. On the contrary, Rohault insisted, "every useful Science ought to descend immediately to Particulars."[33] According to him, the properties of chemical bodies had to be established through laboratory experiment, and not, as with Descartes, by means of a deduction of their process of formation in the earth. No doubt he wrote that "all that which is produced by art in the Laboratories of Chymists, is done naturally in the Bowels of the Earth,"[34] but this parallelism, frequent in the works of alchemists, resulted in nothing but a greater importance given to laboratory work, thus considered as a natural process. Thus, it became possible to transmute Salts into acid liquids, "which the Chymists call *Oyl*, or *Spirit of Salt*, or *Aqua-Fortis* which is used to dissolve Metals with."[35] Rohault then gave the recipe for this operation, which consisted in distilling salt mixed up with piled up clay. A little further on, he saw "the Experiments of Chymists, who by the Resolution of Metals, can draw Salt and Sulphur out of them,"[36] as a confirmation of his theses on the constitution of metals, which accounted for his former assertion that the transmutation of metals was a "moral impossibility," but not an "absolute one."[37]

In the first part of his work, Rohault had fully recognized the importance of the works of chemists, granting them praises such as could not be found in Descartes:

> Without doubt the whole World, and the Philosophers particularly, are very much obliged to them for the Pains they have taken, and which they continue to take, to make a great Number of Experiments, whereby they come to the Knowledge of diverse Properties of many different Things. This gives them opportunity to find out and discover the Nature of Things…[38]

---

[33] Rohault 1671, unpaginated preface: "une science d'usage doit bientôt descendre dans le particulier." An English translation was made available in the early eighteenth century and reprinted as Rohault 1987. For Rohault and his Cartesian experimental physics, see Chap. 9 by Dobre.

[34] Rohault 1671, II, 177: "tout ce que l'artifice produit dans les laboratoires des Chymistes se fait naturellement dans les entrailles de la Terre." Rohault 1987, II, 148.

[35] Rohault 1987, II, 148.

[36] Rohault 1987, II, 155. Clarke's English translation misses the reference to mercury in the French text. For the original French text, see Rohault 1671, II, 188: "l'expérience des Chymistes, qui par la résolution des métaux en peuvent tirer leur sel et leur soufre, et même, si l'on en croit quelques-uns, leur Mercure." The experiments usually put forward as evidence for validating their theses, were aimed at distilling vegetals and particularly wood. It is noteworthy that Rohault took up, for his own account, the distillation of a metal, which could hardly be effected, and which is one example of the extrapolations of which alchemists were often guilty.

[37] Rohault 1671, II, 186. Quite strangely the expression "moralement impossible" is not present in the English translation, Rohault 1987, II, 154.

[38] Rohault 1987, I, 109. Rohault 1671, I, 137: "Tout le monde sans doute, et les Philosophes en particulier, leur sont fort obligés de la peine qu'ils se sont donnée, et qu'il se donnent encore tous les jours, à faire un très grand nombre d'expériences, par le moyen desquelles ils leur font connaître les diverses propriétés de plusieurs Êtres différents. Ce qui leur donne la commodité de rechercher et découvrir la Nature des choses."

But he accompanied this recognition of the importance of the chemists' experimental research for natural philosophy with strong reservations about both their method and their principles. They were wrong in their principles, whose number it was impossible to establish, and even more in their method, because, limiting themselves to the knowledge of the "sensible Parts of which a Body is composed,"[39] they prevented themselves from understanding the deep structure of matter, and notably the part played by subtle matter, also called matter of the first element. Far from promoting a Cartesian chemistry, Rohault situated his chemical observations within the framework of physics, a species of physics which remained perfectly compatible with the spirit of Cartesian thought, insofar as it accounted for all the operations and properties of bodies by the form and movement of their constituent particles. Ultimately, there is no doubt that it was Descartes' theory of matter, and not the chemists', which allowed him to explain the multiple operations of the laboratory.

Among all the Cartesian philosophers who showed an interest in chemistry at that period, Pierre-Sylvain Régis (1632–1707) was perhaps the one who went the farthest to take into consideration the specificity of that science. In this respect, the opposition between the fourth and the fifth parts of Book 4 of his physics, which in itself constitutes the second tome of his *Système de la philosophie*, is quite revealing.[40] In the fourth part, entitled "Of the properties which are common to the hard Bodies and to the liquid Bodies," Régis, in fact, dealt mainly with fire, its nature and effects. In terms close to Descartes', he considered fire, not as a specific matter, but only as the situation in which are placed:

> terrestrial bodies whose non-sensitive parts are almost separated from one another, and so much surrounded with the Matter of the first Element, that they receive from it as much of their speed, as the link which still exists between them can allow.[41]

In confirmation of this theory of the nature and properties of fire, he even produced a drawing which mirrored perfectly the one with which Descartes illustrated his point in articles 95–99 of his fourth part of the *Principes de la philosophie*.[42] His framework of thought was undoubtedly Cartesian here, with the development of a theory of fire wholly deduced from the properties of matter, without any resorting to experimentation.

The fifth part, entitled "Of the properties which are discovered to be those of hard Bodies as a result of their resolution by dissolving Liquids," is quite a different trend. It does abound in mechanistic explanations, but these seem to be superadded to the

---

[39] Rohault 1987, I, 110. Rohault 1671, I, 138.

[40] Régis 1690.

[41] Régis 1690, II, 262: "…les corps terrestres dont les parties insensibles sont presque séparées les unes des autres et tellement environnées de la Matière du premier Elément qu'elles en reçoivent autant de sa vitesse que la liaison qui est encore entr'elles le leur peut permettre."

[42] Régis 1690, II, 265; *Principes de la philosophie*. AT IXb 253–255. I here refer to the French edition of Descartes, much more developed than the Latin edition.

accounts of experiments describing in a precise and detailed manner the transformation of salts into "spirits." Thus one can successively witness the transformation of "common Salt" (what we nowadays call sodium chloride) into "spirit of salt" (our hydrochloric acid), that of saltpetre into "Nitre Spirit" (our nitric acid), and that of Vitriol (a sulphate) into "vitriol spirit" (our sulphuric acid). One can then notice that these chemical operations are explained by a reduction of these different salts into their constituent principles, called by Régis their "parties essentielles" (acid, phlegm and earth), and not by any mechanical properties. But one can also notice that Régis insists on the reversibility of these operations: the vitriol spirit poured upon iron "re-embodies itself as Vitriol," while the "nitre spirit poured upon Tartar salt produces Saltpetre."[43] We are here in the very heart of chemistry such as it developed at the turn of the seventeenth and eighteenth century, with operations which became the essential subject of the "Table of the different rapports between the different substances" presented by Etienne-François Geoffroy in front of the Royal Academy of Sciences in 1718. Régis' whole analysis is here founded on experimental work, and not on the presuppositions of Cartesian physics.

At the end of Book 4, which reads like some kind of "chemistry course" encapsulated in his work, Régis does take some distance with the chemical concept of principle, which must not, he says, "be taken quite in a literal way when used by Chemists." Chemists, as a matter-of-fact, "only consider things in a palpable way," from which it follows that:

> The substances which they call by that name are Principles only with regard to us, and to the extent that Arts cannot go any further in the division of Bodies, even though we should know as a certainty that those Principles themselves are made of an infinite number of parts belonging to the first Element, which could with better justice claim the name of *Principles*.[44]

This sceptical attitude towards principles and their status was not uncommon in seventeenth-century chemistry; it should by no means be regarded as an effect of Cartesian criticism: it is already present, argued in strictly chemical terms, in 1641 in the *Nouvelle Lumière philosophique* by the chemical physician Etienne de Clave.[45] Thus, this standpoint, which was at the core of the chemists' reflexion at that time, did not induce Régis to debunk the scientific legitimacy of chemistry, in which he saw a genuine natural philosophy. Witness, the fact that at the end of his work, the reader was invited to refer to "Mr Lemery's course, which the Author, not content with executing chemical experiments in an easy and accurate way,

[43] Régis 1690, II, 301–302.

[44] Régis 1690, II, 333: "Les substances qu'ils appellent de ce nom ne sont Principes qu'à notre égard et en tant que l'Art ne peut pas aller plus avant dans la division des Corps, bien que nous soyons assurés que ces principes soient eux-mêmes composés d'une infinité de parties du premier Elément qui pourraient à plus juste titre être appelés *Principes*."

[45] De Clave 1643–2000. For a larger discussion of De Clave's work, see Joly 2001a.

enriched with Remarks worthy of a philosopher who knows Nature perfectly."[46] This is why, outgrowing his criticism of chemical principles, he immediately added:

> However, it must be acknowledged, that if one wishes to get as close as possible to the true Principles of Nature, he has no better choice than following the path of Chemistry; for, even though the division between substances which it operates be but a gross one, nevertheless it gives a really great idea of Nature and of the figure of the insensible particles which enter into the composition of Mixed Bodies, both gross and palpable.[47]

In order to achieve the synthesis between Cartesian physics and chemical theories, Régis had, from the very first part of Book 4, distinguished three levels in the composition of matter:

> Just as Water, Air, and Oil are made of particles that move in all directions, these same particles of Water, Air, and Oil are made of the first Element that rest against one another [sic].[48] This is why, in order to distinguish between those different compositions, we shall name the particles of which Water, Air, and Oil are composed, *essential parts*. We shall name the parts of the first Element of which the essential parts are made, *radical parts*; last, we shall name the small drops of Water, Air, or Oil of which larger drops are formed, *integrating parts*. Thus, all the kinds of parts that can be associated with Water, Air, Oil, and Salt, will fall into *radical parts*, *essential parts*, and *integrating parts*.[49]

This distinction introduces between the corpuscular structure of matter and the sensitive bodies an intermediary layer that was missing from Descartes. It also ensured that Cartesian theory of matter could be coupled with the chemists', since the parts of matter belonging to the first (or the third) element have no physical manifestation to the eyes of chemists. It is in the very nature of Cartesian theory of matter to escape sensible apprehension, and yet it provides the whole background against which all the operations and activities of matter can be conceived. In this respect, it belongs to the "metaphysical physics," according to Daniel Garber's expression, a species of physics whose principles cannot be grasped by the senses

---

[46] Régis 1690, II, 336.

[47] Régis 1690, II, 333–334: "Il faut avouer pourtant que si l'on veut approcher autant qu'il se pourra des véritables Principes de la Nature, on ne peut prendre une voie plus assurée que celle de la Chymie; car quoique la division qu'elle fait des substances soit grossière, elle donne néanmoins une fort grande idée de la Nature et de la figure des particules insensibles qui entrent dans la composition des Corps Mixtes, grossiers et palpables."

[48] I suggest the correction: "are made *of parts* of the first Element…" It is strange to observe that Régis speaks here about the first element of matter, and not about the third. For indeed, according to Descartes, the most elementary parts of terrestrial bodies belong to the third element.

[49] Régis 1690, II, 135: "Comme l'Eau, l'Air et l'Huile sont composés de particules qui se meuvent en tous sens, ces mêmes particules d'Eau, d'Air et d'Huile sont composées [des parties] du premier Elément qui sont en repos les unes auprès des autres. C'est pourquoi pour distinguer ces différentes compositions, nous nommerons les particules dont l'Eau, l'Air et l'Huile sont composés des *parties essentielles*. Nous appellerons les parties du premier Elément dont les parties essentielles sont faites, des *parties radicales*; et nous donnerons enfin le nom de *parties intégrantes* aux petites gouttes d'Eau, d'Air ou d'Huile dont d'autres plus grosses gouttes se seront formées. Ainsi l'on pourra réduire toutes les sortes de parties que l'on peut concevoir dans l'Eau, dans l'Air, dans l'Huile, et dans le Sel, aux *parties radicales*, aux *parties essentielles*, et aux *parties intégrantes*." Régis' emphasis.

and whose objects are not to be met in the laboratory, but only in thought.[50] At the other end, there are "integrating parts," such as tiny drops of water, air, or oil, which can be observed by anyone, even though their properties remain unexplained. Thus, between the theoretical physics which speculates on a matter inaccessible to the senses, and the sensory observation which does not understand the properties of the observed objects, there is a science of chemistry which deals with the essential parts of bodies, or in the chemists' own words, their principles.

To conclude with this subject, Pierre-Sylvain Régis gave chemistry the place it had to occupy at the end of seventeenth century, that of an empirical science, which certainly did not pretend to have the last word on the intimate structure of matter, but which provided a theoretical framework allowing one, not only to understand chemical operations as a whole, but also to foresee their results. The distinction between the "rooting" and the "essential" parts found its equivalent in the distinction between Cartesian principles, whose reality chemists admitted without having to resort to them, and chemical "principles," which, it must be acknowledged, did not really have the rank of principles, but allowed for the understanding of laboratory operations. Thus, Régis' works made it possible to envision at the same time the possibility of a chemistry of Cartesian inspiration, and the integration within chemistry of an empiricist conception of science.

6.3 Chemistry as a Demand for Empiricism

As we shall now see, Cartesianism was again associated, in different ways, with an empirical approach of science in the works of Robert Boyle (1627–1691), Nicolas Lémery (1645–1715) and his son Louis Lémery (1677–1743), three chemists of that period who were generally regarded as Cartesians. Robert Boyle expressed his gratitude towards the work of Descartes on many occasions, and showed a precise and detailed knowledge of his writings.[51] Thus, in *The Excellency of Theology, Compar'd with Natural Philosophy* (1674), he gave an account of the Cartesian theses on the attributes of God, the immortality of the soul and its distinction from the body.[52] In that work, as in *Some Considerations about the Reconcileableness of Reason and Religion* (1675), he showed the methodological difference between the second part of the *Principia philosophiae*, based on metaphysical deduction, and the third and fourth parts which are hypothetical.[53] In a general way, Boyle referred on numerous occasions to Descartes' metaphysical texts, published works or extracts from his correspondence, such as the latter could be known at the time. Yet he never quoted Descartes on the topic of chemistry. This could seem all the more surprising as Boyle asserted at the same time that his natural philosophy belonged to the framework of

---

[50] See Garber 1992.

[51] I summarize here an analysis developed in Joly 2009. See also Joly 2011, Chap. V, 193–202.

[52] Boyle 1999–2000, VIII, 3–98.

[53] Boyle 1999–2000, VIII, 233–294.

mechanical philosophy, and that it was in chemistry that it was first manifested. As the "Bologna stone" (a phosphorescent stone) becomes luminous only after a chemical preparation, natural bodies can only produce light if they have previously been studied by chemistry.[54] In an undated letter, he wrote:

> But as I cultivated Chymistry not so much for itself, as for the sake of Natural Philosophy, & in order to it; so most of the Experiments I devis'd & pursu'd, were generally such as tended not to multiply Chymical Processes or gain the Reputation of having store of difficult & Elaborate ones; but to serve for Foundations, & other useful Materials for an Experimental History of Nature.[55]

Yet it was not in Descartes that Boyle found the sources of these chemical experiments which became the foundation for his natural philosophy inspired by mechanism, but rather in his reading of authors like Daniel Sennert (1572–1637), Johann Baptista Van Helmont (1579–1644), Pierre Gassendi (1592–1655) and, in a general way, in the works of the alchemical tradition with which he was perfectly acquainted.[56] It can even be noticed that, when he quoted from the fourth part of *Principia philosophiae*, he took pains to avoid the passages referring to chemistry. This happened notably when, in *New Experiments Physico-Mechanical, Touching the Spring of the Air and its Effects* (1660)[57] or *An examen of Mr. T. Hobbes his Dialogus Physicus De Natura Aeris* (1662),[58] he quoted Descartes to criticize his explanation of tides, or to argue on the basis of his conception of subtle matter, which seemed to him liable to justify his position on void. In fact, contrary to Descartes, Boyle was careful not to directly invoke the geometrical figure of the most elementary particles of matter to account for their chemical properties. This appears clearly from the comparison between two passages in which the two authors stated with precision their conceptions of chemical principles. In the fourth part of *Principia philosophiae*, Descartes wrote:

> And I have here given an account of three kinds of bodies which, to my mind, seem to have very strong analogies with those that chemists commonly hold as their three principles, and which they name salt, sulfur, and mercury. For these corroding saps can be regarded as their salt, these small branches which form an oily matter as their sulfur, and quicksilver as their mercury.[59]

---

[54] This comparison is developed in the preface of *Experiments and Notes about the Producibleness of Chymicall Principles* (1680). See Boyle 1999–2000, IX, 27.

[55] Boyle 1999–2000, XII, 365.

[56] As proved by Clericuzio 1990, 1993, 2000, 2000–2003; Newman 1996; Newman and Principe 2002; Principe 1998.

[57] Boyle 1999–2000, I, 166.

[58] Boyle 1999–2000, III, 125.

[59] *Principes de la philosophie* IV 63. AT IXb 235: "Et j'ai ici expliqué trois sortes de corps qui me semblent avoir beaucoup de rapports avec ceux que les chymistes ont coutume de prendre pour leurs trois principes, et qu'ils nomment le sel, le soufre et le mercure. Car on peut prendre ces sucs corrosifs pour leur sel, ces petites branches qui composent une matière huileuse pour leur soufre, et le vif argent pour leur mercure."

Salt, sulphur and mercury are not really principles, but chemical substances whose properties are deduced exclusively from the configuration of their constituent particles, as was explained in his previous articles: mechanistic considerations are sufficient account for their properties, hence, there is no need for chemical ones. It can be noticed that Descartes did not need to introduce an intermediary layer, that of molecules, between the invisible corpuscular structure of matter and the different chemical bodies whose properties can be observed, since the differences between substances are rooted in those owing to which the very particles of what he calls the third element of matter can be distinguished.

Boyle was of a very different opinion. He did object, as Descartes had done, to the fact that the three chemical "principles" were actually principles, but it was precisely by means of that criticism that he could maintain their chemical specificity; at a lower level, that of mechanical properties, the organization of matter resulted from the interplay between the bulk, form, and movement of particles. To quote from *Of the Imperfection of the Chymist's Doctrine of Qualities*:

> The Chymist's Salt, Sulphur and Mercury themselves are not the first and most simple Principles of Bodies, but rather primary Concretions of Corpuscles or Particles more simple than they, as being endowed only with the first, or most radical (if I may so speak) and most Catholick Affections of simple Bodies, namely Bulk, Shape, and Motion, or Rest; by the different Conventions or Coalitions of which minutest portions of matter are made those differing Concretions that Chymists name Salt, Sulphur and Mercury.[60]

For Boyle then—he expatiated on the subject in several works[61]—there are two levels of organization in matter: a mechanical level, inaccessible to chemical analysis, and a superior level, made of the aggregation ("clusters," "concretions") of elementary particles of different shape and bulk, at which the chemical properties of bodies manifest themselves. This conception had this in common with Descartes, that it avoided having to resort to the hypothesis of the substantial properties of principles, as traditional alchemy had done, but it was distinct from that of the French philosopher by introducing an intermediary molecular layer inspired by the works of Gassendi and by a corpuscular alchemical tradition. This allowed him to restore the specific field of chemistry, one in which explanations essentially derive from laboratory work and not from hypothetico-deductive considerations[62]—which did not mean that chemistry could escape developing on the basis of a mechanical philosophy, ruled by the Cartesian principles of the organization of matter

---

[60] *Of the Imperfection of the Chymist's Doctrine of Qualities*, in *Experiments, Notes, &c., about the Mechanical Origin of Qualities* (1675–1676), in Boyle 1999–2000, VIII, 401.

[61] See Clericuzio 2000.

[62] On the part played by Gassendi, see Clericuzio 1997, 2000, 2003. Nevertheless, one should keep in mind the fact that Gassendi called molecules the "seeds of things," an expression which had been borrowed from the Danish Paracelsian physician Petrus Severinus, whose work *Idea medicinae philosophicae* (Severinus 1591), Boyle had carefully read. See Hirai 2005. Boyle's mention of Gassendi thus refers both to the atomistic tradition and to the alchemical tradition, one branch of which emphasized the part played by the *minima partes*. See Newman 1996.

according to bulk, shape and movement. Indeed, as Boyle wrote in *About the Excellency and Grounds of the Mechanical Hypothesis* (1674):

> Though Chymical Explications be sometimes the most obvious and ready, yet they are not the most fundamental and satisfactory: For, the Chymical Ingredient it self, whether Sulphur or any other, must owe its nature and other qualities to the union of insensible particles in a convenient Size, Shape, Motion or Rest, and Contexture; all which are but Mechanical Affections of convening Corpuscles.[63]

The mechanical explanation was not substituted to the chemical one, but became its underlying assumption, by its supposition of the size, shape and movement of non-sensitive particles escaping chemical analysis. So doing, mechanical philosophy was presenting chemistry with a permanent and universal basis, a background against which chemical hypotheses could develop at pace with the progress of laboratory work:

> Though the further Sagacity and Industry of Chymists (which I would by no means discourage) should be able to obtain from mixt Bodies homogeneous substances differing in number, or nature, or both, from their vulgar Salt, Sulphur, and Mercury; yet the Corpuscular Philosophy is so *general* and *fertile*, as to be fairly reconciliable to such a Discovery; and also *useful*, that these new material Principles will, as well as the old *Tria Prima*, stand in need of the more Catholick Principles or *Corpuscularians*, especially Local Motion.[64]

Thus Boyle had no need to refer to the mechanistic reductions of chemistry, which Descartes had resorted to; as a matter-of-fact, they might have imperilled the very existence of chemistry, the theoretical and practical interest of which Boyle fully recognized. Yet, Boyle remained Cartesian insofar as he recognized the ultimate character of mechanistic explanations according to size, shape and movement. It was the distinction he made between these two levels that allowed him to give an essential place to an empiricist conception of chemistry in the very framework of Cartesianism. Free of any obligation to provide its own principles by itself, chemistry could unfold, inside the very framework of a mechanistic conception referring to the work of Descartes, as a specific science deriving its knowledge from experience.

Less sophisticated were the speculations of Nicolas Lémery.[65] When one has a look at the *Cours de chymie* he published in 1675,[66] one can even wonder how he could cast a figure as one of the prominent representatives of Cartesian chemistry, to the extent of being nicknamed the "Descartes of chemistry" by Dortous de Mairan in the eulogy of his son Louis.[67] This work, both in form and substance, follows in a long tradition of "chemistry courses" which had developed in France since the publication of *Elemens de chymie* by Jean Beguin (1550–1620) in 1612.

[63] Boyle 1999–2000, VIII, 110–111.

[64] Boyle 1999–2000, VIII, 112. Emphasized by Boyle.

[65] For Nicolas Lémery's work, see Bougard 1998.

[66] Lémery 1690.

[67] Dortous de Mairan 1747.

These were nothing but collections of recipes intended for apothecaries. Contrary to works of Galenic inspiration, they promoted the use of mineral substances and the massive resorting to the multiple operations made possible by chemistry at that time. These "chemistry courses" were one of the channels of development of a new conception of medicine, founded on chemistry, of Paracelsian inspiration. As a matter-of-fact, the title of Lémery's work evokes operations such as were "in use in medicine." The book was widely acclaimed in its time, not only because it introduced new recipes, but also because it developed a theory of the points of acids and the pores of alkalis, which sounded perfectly original.[68] It was that reference to a mechanistic doctrine, instead of the traditional theory of principles, which allowed Lémery to be considered, by some, as the founder of modern chemistry.

In spite of its mechanistic outlook, that doctrine owed little to Cartesianism. The spread of Johann Baptista Van Helmont's ideas, from the publication of *Ortus medicinae* in 1648, 4 years after his death, had familiarized the public with the idea that acids and alkalis were agents of digestion. Otto Tachenius (1610–1680), in *Hippocates chimicus* (1666) generalized that doctrine by making acid and alkali the two constituent principles of all bodies.[69] François André (fl. 1670–1690), a French physician from Caen, seems to have been the first to assert, in his *Entretiens sur l'Acide et l'Alcali* published in 1672, that "acid salt…is composed of small pointed parts which insinuate themselves into the pores which they encounter, thus bringing about either the breaking apart of these parts, or their coagulation,"[70] whereas alkali salt is an "extremely porous body, empty and rough in its surface, whose parts are unequal," which accounts for the fact that it takes off the dirt from linen and stuff.[71] He considered these two salts as the genuine chemical principles, since they could solve all existing bodies, without being converted back to any other.

Nicolas Lémery took up that doctrine and put it to work to account for most of the chemical operations whose recipes he was giving. His view of things was very clearly expounded in a long introduction in which he defined acid in those terms:

> As the nature of such a secret thing as salt cannot be better explained than by allowing all its constituent parts to be made of figures which suit all the effects it produces, I here maintain that the acidity of a liquor consists in pointed particles of salt, which are agitated; and I do not think it can be contested that acid has points, since all experiments show it, and one has only to taste it to be convinced of that; for one feels a tickling on our tongue similar or very close to that which could be produced by some matter sharpened into very thin

[68] The idea according to which the pungent nature of acid or fire results from the pointed shape of their corpuscles can be found in Democritus and in Plato's *Timaeus*. It was taken up again in the medical tradition to account for some workings of sensation. *Timaeus*, followed by book 4 of Aristotle's *Meteorologica*, put forward the filling in of the pores of some bodies by air, water, or fire, to account for processes such as permeability, fusibility, or combustibility. But using the penetration of pores by points as the key for interpreting all chemical processes was indeed a novelty.

[69] Tachenius 1666.

[70] André 1672. I quote from the second edition, André 1677, 15.

[71] André 1677, 31.

points. That acid is made of pointed parts is eloquently and convincingly demonstrated by the fact that, not only do all acid salts crystallise in the shape of points, but also all the dissolutions of different matters produced by acid liquors take on that very same shape in their crystallisation.[72]

Endow the parts of salt with figures that are fit to answer for the produced effects: one can see how, with Lémery, the mechanistic scheme was deduced from the sensitive properties of bodies, whereas with Descartes it resulted from deductions drawn from metaphysical principles, to which the particular properties of bodies were added at the end. In this respect, one could say that Lémery, by giving up any metaphysical reference, built up a kind of "empirical mechanism" which drew sensibly apart from Descartes' method used in the *Principia philosophiae*.

The same procedure was used to define alkali, whose description of the porous structure is both introduced by one observation, and confirmed by another:

As far as alkalis are concerned, they can be recognized as such when one pours acid over them; for instantaneously or a short while later a violent effervescence is produced, which lasts till acid does not find any more matter to rarefy. Such an effect affords reasonable certainty that alkali is a matter composed of stiff, brittle parts, whose pores are shaped in such a way that, once they have been penetrated by the points of acids, the latter break and push aside everything that impedes their movement, and that, depending on whether the parts which compose this matter are more or less solid, the acids meet with more or less resistance and the effervescence they produce is more or less powerful.[73]

What Lémery called the "struggle" between acid and alkali[74] then became the interpretive key for most chemical operations, all through the 58 chapters of his *Cours de chymie* whose recipes fall into three traditional groups: minerals, vegetals, and, much more briefly, animals. The mechanical explanations go together with the detailed account of the different operations, they adapt themselves to them

---

[72] These developments cannot be found in the first edition of the work (1675). They appeared only from the fourth edition (1681) onward. Page numbers are here taken from the seventh edition (1690): Lémery 1690, 24–25: "Comme on ne peut pas mieux expliquer la nature d'une chose aussi cachée qu'est celle du sel, qu'en admettant aux parties qui le composent des figures qui répondent à tous les effets qu'il produit, je dirai que l'acidité d'une liqueur consiste dans des particules de sel pointues, lesquelles sont en agitation; et je ne crois pas qu'on me conteste que l'acide n'ait des pointes, puisque toutes les expériences le montrent, il ne faut que le goûter pour tomber dans ce sentiment; car il se fait des picotements sur la langue semblables ou fort approchants de ceux qu'on recevrait de quelque matière taillée en pointes très fines. Mais une preuve démonstrative et très convaincante que l'acide est composé de parties pointues, c'est que non seulement tous les sels acides se cristallisent en pointes, mais toutes les dissolutions de matières différentes faites par les liqueurs acides prennent cette figure dans leur cristallisation."

[73] Lémery 1690, 25: "Pour ce qui est des alkali, on les reconnaît quand on verse de l'acide dessus; car aussitôt ou peu de temps après il se fait une effervescence violente qui dure jusqu'à ce que l'acide ne trouve plus de corps à raréfier. Cet effet peut faire raisonnablement conjecturer que l'alkali est une matière composée de parties raides et cassantes, dont les pores sont figurés de façon que les pointes des acides y étant entrées, elles brisent & écartent tout ce qui s'oppose à leur mouvement, et que selon que les parties qui composent cette matière sont plus ou moins solides, les acides trouvant plus ou moins de résistance, ils font une plus forte ou plus petite effervescence."

[74] Lémery 1690, 27.

and modify themselves as they are being objected to. Lémery aims to reveal the continuity between what can be observed and what cannot be seen, but whose reality is sometimes perceptible, as when one observes crystal salt aggregates, or uses the microscope.

The points-and-pores hypothesis was thus supported by empirical observations and not by deductions drawn from a theoretical conception of matter. This hypothesis translates data provided by sensations into the field of the invisible, but is eventually deprived of any heuristic dimension. It goes alongside with the description of the properties of different chemical bodies, but does not allow one to predict any new ones. Far from being, as in Boyle, the underlying assumption of chemical theory, the mechanical hypothesis is nothing but its ornament. Not surprisingly, Boyle had criticized André's doctrine in his *Reflexions upon the hypothesis of alkali and acidum* of 1675.[75] According to Boyle, some chemical substances could neither be ranked among acids, nor among alkalis, and the new theory brought nothing more than that of the three chemical principles.[76]

Nicolas Lémery had been appointed a member of the Royal Academy of Sciences in 1699. At his death in 1715, his son Louis, who had been a élève in the Academy since 1700 and an membre associé since 1712, became in his turn a membre pensionnaire. His chemical works and theories were known through the numerous memoirs he presented in front of the Academy, as well as his works on the anatomy of monsters, which made him famous. Louis Lémery took up his father's mechanistic hypotheses, but did not limit the point-and-pore structure of matter to acids and alkalis. He also showed how the movements of corpuscles which differ in shape and size, gave rise to the whole activity of observable bodies. This appeared, for instance, in the quarrel which opposed him, from 1706, to his colleague Etienne-François Geoffroy (1672–1731), about the artificial making of iron.[77] Harping on a recipe borrowed from the alchemist Johann-Joachim Becher (1635–1682), Geoffroy thought he could demonstrate that the combustion of plants produces iron by favouring the blending of its constituent elements, namely earth, an acid salt, and the oil of a vegetal.[78] Witness the fact that a magnetized blade can attract particles extracted from vegetal ashes.[79] On the contrary, Louis Lémery, was convinced that iron was already present in the plant's marrow, and had been drawn out of the earth by means of its roots. He explained that acids, mixing with oily parts of iron, form vitriols, which obtrudes the pores of metal and impedes the flow of magnetic matter, which results in iron being undetected by a magnet. When a violent fire ruins this vitriolic union, the pores allow magnetic matter to flow again and iron can be detected.

[75] This short text belongs to *Experiments, Notes, &c., about the Mechanical Origin of Qualities*, Boyle 1999–2000, VIII, 407–720.

[76] It was to counter these objections that François André published a considerably enlarged second edition of his work in 1677. The main elements of the dispute appear again in Bertrand 1683.

[77] I have developed the analysis of this quarrel in Joly 2007b; see also Joly 2008.

[78] Geoffroy 1704.

[79] Geoffroy 1705.

This theory of "hidden iron" provided an explanation for the fact that a magnet could detect iron only after combustion.[80]

Later on, Louis Lémery explained how, in spite of gravity, iron can rise in the fibres of vegetals, provided it is dissolved in an acid: "The iron dissolved by acids can easily be reduced into particles small and light enough to be able to run up the smallest and highest pipes of plants."[81]

Whereas Geoffroy drew on the works of chemists and their theory of principles for his argumentation, Lémery strongly opposed him by invoking what he called the "*mécanique*" of chemical operations, the "*mécanique* owing to which vitriolic acid rises up along the stems of plants,"[82] or that which results in acid being discomposed during combustion, allowing the iron which had been hidden inside it to appear again.[83]

To bring the quarrel to an end, Louis Lémery presented a memoir in front of the Academy, in 1709, entitled "Conjectures and reflections on the matter of fire or light."[84] Contrary to Descartes, for whom fire was nothing but a particular form that could be assumed by all terrestrial corpuscles in certain conditions of movement,[85] Lémery considered that "the matter of fire should be regarded as a fluid of a certain nature, and endowed with specific properties which make it different from any other fluid."[86] And when he wanted to tackle the question of the increase in the weight of a metal after it has been burnt, he asserted that fire is made of particles which increase the weight of bodies by creeping themselves inside their pores, where they remain enclosed as in a prison. He thus comes back to a conception of fire centring on its substance, which was discarded only by the works of Lavoisier.[87]

Lémery's constant references to the "mécanique" of chemical operations are a sure sign of his determination to "spread the light of modern philosophy," in other words "Cartesianism," over the science of chemistry[88]: thus, in 1707, he expounded the "mécanique" owing to which nut-gall, by freeing the iron hidden in vitriol, allows the making of ink[89]; in 1711, he expressed his determination to have the "mécanique" at work in metallic precipitations by alkalis "properly understood"[90]; in 1716, he put

[80] Lémery 1706a.

[81] Lémery 1706b, 416: "Le fer dissous par les acides peut être aisément réduit en des particules assez petites & d'une assez grande légèreté pour pouvoir pénétrer les tuyaux les plus petits & les plus élevés des plantes."

[82] Lémery 1708, 394.

[83] Lémery 1708, 400.

[84] Lémery 1709.

[85] Descartes, *Principes de la philosophie* IV 80, AT IXb 243.

[86] Lémery 1709, 405.

[87] Lavoisier later showed that the increase in the weight of burnt out metal results from the presence of oxygen particles: combustion is an oxidation.

[88] Quoted from Fontenelle's presentation of a memoir by Louis Lémery in 1711; see Fontenelle 1711, 33.

[89] Lémery 1707.

[90] Lémery 1711.

into relief the "mécanique" of the dissolution of salt in water.[91] However, far from causing chemistry to lose its identity in what he called the système du mécanisme,[92] he insisted that the latter should be integrated to chemistry which, better than any other science, allows to deepen its results. All things being considered, he brought back the theoretical foundations of such an interpretation within the field of an experimental practice of chemistry, through which, he conjectured, the hidden structure of matter could be exhibited.

6.4 Conclusion

Descartes had no contempt for chemical experiments, but he thought that they could not become the foundation stone of a science of chemistry. Yet chemistry had such a long-standing tradition and such a practical interest, that it did not suffer any damages from the Cartesian critique. The example of the so-called Cartesian chemists shows that it was possible to build up connections between Cartesian thought and chemistry through an empirical approach to natural philosophy. Though Descartes' standpoint led him to favour a reduction of chemistry to the principles of mechanism, he nevertheless remained dependent upon laboratory work, without which the very objects of chemistry, as well as its operations, could have no existence. Thus it was not surprising that his successors, and more particularly those who remained convinced of the specificity of the science of chemistry, endeavoured to keep up with an empirical approach of these objects and operations, So doing, they promoted an empirical dimension of Cartesianism, which a philosophical system intent on interpreting the whole of material reality, could not wholly dismiss. These chemists thus brought into relief what Descartes had constantly seemed to neglect or ignore, namely the necessity of submitting the chemical theories to the test of experimentation.

Yet it can be wondered whether a Cartesian conception of chemical matter could really have a heuristic interest.[93] In what respect did it benefit the progress of chemistry? Eventually one feels that it is by discarding their Cartesianism, by making it a background without any direct link with their practice, that Cartesian chemists made any improvements. In fact, the one genuine empiricist among French chemists at the beginning of the eighteenth century, Etienne-François Geoffroy, was neither a Cartesian nor a Newtonian.[94] It was by his rejection of any pre-established creed, and by limiting himself to the data provided by laboratory work, that he built up his "Table of the different relationships between different substances," which was to become one of the foundation stones of eighteenth century chemistry.

---

[91] Lémery 1716.

[92] See Lémery 1712, 53.

[93] It could also be wondered how the criticism of Cartesianism by chemists who would not be called Cartesians influenced their doctrines and practices. But this would deserve another enquiry.

[94] See Joly 2013.

Acknowledgments I am exceedingly grateful to Michel Maillard for his translation of the present chapter. All translations are Maillard's except where otherwise noted.

References

André, François. 1672. *Entretiens sur l'acide et l'alcali*. Paris.

André, François. 1677. *Entretiens sur l'Acide et sur l'Alkali où sont examinées les objections de Mr Boyle contre ces principes, avec une replique à la lettre de Mr S... Docteur en Medecine aggrégé au College de ... touchant la nature de ces deux sels*. Paris.

Bertrand. 1683. *Reflexions nouvelles sur l'Acide et sur l'Alcali ou après avoir demontré que ces deux sels ne peuvent pas être les principes des Mixtes, on fait voir le véritable usage qu'on en peux faire dans la Physique et la Médecine*. Lyon: Thomas Amaulry.

Bougard, Michel. 1998. *La chimie de Nicolas Lémery, apothicaire et médecin (1645–1715)*. Turnhout: Brepols.

Boyle, Robert. 1999–2000. *The works of Robert Boyle*, 14 vols., ed. Michael Hunter and Edward B. Davis. London: Pickering & Chatto.

Clericuzio, Antonio. 1990. A redefinition of Boyle's chemistry and corpuscular philosophy. *Annals of Science* 47(6): 561–589.

Clericuzio, Antonio. 1993. From Van Helmont to Boyle: A study of the transmission of Helmontian chemical and medical theories in seventeenth-century England. *British Journal for the History of Science* 26(3): 303–334.

Clericuzio, Antonio. 1997. L'atomisme de Gassendi et la philosophie corpusculaire de Boyle. In *Gassendi et l'Europe (1592–1792)*, ed. Sylvia Murr, 227–235. Paris: Vrin.

Clericuzio, Antonio. 2000. *Elements, principles and corpuscles. A study of atomism and chemistry in the seventeenth century*. Dordrecht: Kluwer.

Clericuzio, Antonio. 2000–2003. Philosophie de la nature, chimie et alchimie au XVIIe siècle: Jean-Baptiste van Helmont et Robert Boyle. *Chrysopœia* 7: 315–326.

De Clave, Etienne. (1643). *Nouvelle lumière philosophique des vrais principes et elemens de nature*. (Reprint, Paris: Fayard, Corpus des œuvres de philosophie en langue française, 2000.)

Descartes, René. 1970. *Oeuvres de Descartes*, ed. Charles Adam and Paul Tannery. Paris: J. Vrin.

Descartes, René. 1985–1991. *The philosophical writings of Descartes*, 3 vols., ed. John Cottingham, Robert Stoothoff, and Dugald Murdoch. Cambridge: Cambridge University Press.

Dobre, Mihnea. 2011. Cartesianism and chymistry. *Society and Politics* 5(10): 121–136.

Dortous de Mairan, Jean-Jacques. 1747. Eloge de Louis Lémery. In *Eloges des Académiciens de l'Académie royale des sciences morts dans les années 1741, 1742 et 1743*. Paris: Durand.

Fontenelle, Bernard le Bouyer de. 1711. *Mémoires de l'Académie royale des sciences*. Paris.

Garber, Daniel. 1992. *Descartes' metaphysical physics*. Chicago: The University of Chicago Press.

Geoffroy, Etienne-François. 1704. Manière de recomposer le souffre commun par la réunion de ses principes, et d'en composer de nouveau par le mélange de semblables substances, avec quelques conjectures sur la composition des métaux. In *Mémoires de l'Académie royale des sciences*, 278–286. Paris.

Geoffroy, Etienne-François. 1705. Problème de chimie. Trouver des cendres qui ne contiennent aucunes parcelles de fer. In *Mémoires de l'Académie royale des sciences*, 362–363. Paris.

Gouhier, Henri. 1958. *Les premières pensées de Descartes. Contribution à l'histoire de l'anti-renaissance*. Vrin: Paris.

Halleux, Robert. 1979. *Les textes alchimiques*. Turnhout: Brepols.

Hirai, Hiro. 2005. *Le concept de semence dans les théories de la matière à la Renaissance de Marsile Ficin à Pierre Gassendi*. Turnhout: Brepols.

Joly, Bernard. 2001a. La théorie des cinq éléments d'Etienne de Clave dans La Nouvelle Lumière philosophique. *Corpus* 39: 9–44.

Joly, Bernard. 2001b. La figure du sceptique. In *La vérité des sciences de Marin Mersenne, Le scepticisme aux XVIe et XVIIe siècles*, ed. Pierre-François Moreau, 257–276. Paris: Albin Michel.

Joly, Bernard. 2002. La chimie contre Aristote. La distillation du bois et la doctrine des cinq éléments au XVIIe siècle en France. In *Proceedings of the XXth International Congress of History of Science* (Liège, July 20–26, 1997), vol. XVIII, Alchemy, chemistry and pharmacy, ed. Michel Bougard, 67–75. Turnhout: Brepols.

Joly, Bernard. 2003. Francis Bacon réformateur de l'alchimie: tradition alchimique et invention scientifique au début du XVIIe siècle. *Revue philosophique de la France et de l'étranger* 193(1): 23–40. doi:10.3917/rphi.031.0023.

Joly, Bernard. 2007a. A propos d'une prétendue distinction entre la chimie et l'alchimie au XVIIe siècle: questions d'histoire et de méthode. *Revue d'histoire des sciences* 60–61: 167–183.

Joly, Bernard. 2007b. Quarrels between E. F. Geoffroy and Louis Lémery at the Académie Royale des sciences in the early eighteenth century: Mechanism and alchemy. In *Chymists and chymistry. Studies in the history of alchemy and early modern chemistry*, ed. Lawrence Principe, 203–214. Philadelphia/Sagamore Beach: Chemical Heritage Foundation-Science History Publications.

Joly, Bernard. 2008. Le mécanisme et la chimie dans la nouvelle Académie royale des sciences: les débats entre Louis Lémery et Etienne-François Geoffroy. *Methodos* 8. doi:10.4000/methodos.1403.

Joly, Bernard. 2009. Le cartésianisme de Boyle du point de vue de la chimie. In *La philosophie naturelle de Boyle*, ed. Myriam Dennehy and Charles Ramond, 139–155. Paris: Vrin.

Joly, Bernard. 2011. *Descartes et la chimie*. Paris: Vrin.

Joly, Bernard. 2013. Was early eighteenth century's chemistry an empirical science? In *What does it mean to be an 18th century empirist?*, ed. Anne-Lise Rey and Siegfried Bodenmann. Dordrecht: Springer.

Lémery, Nicolas. 1690. *Cours de chymie contenant la manière de faire les opérations qui sont en usage dans la Médecine, par une méthode facile. Avec des raisonnements sur chaque opération, pour l'instruction de ceux qui veulent s'appliquer à cette science*, 7th ed. (First edition 1675). Paris.

Lémery, Louis. 1706a. Diverses expériences et observations chimiques et physiques sur le fer et sur l'aimant. In *Mémoires de l'Académie royale des sciences*, 119–135. Paris.

Lémery, Louis. 1706b. Que les plantes contiennent réellement du fer, et que le métal entre nécessairement dans leur composition naturelle. In *Mémoires de l'Académie royale des sciences*, 411–418. Paris.

Lémery, Louis. 1707. Eclaircissement sur la composition des différentes espèces de vitriols naturels & explication physique & sensible de la manière dont se forment les Ancres [sic] vitrioliques. In *Mémoires de l'Académie royale des sciences*, 538–549. Paris.

Lémery, Louis. 1708. Nouvel éclaircissement sur la prétendue production artificielle du fer, publiée par Becher, & soutenue par M. Geoffroy. In *Mémoires de l'Académie royale des sciences*, 376–402. Paris.

Lémery, Louis. 1709. Conjectures et réflexions sur la matière du feu ou de la lumière. In *Mémoires de l'Académie royale des sciences*, 400–417. Paris.

Lémery, Louis. 1711. Mémoire sur les précipitations chimiques; ou l'on examine par occasion la dissolution de l'or & de l'argent, la nature particulière des esprits acides, & la manière dont l'esprit de nitre agit sur celui de sel dans la formation de l'eau régale ordinaire. In *Mémoires de l'Académie royale des sciences*, 56–79. Paris.

Lémery, Louis. 1712. Conjectures sur les couleurs différentes des précipités de Mercure. In *Mémoires de l'Académie royale des sciences*, 51–69. Paris.

Lémery, Louis. 1716. Explication mécanique de quelques différences assez curieuses qui résultent de la dissolution de différents sels dans l'eau commune. In *Mémoires de l'Académie royale des sciences*, 154–172. Paris.

Maillard, Jean-François. 1998. Descartes et l'alchimie: une tentation conjurée? In *Aspects de la tradition alchimique au XVIIe siècle*, ed. Frank Greiner, 95–109. Paris/Milan: Séha/Archè.

Mandosio, Jean-Marc. 1990–1991. La place de l'alchimie dans les classifications des sciences et des arts à la Renaissance. *Chrysopœia* 4: 199–282.
Mandosio, Jean-Marc. 1998. Quelques aspects de l'alchimie dans les classifications des sciences et des arts au XVIIe siècle. In *Aspects de la tradition alchimique au XVIIe siècle*, ed. Frank Greiner, 19–61. Paris/Milan: Séha/Archè.
Matton, Sylvain. 1998. Cartésianisme et alchimie: à propos d'un témoignage ignoré sur les travaux alchimiques de Descartes. Avec une note sur Descartes et Gómez Pereira. In *Aspects de la tradition alchimique au XVIIe siècle*, ed. Frank Greiner, 111–184. Paris/Milan: Séha/Archè.
Mehl, Edouard. 2001. *Descartes en Allemagne. 1619–1620. Le contexte allemand de l'élaboration de la science cartésienne*. Strasbourg: Presses universitaires de Strasbourg.
Morhof, Daniel Georg. 1673. *De metallorum transmutatione*. Hamburg.
Newman, William. 1996. The alchemical sources of Robert Boyle's corpuscular philosophy. *Annals of Science* 53(6): 567–585.
Newman, William, and Lawrence Principe. 2002. *Alchemy tried in the fire. Starkey, Boyle and the fate of Helmontian chymistry*. Chicago: The University of Chicago Press.
Peterschmitt, Luc. 2007. The Cartesians and chemistry. Cordemoy, Rohault, Régis. In *Chymists and chymistry. Studies in the history of alchemy and early modern chemistry*, ed. Lawrence Principe, 193–202. Sagamore Beach: Science History Publications.
Principe, Lawrence. 1998. *The aspiring adept. Robert Boyle and his alchemical quest*. Princeton: Princeton University Press.
Principe, Lawrence, and William Newman. 1998. Alchemy vs. chemistry, the etymological origins of a historiographical mistake. *Early Science and Medicine* 3(1): 32–65.
Principe, Lawrence, and William Newman. 2001. Some problems with the historiography of alchemy. In *Secrets of nature: Astrology and alchemy in early modern Europe*, ed. William Newman and Anthony Grafton, 385–431. Cambridge, MA: MIT Press.
Régis, Pierre-Sylvain. 1690. *Système de philosophie contenant la logique, la métaphysique, la physique et la morale*. Paris.
Rohault, Jacques. 1671. *Traité de physique*. Paris.
Rohault, Jacques. 1987. *System of natural philosophy, illustrated with Dr. Samuel Clarke's notes, taken mostly out of Sir Isaac Newton's philosophy*. New York/London: Garland Publishing.
Severinus, Petrus. 1591. *Idea medicinae philosophicae*, Basel, 1571.
Tachenius, Otto. 1666. *Hippocrates Chimicus, per ignem et aquam methodo inaudita novissimi salis viperini antiquissima fundamenta ostendens*. Venice.

Part II
Cartesian Natural Philosophers

Chapter 7
Empiricism Without Metaphysics: Regius' Cartesian Natural Philosophy

Delphine Bellis

Abstract This chapter is devoted to the philosophy of Henricus Regius, a Dutch philosopher and one of the first followers of Descartes. Regius' philosophy presents an original version of Cartesianism insofar as it relies on a certain number of Cartesian principles and on many particular Cartesian explanations in natural philosophy, while at the same time rejecting Descartes' metaphysics. Regius' empiricist theory of knowledge is precisely intended to replace Descartes' metaphysics. I first explore this original empiricist theory by relying on a systematic comparison of the three editions of Regius' main work. The expression of this empiricism becomes more and more radical and goes hand in hand with a limited skepticism. I show that it leads Regius to a new conception of vision that, although close to that of Descartes, has to account for the visual perception of the geometrical and spatial properties of objects without any innate ideas. Then I present the consequences of Regius' empiricist theory of knowledge on the way the principles of natural philosophy can be grasped and on the role that can be attributed to experience in the explanation of natural phenomena.

Henricus Regius (1598–1679), Professor of medicine at the University of Utrecht since 1638, was one of the first followers of Descartes in natural philosophy. At first, Regius was an enthusiastic disciple of Descartes and the French philosopher was flattered by his Dutch friend's admiration. Descartes supported Regius in the Utrecht dispute against Gisbertus Voetius (1589–1676) and gave him advice on the courses he taught at the University. But after several exchanges by letter, the collaboration between the Dutch and the French philosophers ended in 1646 when Regius decided

D. Bellis (✉)
Department of Philosophy and Moral Sciences, Ghent University, Ghent, Belgium
e-mail: delphine.bellis@gmail.com

M. Dobre and T. Nyden (eds.), *Cartesian Empiricisms*, Studies in History and Philosophy of Science 31, DOI 10.1007/978-94-007-7690-6_7,
© Springer Science+Business Media Dordrecht 2013

to publish his *Fundamenta physices* with which Descartes disagreed, particularly in regard to metaphysics. Beyond the light this dispute sheds on Descartes' own philosophy, Regius is an interesting character for the history of the Cartesian approach to natural philosophy, since his works tend to relegate metaphysics to the background and give prominence to natural philosophy itself. Some commentators, like Charles Adam, even praise Regius for having gone in that direction.[1] Should we therefore contrast a speculative Descartes with an experimental Regius who only applies operative aspects of Descartes' physics, discarding his metaphysics?[2] Some commentators have even gone as far as to depict Regius as "a crude empiricist who was insufficiently clever to understand Descartes's metaphysics."[3] On the other hand, Paul Mouy presents a very different image of Cartesian philosophers in general:

> Now, in a singular turn of events, no experimenters are found in the Cartesian school proper, both in Paris and in Holland. The only true physicist, Rohault, died too young to give his full measure. Indeed, the originality of the disciples will be much more apparent in their theories pertaining to the metaphysics of Descartes' system, or their solutions to the mind-body problem than in their experimental research....the disciples are always dependent on the experimental work of others, and very concerned with integrating it, and, dare one say, digesting it.[4]

In a way, Theo Verbeek summarizes these divergent interpretations, which can be applied to Regius by acknowledging an intrinsic tension in Regius' thought and emphasizing that there is in Regius "a combination—at times unsteady—between a mainly rationalist physics and an empiricist epistemology."[5] A study of Regius' natural philosophy and epistemology will provide us with an early interpretation of Cartesian philosophy and enable us to situate Regius' philosophy in relation to these extreme theoretical positions. The issue is to understand the reasons why

---

[1] AT XI 675: "À la distance où nous sommes aujourd'hui, nous pouvons ne point penser que Regius ait eu tellement tort, théoriquement: la partie solide, celle qui subsiste, de l'œuvre de Descartes, est bien la physique telle qu'il l'entendait, c'est-à-dire l'application de la mathématique à la physique; et sans doute il n'était pas besoin pour cela de tant de métaphysique, ni surtout d'une métaphysique comme celle de Descartes" ("Seen from a distance today, Regius does not seem to have been so mistaken, in theoretical terms: the solid part of Descartes' work, the enduring part, is indeed physics as he intended it, that is, the application of mathematics to physics; this didn't really require so much metaphysics, especially a metaphysics such as Descartes'").

[2] For a discussion of how the Rationalist-Empiricist distinction can cause us to misunderstand or underestimate seventeenth-century philosophers, see Chap. 1 by Dobre and Nyden.

[3] Verbeek 1994, 533–551. This is an interpretation that Theo Verbeek precisely intends to challenge. See, for example, Damiron 1846, II, 103–108; Bouillier 1868, I, 266–267.

[4] Mouy 1934, 71: "Or, par une singulière destinée, dans l'école cartésienne proprement dite, aussi bien à Paris qu'en Hollande, il n'y aura pas d'expérimentateur. Le seul vrai physicien, Rohault, mourra trop jeune pour donner sa mesure. Et précisément l'originalité des disciples se marquera bien plus par leurs théories relatives à la métaphysique du système cartésien ou par leurs solutions du problème psycho-physiologique que par leurs recherches expérimentales....nous verrons les disciples toujours à la remorque des recherches expérimentales d'autrui, et très préoccupés de les intégrer, et, si l'on ose dire, de les digérer."

[5] Verbeek 1993a, viii: "un mariage, parfois assez mal assorti, entre une physique essentiellement rationaliste et une épistémologie empiriste."

Regius became so quickly interested in Descartes' 1637 *Discours de la méthode* and *Essais* and why the two philosophers eventually realized the incompatibility between their respective philosophies. In other words, how far can Regius be considered a Cartesian?

It is noteworthy that very few studies of Regius' natural philosophy are available,[6] except on his physiology and his therapeutics.[7] Now, it is known that Regius taught various topics in natural philosophy, even before Descartes published his *Principia philosophiae*.[8] Descartes himself admits that, after having read only the *Dioptrique* and the *Météores*, Regius wrote a whole *Physiologia*.[9] Even if this natural philosophy is largely inspired from that of Descartes, we should not go as far as Paul Mouy does when considering that the differences between Descartes and Regius are insignificant.[10] Admittedly, the explanations of particular phenomena can often be similar in Descartes and Regius. Regius remains Cartesian because he adopts most of Descartes' explanations in cosmology, meteorology, optics and because he relies on the rejection of vacuum and of substantial forms, as well as on the reduction of matter to extension and the subsequent elaboration of a corpuscular natural philosophy. But relevant and deeply irreconcilable differences in Regius' and Descartes' natural philosophies can be pointed out if one understands the articulation between the particular explanations and the underlying epistemology and theory of knowledge that account for the way the principles of physics are deduced, as well as the subsequent role experience can play in their physics. Experience is thus an illuminating topic at the crossroads of natural philosophy, psychology and metaphysics which can enable us to understand the peculiarities and possibilities of the evolution of Cartesian philosophy in the seventeenth century.

---

[6] De Vrijer himself acknowledged his lack of competence in this area, when it came to undertaking a thorough analysis of Regius' physics and medicine: "Het is mij dus niet mogelijk dan zeer bescheiden te handelen over Regius als physicus en medicus" (De Vrijer 1917, 201: "It is therefore not possible for me to deal with Regius as physicist and physician more than in a very modest way"). On the contrary, there are a certain number of studies on the nature of the soul and its relation with the body in Regius: Rodis-Lewis 1993; Wilson 2000; Clarke 2010; Kolesnik-Antoine 2012. For the historical context, see Verbeek 1988, 1992, 1993b.

[7] See Dechange 1966; Rothschuh 1968; Farina 1975; Gariepy 1990; Bitbol-Hespériès 1993; Caps 2010, 87–116; Kolesnik-Antoine 2010, forthcoming a, unpublished manuscript 2013 (thanks to Delphine Kolesnik-Antoine for having shared with me the unpublished manuscript of these articles).

[8] Theo Verbeek quotes a passage of the *Brevis explicatio*, written by Petrus Wassenaer, in which are listed a variety of topics in natural philosophy, including the human mind, the laws of motion, the motions of the animals, the forces of machines, the vortexes of the skies, the Sun, the fixed stars, the daily and annual motion of the planets, the sea tides, comets, magnets, the nature of meteors, minerals, plants, animals, men, and a lot of other things regarding physiology and medicine (see Verbeek 1994, 541, n46). But Theo Verbeek points to the fact that the *Physiologia* is still "very sketchy on cosmological and purely physical issues" (Verbeek 1994, 544).

[9] Letter to Dinet, AT VII 582–583.

[10] See Mouy 1934, 84.

7.1 Regius' Empiricist Psychology and Theory of Knowledge

We find many references throughout Regius' works to experience. But one can only understand the status experience plays in his natural philosophy if one grasps the articulation between the various levels at which experience takes on a cognitive role for him. One methodological principle I will follow here is to read his natural philosophy backwards. Indeed, Regius gives his reader the key principle of his epistemology only at the end of the *Fundamenta physices*. His epistemology is fundamentally empiricist in the sense that he considers that all our ideas (apart from what pertains to Revelation) ultimately come from sensation. This epistemology has important consequences on the way the principles of natural philosophy can be grasped and on the role that can be attributed to experience in the explanation of natural phenomena.

Before entering into the details of Regius' empiricist psychology, a few remarks are in order to locate it in its historical context. Indeed, several possible sources can account, at least partly, for Regius' empiricism. First of all, we have to acknowledge that the principle according to which all knowledge comes from the senses is a traditional Scholastic principle that Regius shares with his enemy Voetius.[11] In the 1661 edition of the *Philosophia naturalis*, Regius will recall Aristotle's authority when considering the mind as a *tabula rasa* which will receive inscriptions from sense experience.[12] From that point of view, Regius could be considered as an *arrière-garde* Cartesian who failed to fully endorse Descartes' rationalism, especially insofar as it impacts the theory of knowledge and the principles of natural philosophy. But other more modern sources also come into play and are important to understand Regius' empiricism as being different from that of the Scholastics. Paolo Farina has indicated that Regius was influenced by Santorio Santorio (1561–1636), his professor of medicine in Padua.[13] This influence can account, at least in part, for the role Regius primarily attributed to experience in his physiology, even if Regius' own

[11] Geneviève Rodis-Lewis therefore rightly notes: "Regius and Voetius reduce intellectual thought to an abstraction from the sensible" (Rodis-Lewis 1993, 44: "Regius et Voetius réduisent la pensée intellectuelle à une abstraction à partir du sensible"). His opponents will not reproach Regius for his adherence to empiricism. On the commitment to sense experience in Dutch universities, see Verbeek 1992, 6–9, 21–23, 35. Theo Verbeek emphasizes "a general tendency to stress the empiricist elements in Peripatetic doctrine" in the Low Countries at the time. Voetius precisely criticized the rejection of the senses by Descartes: see Verbeek 1992, 56.

[12] Regius 1661, 419: "Atque hinc recte olim dixit Aristoteles, quod mens hominis recens nati sit instar tabulae rasae, cui nihil inscriptum, sed quaevis inscribi possunt" ("Hence Aristotle once rightly said that the soul of a newly born man is like a blank slate on which nothing has been written, but on which one can write anything"). This late addition might correspond to Regius' wish to give an authoritative support to his empiricism, even if his is far from being identical to Aristotle's. In *De anima*, III, 4, 430a, Aristotle compares the intellect to a blank writing tablet.

[13] See Farina 1975, 398. Farina also points to the possible influence of Francis Bacon through Reneri: see Farina 1975, 399.

training in medicine might also have played a role.[14] One could also mention the influence of Bacon in the Low Countries at the time, an influence which was particularly perceptible on Henricus Reneri (1593–1639), Regius' friend.[15] For Reneri, if we can stick to the Scholastic tradition as far as the general principles of physics are concerned, when we turn to *physica specialis*, observation becomes the most important of the constitutive elements.[16]

However important these sources might be, Regius' empiricist psychology and theory of knowledge is largely original. For Regius, all our ideas come from experience or sensation (apart from what concerns Revelation) and sensation (not the cogito) is the principle of all knowledge.[17] As a consequence, and in overt opposition to Descartes, Regius considers that there are no innate ideas. In the *Fundamenta physices*, Regius appeals to a principle of economy: there is no need of innate ideas, as is manifest in our perception of pain, color, taste, and of other

[14] Examining the fundamental concepts constituting the core of Regius' general physiology in his 1641 *Physiologia*, Klaus Dechange rightly notes that the representation of the senses served as a criterion of organization (Dechange 1966, 24). At the time, "*physiologia*" was often taken, in accordance with its etymological meaning, as a synonym of "natural philosophy": see Des Chene 1996. But Regius considers the term to refer to the "knowledge of health," as the subtitle of his work *Physiologia* indicates (*Physiologia sive cognitio sanitatis*).

[15] See Dibon 1990.

[16] Dibon 1990, 213–214. On Rohault's passage from general principles to particulars, see Chap. 9 by Dobre.

[17] Regius 1654, 335: "[Actiones cogitativae] omnes, quae non sunt ex revelatione divinâ, sunt sensationes, vel à sensatione originem ducunt. Nam nihil possumus velle, dijudicare, reminisci, nec de quoquam imaginari, nec quicquam aliter percipere, nisi ejus idea per sensationem, mediatè, vel immediatè, in nobis antea producta, vel postea excitata, & menti oblata fuerit" ("All mental acts that do not come from divine revelation are sensations or originate from sensation. Indeed we cannot wish, judge or remember anything, nor imagine, nor otherwise perceive anything, unless the idea of that thing has previously been produced in us, either indirectly or directly, or afterwards excited and brought before the mind, by sensation"). The 1661 edition adds: "Atque hinc patet sensum aliquem omnis cognitionis, reliquarumque actionum cogitativarum esse principium: Ac proinde non esse omnis cognitionis principium, sive primum cognitum, *Cogito*; nedum, *Cogito, ergo sum*. hi enim sunt conceptus generales, qui ex speciali aliquo sensu primam originem duxerunt" (Regius 1661, 399: "And so it appears that some sense or other is the principle of all knowledge as well as of the remaining mental acts: and hence *I think* is not the principle of all knowledge or the first thing known; and even less so *I think, therefore I am*. These are indeed general notions which have drawn their first origin from some specific sense"). This passage is summed up as follows in the table of contents (unpaginated): "[Actiones cogitativae] sunt sensationes, vel ab iis ortae. Hic vel ille sensus est omnis cognitionis principium: non vero cogito; vel cogito ergo sum" ("[Mental acts] are sensations or originate from them. This or that sense is the principle of all knowledge: and not in fact I think or I think therefore I am"). The critical mention of the cogito is absent in the *Fundamenta physices* and in the 1654 edition of the *Philosophia naturalis*. Regius 1654, 343 (about the necessary role of the brain in the activity of thinking): "Atque hoc ex eo patet, quod ut de corporeis, ita etiam de divinis & spiritualibus rebus non nisi corporeâ sensatione & imaginatione de humanis & corporeis rebus praecedente…" ("And this appears from the fact that, as with corporeal things, we can only think something about the divine and spiritual things insofar as we have previously had a corporeal sensation and imagination of human and corporeal things…"); Regius 1661, 407.

similar things which are correctly (*rectè*) perceived by the mind, although none of these ideas are innate to the mind. And there is no reason why some ideas would be more innate by nature than others.[18] But after the polemics with Descartes and the publication of the *Notae in programma quoddam* (1648) in which Descartes goes as far as to declare that all our ideas, even the sensible ones, are in a sense innate,[19] Regius feels the need to reply. Therefore in the 1654 and 1661 editions of the *Philosophia naturalis*, Regius writes:

> Therefore we need not invent in us any innate ideas of light, colors, sounds, smells, tastes, or shape: since, according to the reason already given, these ideas are either newly produced in the mind by the objects, or have been previously produced by them in the mind, they are either excited by the objects or by their traces.[20]

For Regius, the first data of all our knowledge are the images that are formed from sense perception and on which the mind operates. This means that sense data are not really constitutive of the human mind as a raw material, but are elaborated upon by the mind's activity[21]—that is to say, as we shall see, that of imagination and judgment[22]—which is the sole innate component of our knowledge.[23] Regius then explains the genesis of all our ideas, including the more general ones, from the sense perception of individuals. This is realized by a process of induction.[24]

---

[18] Regius 1646, 251: "Nullis videtur menti ad cogitandum opus esse ideis, imaginibus, notionibus, vel axiomatis innatis; sed sola innata cogitandi facultas ipsi ad omnes actiones cogitativas peragendas sufficit; quod in doloris, coloris, saporis, aliorumque similium perceptione est manifestum, quae à mente rectè percipiuntur, quamvis nullae eorum ideae menti sint innatae. Nec est ulla ratio, cur unae ideae magis à natura sint insitae, quàm aliae" ("It seems that, in order to think, the mind does not need any innate ideas, imaginations, notions or axioms; but the innate faculty of thinking itself alone suffices to perform all mental acts; this is manifest in pains, colors, tastes and in the perception of other similar things, which are perceived correctly by the mind, although none of their ideas are innate to the mind. Nor is there a reason why some ideas would be more innate by nature than some others"). The paragraph is entitled "Mens non indiget ideis innatis" ("The mind does not need any innate ideas").

[19] CSM I 304: "So much so that there is nothing in our ideas which is not innate to the mind or the faculty of thinking, with the sole exceptions of those circumstances which relate to experience, such as the fact that we judge that this or that idea which we now have immediately before our mind refers to a certain thing situated outside us....Hence it follows that the very ideas of the motion themselves and of the figures are innate in us. The ideas of pain, colours, sounds and the like must be all the more innate if, on the occasion of certain corporeal motions, our mind is to be capable of representing them to itself, for there is no similarity between these ideas and the corporeal motions" (AT VIIIb 358–359).

[20] Regius 1654, 354: "Non est itaque, quod ullas luminis, colorum, sonorum, odorum, saporum, vel figurae ideas nobis innatas esse fingamus: cum illae in mente, ratione jam explicatâ, ab objectis recens producantur, vel antea ab iis in ea productae, ab objectis vel eorum notis, excitentur." Regius 1661, 419–420.

[21] This is only made explicit in the 1654 *Philosophia naturalis*.

[22] This excludes any purely intellectual activity as, for example, the cogito.

[23] See Regius 1646, 251.

[24] Regius 1654, 355: "Illas autem omnes, similesque alias quaslibet, ex observationibus rerum, primò per singularem individuorum perceptionem, & deinde per multorum singularium collectionem, & inde factam inductionem, universales notiones inferentem, ipsi formavimus, vel ex alienâ

Now, induction is carried out mainly thanks to imagination. Imagination appears as the mental power that can operate on the sense perceptions and modify them.[25] This enables the mind to form diversified[26] or more general ideas from particular sense perceptions.[27] The process actually takes place in three steps: first the perception from the particulars, then *collectio* (that is to say the gathering of some particulars), and finally induction. This process of abstraction from the sensible is based on the action of the imagination, which removes the particular aspects of the things perceived,

---

traditione illas ab aliis accepimus. Atque hoc cuivis, qui primam notionum istarum in animis nostris nascentium productionem rectè mecum consideraverit, est manifestum" ("Regarding all these [ideas] and any other similar ones, we ourselves have formed them from observations of things, at first through the singular perception of individuals, and then through the collection of many particulars, and thence, by performing an induction and inferring universal notions; or we have received them through a foreign tradition from others. And this is manifest to whoever has rightly considered with me the first production of these notions that are born in our souls"). The 1661 edition introduces a few modifications, specifying that the observations are "probable," that an innate faculty of the mind is involved in the process, and that the general notions are useful in various ways in life: see Regius 1661, 420–421.

[25] Regius 1654, 355, 1661, 421: "Nam puer ille non de communi & universali notione, sed re singulari interrogatur, eaque sensibus ipsius praesens cognoscenda proponitur, iisque ab ipso percipitur, & deinde imaginatione ejus augetur, & imminuitur, aliterque mutatur" ("For that child is not questioned about the common and universal notion, but about the particular thing, and one places before him this [object of knowledge], which is present to his senses, which is perceived by him through the senses, and then is augmented, diminished and changed in other ways by his imagination"). The child referred to here is the *παῖς* in Plato's *Meno* (82b–86c). The passage is not to be found in the *Fundamenta physices*.

[26] Regius 1654, 353–354, 1661, 419: "[imagines] mens sive cogitandi facultas deinde considerat, perpendit, examinat, componit, dividit, inter se confert, aliisque modis tractat; atque ita sibi alias necessarias ideas & notiones, ad omnes alias posteriores suas cogitationes perficiendas, ita sufficienter ex iis conficit, ut nulla ratio, nec ulla necessitas, ad ullarum talium idearum vel notionum menti innatarum subsidium nobis confugiendum esse, suadeat" ("Then the mind or the faculty of thinking considers [these images], it assesses them carefully, examines, composes, divides them, compares them to each other and handles them in other ways; and thus from them it makes for itself the other ideas and notions that are necessary to carry out all its other later thoughts, so that no reason and no necessity urges us to appeal to the help of any such ideas or notions innate to the mind"). Regius does not give any more detailed genetical account of the diversity of our ideas.

[27] Regius 1646, 285: "Perceptio universalium ad imaginationem pertinet. Universalia enim sunt singularia, in abstracto, sine notis individuationis, hoc, hîc, nunc, ut loquuntur Scholastici, considerata" ("The perception of universals belongs to imagination. Indeed, universals are particulars considered in an abstract way, without any mark of individuation such as *this, here, now*, as the Scholastics say [similar things to which are found, or at least can be found, in many other things]"); Regius 1654, 401, 1661, 473. Regius 1654, 355, 1661, 421: "omnis cognitio singularium ex sensibus oritur, & quaecumque alia ibi fuerit, ea imaginatione sensus sequente recens formatur" ("All knowledge of particulars originates from the senses; and any other knowledge there is, it is newly formed by the imagination drawing on the senses"). The passage is not to be found in the *Fundamenta physices*. One could see Regius' conception of imagination in the line of the Scholastic theories of sensation in which the imagination plays a role of generalization in a whole process of abstraction from the impressions received by the sense organs. But the process, as described by Scholastic psychology in its various forms, ends up by a specific operation assigned to the intellect. See Tachau 1988. Regius does not endorse this final step of the process, but rather displaces it and attributes it to imagination and judgment. This distinguishes him clearly from Scholastic psychology.

and on the resemblance (*similitudo*) the mind perceives by judgment between the various items of particulars that have been collected.[28] As we shall see, Regius frequently appeals to similarities in his natural philosophy (for example between visible and invisible processes or between mechanical and physiological phenomena). Imagination plays a central role in constructing explanatory comparisons in natural philosophy. Regius here provides the epistemological foundation of such methodical tools, since imagination is conceived as *drawing on* the senses when operating on the sense data.[29] There is a continuity between imagination and the senses, as they are simply two different kinds of perception.[30] As a consequence, it appears that imagination is double-sided. On the one hand, it is active since it operates on the sense impressions.[31] But on the other hand, it is a passive faculty by which the mind represents

[28] Regius 1654, 355–356: "Neque mirum est, quòd, ex paucorum singularium à nobis observatorum collectione, universales notiones inferri possint. Nam notiones illae, ex singularium quorundam observatione per inductionem collectae & acquisitae, eatenus tantum sunt universales, quatenus illae propter similitudinem, quam habere creduntur cum aliis quibusvis singularibus, quae à nobis per sensus non fuerunt observata, universis, hoc est, quibusvis aliis similibus singularibus, competere judicantur. Neque hoc ullam etiam admirationem meretur. Vniversalia enim nihil aliud sunt quam singularia quaelibet, absque notis individuationis, *hoc*, *hic*, *nunc*, per imaginationem detractoriam, considerata, quorum similia in aliis inveniuntur, vel saltem inveniri possunt....Nam singulares immediate à sensibus; universales verò mediante, ut jam dixi, imaginatione & judicio ab iisdem originem ducunt" ("And one should not be surprised that universal notions can be inferred from the collection of a few particulars observed by us. For those notions, which have been collected and acquired through induction from the observation of some particulars, are universal insofar as we judge them to agree with all the other similar particulars because of a similarity that we believe they have with all the other particulars that have not been observed by us through the senses. And this does not even deserve any admiration. Universals are indeed nothing other than particulars considered, through a subtraction of the imagination, without any mark of individuation such as *this*, *here*, *now*, similar things to which are found, or at least can be found, in other things.... Indeed particulars immediately draw their origin from the senses; while universals, as I have already said, also draw their origin from the senses, but indirectly, by means of imagination and judgment"); Regius 1661, 421–422.

[29] The knowledge which does not originate in the senses is formed "by the imagination drawing on the senses" (Regius 1654, 355, 1661, 421: "imaginatione sensus sequente").

[30] Regius 1646, 252, 1654, 361, 1661, 428: "[Perceptio] triplex est; sensus cogitativus, reminiscentia, & imaginatio" ("Perception is threefold: cogitative sense, reminiscence, and imagination"). Regius 1646, 285: "Imaginatio est perceptio, qua è vestigiorum cerebri varia mutatione, vel spirituum animalium certa dispositione & motu, novae imagines gignuntur, animaeque offeruntur" ("Imagination is a perception by which, from the modification of the traces of the brain or from a certain disposition and motion of the animal spirits, new images arise and are brought before the soul"). Regius 1654, 399–400: "Imaginatio est perceptio, quâ novae imagines & ideae, è vestigiorum cerebri variâ mutatione, vel spirituum animalium certâ dispositione & motu, vel aliâ novâ imaginum & idearum antea perceptarum oblatione, genitae vel productae, menti offeruntur" ("Imagination is a perception by which new images and ideas, from the modification of the traces of the brain or from a certain disposition and motion of the animal spirits, or from another new presentation of images and ideas which had previously been perceived, arise, are produced and brought before the mind"); Regius 1661, 471–472.

[31] Regius 1646, 285: "Mutatio illa vestigiorum cerebri sit, dum vestigia ista vel composita, vel separata, vel detorta menti objiciuntur" ("This modification of the traces of the brain occurs while those traces are presented to the mind either united or separated or distorted"); Regius 1654, 400, 1661, 472: "Mutatio illa vestigiorum cerebri sit, dum vestigia ista vel composita, vel ampliata, vel imminuta, vel separata, vel detorta, vel inter se comparata aut collata,

to itself some ideas, which are nothing else than images[32] caused by motions of the animal spirits in the brain.[33] This passive faculty of the mind is common to sensation and imagination; it is a natural disposition that does not require any more explanation on its nature and origin than the understanding or the will. One could then ask: what function does the intellect play in Regius' epistemology? The most obvious reply is certainly: none. Indeed Regius ends by simply replacing the pure intellect with imagination and judgment.[34] This leads him to formulate a radical empiricism.[35]

That being said, Regius' empiricism is not tantamount to a foundation of certain knowledge on sense data. It rather means that certainty, if any exists in the realm of rational knowledge, is only concerned with appearances.[36] Regius expresses a kind

---

menti objiciuntur" ("This modification of the traces of the brain occurs while those traces are presented to the mind, either united or enlarged or diminished or separated or distorted or compared or opposed to one another").

[32] Regius 1661, 426: "Nulla enim alia alicujus rei idea unquam in mente nostra datur, quam quae est imago, simulacrum, vel similitudinis quaedam nota, in mente hominis existens…" ("For there is never any other idea of any thing in our mind than an image, a likeness or a certain sign of resemblance that exists in the mind of man").

[33] See Regius 1654, 400, 1661, 472–473. Only in the 1661 edition does Regius evoke in this passage "innatam quandam peculiarem mentis facultatem passivam" ("a certain particular passive innate faculty of the mind"). The 1661 edition adds the following paragraph: "Haec autem innata peculiaris mentis facultas passiva menti, ut facultas intelligendi, volendi, spirituumque motum in hanc vel illam partem determinandi, per ejus essentiam inest: ita ut de unius origine & natura non magis, quam de cujusvis alterius facultatis mentis, ipsi per se competentis, ortu & essentia hic sit quaerendum" ("Now, this particular passive innate faculty of the mind belongs to the mind by its own essence, as does the faculty of understanding, willing, and determining the motion of the spirits towards this or that part: so that one does not have to search more for the origin and nature of the former than for the source and essence of the latter faculty of the mind, which is by itself proper to the mind").

[34] Regius 1654, 404, 1661, 477: "Per intellectum autem purum quorundam, est imaginatio & judicio nostrum intelligendum" ("But, by the pure intellect that some speak about, it is to be understood our imagination and judgment"). Such a radical statement is not present in the *Fundamenta physices*.

[35] On the apparent variations in Regius' statements of a thoroughgoing empiricism, see Bos 2013. Bos convincingly shows how in his 1641 *Physiologia*, Regius reintroduced, under Descartes' pressure, the category of inorganic perception through which the soul can perceive incorporeal things, like God, without being aided by the body. In his *Explicatio mentis humanae*, this category of inorganic perception is absent. See Descartes, *Notae in programma quoddam* (AT VIIIb 363–364, CSM I 307), where Descartes identifies Regius' theory of knowledge as a wholehearted empiricism which does not open up the possibility for the soul to conceive by itself immaterial objects. Even if Regius acknowledges that the soul is a substance in itself, and thus is distinct from the body, as Revelation teaches us, it does not mean that this ontology has any impact on Regius' psychology and theory of knowledge. On the contrary, in his *Philosophia naturalis* where he replies to Descartes' criticisms as formulated in the *Notae in programma quoddam*, Regius claims that, as long as the soul is united to the body during this life, it cannot perform any operation without being assisted therein by the body (see Regius 1654, 342–343, 356–357). On Descartes' letter to Regius [first half of May 1641], see also Bos 2002, 71n5.

[36] Regius 1646, 287, 1654, 403, 1661, 476: "An autem satis clarè & distinctè rem perceperimus & examinaverimus, mens secundùm apparentiam tantùm dijudicat" ("The mind judges only according to the appearance whether we have perceived and examined the thing clearly and distinctly enough"). See Descartes to Mersenne, November 23, 1646; CSMK 301: "You are right in supposing that I do not share Regius' opinion…that we know nothing except by appearance; for in my writings I have said exactly the opposite" (AT IV 566). On the question of moral certainty, see Chap. 9 by Dobre, Chap. 10 by Nyden, Chap. 11 by Hatfield.

of limited skepticism,[37] saying that our mind can receive impressions from imaginary causes, as well as from real causes. Therefore we can doubt whether we conceive real or imaginary things and, through rational reasoning (so apart from Revelation) we can only have a probable knowledge of the existence of external things.[38] Admittedly God is not a deceiver for Regius. But he gives no metaphysical justification of this and this can only provide us with the guarantee that we do have the perceptions we have, and not that these appearances correspond to something similar in the world.[39] Eventually, in the *Philosophia naturalis*, the confidence in the sense data is justified by an *a fortiori* argument: since we have to trust testimonies for Revelation, then *a fortiori* we have to trust our senses and what we can deduce from them.[40] The epistemological justification of the appeal to experience in Regius' natural philosophy mostly relies on this minimalist argument. For Regius, our knowledge of the world is a kind of testimony on the world. The senses are not so much direct witnesses of the world as they are one source among others, but certainly the most important one, in the constitution of testimony on the world.[41] Thus, Biblical Revelation allows us to think that there is a material world outside us in general and that we have to trust the testimony of our senses. But Revelation does not enable us to discriminate between our representations and philosophical reasoning does not allow us to remove the residual doubt about the existence of any particular body represented by particular ideas. Even within a kind of natural-theological reasoning on the nature of God and the possibility that he could be a deceiver, it remains impossible to determine, on the basis of our representations, if the latter are produced by some external bodies or if they are imaginary.[42] Therefore, philosophical reasoning can in no way allow us to go beyond the mere level of phenomenalism. Only Revelation can make us access the level of realism, but without really impacting

---

[37] On Regius' skepticism, see Kolesnik-Antoine forthcoming b (thanks to Delphine Kolesnik-Antoine for having shared with me the unpublished manuscript of her article).

[38] See Regius 1646, 249. The section is entitled "Quia mens nostra non tantum a veris, sed etiam ab imaginaris potest affici, ideo dependet certitudo et veritas nostrarum cogitationum a revelatione in Verbo Dei facta" ("Because our mind can be affected not only by true things, but also by imaginary ones, the certainty and truth of our thoughts depend on the Revelation made in the Word of God"). The argument and the tendency towards a natural doubt on the existence of material things is developed and reinforced in the *Philosophia naturalis*: see Regius 1654, 346–351, 1661, 411–416.

[39] See Regius 1654, 349–350, 1661, 414–415.

[40] Regius 1654, 351, 1661, 416, summed up as follows in the table of contents (unpaginated): "Cum fides sit testibus adhibenda; idcirco ea magis nostris sensibus, & iis quae inde recte deducta sunt, debetur" ("Since one has to give credit to witnesses, it follows that one has to give credit all the more to our senses and to the things that have thence been rightly deduced from them").

[41] Regius 1654, 351, 1661, 417: "hinc jam facile intelligimus, qualem fidem circumspecta nostra, quantum fieri potest, judicia, à nostris sensibus & diligentibus rerum observationibus ac traditionibus petita, mereantur.…Certè, cum nostrum ipsorum testimonium apud nos ipsos multo pluris sit, quam alienum, cui fidem tamen adhibendam jam probavi…" ("Hence we also easily understand what credit our judgments deserve, as circumspect as they can be, insofar as we have reached them through our senses and through the careful observations of things and traditions.…Certainly, since our own testimony is of much higher value for us than that of someone else, to which nevertheless we have to give credit, as I have already shown…").

[42] See Regius 1654, 349–350, 1661, 414–415.

Regius' epistemology. Regius' epistemology remains concerned with appearances. And not only are the senses the sole source of knowledge apart from Revelation; they are also the criterion against which to judge the validity of a conception, even of our most abstract ones (like circles, centers, triangles, essence and existence). Indeed the knowledge of all those things depends only on the verisimilitude of the senses or, in other words, of what appears true to the senses.[43]

But since the 1641 *Physiologia*, Regius' analysis of sensation was also sustained by an account of vision, clearly inspired by Descartes' *Dioptrique*. Regius was able to make the most of Descartes' physiology and psychology of sense perception as it is presented in the *Dioptrique*. Dechange notices that in his 1641 *Physiologia*, Regius remains very close to Descartes precisely on that point.[44] But his training in physiology and medicine made Regius maybe more deeply aware than Descartes of the soul's dependence on the body to perform its operations.[45] This enables Regius to claim:

> Our senses are mistaken when the organ is damaged, or when the medium is inadequate, or when the object acts on it too violently or too lightly, or when it is not separated by a proper interval or when something similar required for sense is lacking. If none of these conditions is lacking, the senses are not deceptive at all.[46]

---

[43] Regius 1654, 350, 1661, 416: "Cum enim nullum totum, nulla pars, nullus circulus, nullum centrum, nullus triangulus, nulla essentia vel existentia, cogitatione & rerum apparentiâ exceptâ, sit indubitabiliter cognita; cum horum cognitio a sola sensuum verisimilitudine dependeat; nihil etiam quicquam de iis certò & indubitabiliter à quoquam enunciari potest" ("Since indeed we do not know without any doubt any whole, part, circle, center, triangle, essence or existence, except for thought and the appearance of things, and since the knowledge of these things depends only on the verisimilitude of the senses, we cannot assert anything with certainty and indubitability about them").

[44] Dechange 1966, 61.

[45] Regius 1646, 1, 1654, 2, 1661, 2: "Sic mens humana est principium internum, corporeum; quia sensationes, imaginationes, & alia plurima, sine corpore peragere non potest" ("Thus the human mind is an internal bodily principle, because it cannot produce sensations, imaginations and several other things without the body"). Regius 1654, 343: "Atque hoc ex eo patet, quod ut de corporeis, ita etiam de divinis & spiritualibus rebus non nisi corporeâ sensatione & imaginatione de humanis & corporeis rebus praecedente, & corporeis memoriae notis cerebro impressis adjuvantibus, spiritibusque animalibus auxiliantibus, quicquam cogitare possumus./Quid autem...spiritus animales hîc juvent, ex eo est manifestum, quod justâ illorum quantitate deficient, nulla de Deo vel aliâ re, sive corporeâ, sive incorporeâ, à mente fiat cogitatio, ut in somno profundo, apoplexiâ, & magnâ lipothymiâ, passim observatur. Ita ut vel solus spirituum animalium defectus, cogitationes in homine penitus tollens, sufficiens organicae mentis constitutionis, sive ipsius organorum ad cogitandum indigentiae, sit argumentum" ("And this appears from the fact that we cannot think anything whatsoever, neither about corporeal things, nor about divine and spiritual things, unless we have beforehand some corporeal sensation and imagination of human and corporeal things, and unless the corporeal memory traces impressed in the brain contribute and the animal spirits be of assistance./Now, the assistance provided by the animal spirits is manifest from the fact that, when they are below the right quantity, the mind has no thought of God or of anything else, be it corporeal or incorporeal, as it is observed here and there in deep sleep, in apoplexy and in great faintness. So that the mere absence of animal spirits, which removes entirely thoughts in man, is a sufficient proof of the organic constitution of the mind or of the fact that it needs those organs in order to think"); Regius 1661, 407.

[46] Regius 1641, 233: "*Sensus* nostri *falluntur* cum organum est vitiatum, aut medium ineptum; aut objectum nimis vehementer vel leniter agit; aut justo intervallo non est dissitum; aut simile aliquod requisitum sentiendi deest. Si nulla harum conditionum deficit, *nulla* fit *sensuum fallacia*." All the references to this work are to Bos' edition.

Regius will reiterate the same claim, in expanded form, in the *Fundamenta physices* and in the *Philosophia naturalis*.[47] He manages to circumscribe the factors that may threaten the reliability of the senses to access the specific properties of bodies (that is to say the modes of extension according to which they are diversified). But once these conditions are met, it is possible to claim that our senses are reliable to perceive the properties of the external objects themselves. That is why Regius can define vision as "the external sense by which, from the motion of the fibres of the optical nerve conveyed in the brain, the soul perceives the light, color, situation, distance, magnitude and shape of the things which are presented to it."[48] It is noteworthy that, among the properties of the objects seen, Regius includes not only geometrical properties such as position, distance, size, and shape, but also light and color. This can be accounted for by the fact that Regius takes up Descartes' mechanistic explanation of colors that ascribes them to the variation of the rotational speed of the corpuscles of subtle matter.[49] Colors and light can be considered as properties of objects as far as one is able to link their phenomenological appearance with a mechanistic explanation.[50]

To say that Regius' account of vision is largely inspired by Descartes' *Dioptrique* is not to say that the Dutch philosopher could adopt it as it is without his empiricism being threatened by some aspects of Descartes' new optics. This is all the more obvious in two specific aspects. First, Descartes rejected to some extent the notion of resemblance (*verisimilitudo*)[51] on which the empiricist Scholastic theories of sense perception and knowledge relied.[52] But, as Regius wants to show, the deconstruction of the Scholastic model of resemblance in Cartesian optics does not suffice to undermine empiricism in general:

> So that it cannot be objected that we often conceive by our mind some ideas about the shapes of things that are different from their images depicted at the back of our eyes. For this does not arise from innate ideas, but from ideas impressed in the mind through perceptions that have otherwise been made before, as will become more evident in what follows.[53]

---

[47] See Regius 1646, 284, 1654, 398–399, 1661, 470–471.

[48] Regius 1641, 224: "sensus externus, quo ex motu fibrillarum nervi optici in cerebrum delato, anima Lumen, colorem, situm, distantiam, magnitudinem et figuram rerum objectarum percipit."

[49] Regius 1641, 226.

[50] Regius 1641, 204: "Atque ita omnes aliae qualitates sensibiles ex solo motu, figura, magnitudine et situ particularum insensibilium clarissime possent explicari…" ("And thus all the other sensible qualities could be very clearly explained from only the motion, shape, magnitude and situation of insensible particles…").

[51] It is nevertheless possible to show that the notion of resemblance still plays an operative role in Descartes' theory of vision, but in a new way. On that, see Descartes 1997, 335–336; Fichant 1998; Bellis 2010, 347–367. It is very significant that Regius does not even try to take up this refined aspect of Descartes' theory of vision to give support to his empiricism. His solution is more radical, but maybe also more fragile from an epistemological point of view.

[52] On the Scholastic theories of sense perception, see Tachau 1988; Spruit 1994–1995.

[53] Regius 1654, 354; Regius 1661, 420: "Vti nec obest, quod saepe alias de rerum figuris ideas mente nostrâ concipiamus, quam sunt earum imagines, quae in fundo oculorum nostrorum pinguntur.

The fact that the material basis of vision, that is, the retinal picture, does not look perfectly like the ideas we have of external objects could have been an objection addressed in a Scholastic fashion to Regius' empiricism. Indeed, in the Scholastic theories of sense perception, the formal *species* bears the resemblance of the object so that the mind can perceive it as it is. There is thus continuity in resemblance from the external object to the *species* and then to the mental representation.[54] But Regius precisely gives up this continuity: he rejects, as Descartes had done, the *species intentionales* to explain sensation[55] and he deprives the basis of vision (which has become in the meantime corporeal) of the resemblance with the external objects and with our ideas of them. The solution of continuity introduced in the process does not however threaten the probable reliability of the senses in accessing the external world. It is compensated for by the activity of the mind based on previous sense perceptions. Therefore, a resemblance is still guaranteed between the ideas of the mind in their representational dimension and the external objects,[56] but in a way very different from that of the Scholastics or of Descartes. Resemblance is not a kind of *objective* property of the *species sensibiles* and *intelligibiles*,[57] but is the result of the *subjective* activity of the mind (of imagination and judgment, and not

Hoc enim non ex ideis innatis, sed menti, per perceptiones antea aliter factas, impressis, ut posteà magis patebit, contingit."

[54] This is a very simplified presentation of the process. For a more accurate account, see Tachau 1988; Spruit 1994–1995.

[55] See Regius 1646, 253–254, 1654, 362, 1661, 429.

[56] Regius 1661, 426–427: "Nulla enim alia alicujus rei idea unquam in mente nostra datur, quam quae est imago, simulacrum, vel similitudinis quaedam nota, in mente hominis existens, & rem aliquam menti utcunque repraesentare apta. Haec idea cum etiam citra intellectus operationem in mente existat, hinc patet illam pro ipsa intellectus operatione non esse ponendam. Cumque illa idea nihil aliud sit quam imago rem menti repraesentans, vel utcunque repraesentare potens, hinc illam pro re, per operationem intellectus ope istius ideae repraesentata, vel repraesentanda, nequaquam esse sumendam, etiam est manifestum" ("In fact, there is never any other idea of a particular thing in our mind than that which is an image, a likeness or a certain sign of resemblance, which exists in the mind of man and is fit for representing a certain thing to the mind in a certain way. Since this idea also exists in the mind without the operation of the intellect, it appears that this idea should not be regarded as the very operation of the intellect. And since that idea is nothing else than an image representing the thing to the mind, or which can represent it in a certain way, it is also manifest that that idea should by no means be taken for the thing which has been represented, or should be represented, through the operation of the intellect with the help of this idea"). This is an addition of the 1661 edition.

[57] In the 1661 edition, Regius nevertheless claims that there is a confused (*confusa*) resemblance between the motions causing our sensations and the sensible ideas: see Regius 1661, 430. The 1654 edition was already a bit ambiguous, stating that the traces imprinted on the brain by the motions of the animal spirits often (*saepe*) have no resemblance or only a weak (*exiguam*) resemblance with the external objects perceived: see Regius 1654, 399, 1661, 471. But this confused, weak and rare, albeit objective, resemblance (of the motions in the brain with the external objects or with our sensible ideas of them) can in no way provide a solid epistemological foundation for the reliability and for a realist interpretation of sensation.

of the intellect endowed with innate ideas).[58] Therefore, *if* the thing represented by the mental image exists, it resembles the image formed in our mind from the sense perception and imagination. But the mental representation of a thing is in no way a clue to the existence of this thing outside us (since imaginary ideas are similar to sensible ideas from a representational point of view for Regius). This is perfectly coherent with Regius' general theory of ideas since, as we have seen, resemblance is also the basis of the process of formation of universals in induction.[59]

Second, Regius' rejection of innate ideas forces him to reshape Descartes' theory of vision to some extent to comply with the requirements of his empiricist psychology. Indeed, in the *Dioptrique*, Descartes had formulated the framework of a theory of vision in which a natural geometry plays a crucial role. The notion of a natural geometry is mentioned in the context of the perception of depth and distance in vision.[60] It is involved in a calculus by triangulation through which the distance of the object seen is determined in vision. This geometry presupposes that the percipient has knowledge of the distance separating his eyes and of the sensation produced by the rotation of his eyes orientating towards the object seen. This enables the observer to evaluate the two angles and the length of the intermediate side of a triangle formed by his two eyes and a point on the object. By a mental act performed by the imagination, one can therefore determine the length of the other two sides, that is to say the distance between the percipient and the object. This geometry is "natural" insofar as it was instituted in us by God. And more importantly, the very idea of a geometry is revealing in that it suggests that this psychological process is based, as geometry in general is for Descartes, on innate ideas. Descartes' *Reply* to Pierre Gassendi's *Fifth Set of Objections* makes this point very clear.[61] It would have been impossible to conceive the geometrical triangle from the sole vision of a—necessarily imperfect—triangle, if we hadn't had the innate idea of the triangle in our mind. The triangle seen is like a portrait of the geometrical triangle. But for this evocation to be possible, the innate idea of the triangle must be in a way projected unto sensation. Therefore, important as its cognitive role may be for Descartes, he does not consider imagination as having the most fundamental role in the process of

[58] Regius insists on the fact that the absence of a perfect resemblance between our sensible ideas and the motions in the brain that cause them does not imply the need for innate ideas in order to explain sensation: see Regius 1661, 430. This is certainly a reply to Descartes: see *Notae in programma quoddam*, AT VIIIb 358–359, CSM I 304.

[59] See Regius 1654, 355–356, 1661, 421–422.

[60] AT VI 137–138, CSM I 170: "In the second place, we know distance by the relations of the eye to one another. Our blind man holding the two sticks AE and CE (whose length I assume he does not know) and knowing only the distance between his two hands A and C and the size of the angles ACE and CAE, can tell from this knowledge, as if by a natural geometry, where the point E is. And similarly, when our two eyes A and B are turned towards point X, the length of the line AB and the size of the two angles XAB and XBA enable us to know where the point X is.…And this is done by a mental act which, though only a very simple act of the imagination, involves a kind of reasoning quite similar to that used by surveyors when they measure inaccessible places by means of two different vantage points."

[61] See AT VII 381–382.

sense perception. Thanks to the mediation of imagination, innate ideas are what enable us to identify geometrical objects in the sensible. Imagination wouldn't be able to perform anything without the geometrical innate ideas of the intellect. With such explanations of the spatial properties of bodies as perceived by vision, Descartes offers an epistemological analysis of vision through which the circumscribed validity of visual experience for the knowledge of bodies, and so for physics, is ascertained. It enables one to associate some experiential data gained through vision (and bearing on the modes of extension) with the building of a natural philosophy. With vision thus understood, it becomes possible to know the location, distance, and figure of external bodies in nature.

Now, one understands that given his commitment to empiricism, Regius could find it useful to ascertain the validity of vision to grasp the properties of bodies in experience. Very significantly, in Regius' natural philosophy, Cartesian optics is integrated in a study of man, and more precisely of vision.[62] This shows that optics clearly becomes subordinated to a theory of sensation and of human faculties, and ultimately to a theory of knowledge. But how can Regius account for the perceptions of the geometrical properties and of the distances of objects, if he cannot appeal to innate ideas?[63] For Regius, a natural disposition is enough to produce in us different ideas of sensation caused by different motions in the sense organs.[64] On the contrary, for Descartes, innate ideas were required to make this disposition effective. This meant that, in sensation, the intellect had an active dimension by projecting innate ideas onto the perceptions produced by corporeal motions. But, for Regius, the intellect does not come into play in the process.[65] There is only a passive disposition of the mind which receives the sense impressions on which imagination and judgment operate.[66] Regius therefore reverses the Cartesian procedure which

[62] See Regius 1646, Chap. XII, 272–273; *Philosophia naturalis*, Book V De Homine, Chap. III De Visu, 1654, 382–383, 1661, 453.

[63] This also applies to mathematical ideas such as that of the triangle: we know that the three angles of a triangle are equal to two right angles because such a triangle exists in nature. See Regius 1654, 358, 1661, 424. This paragraph echoes the discussion of the *Fifth Objections and Replies* (AT VII 381–382). Here Regius clearly sides with Gassendi.

[64] Regius 1646, 254: "Porrò varii isti motus, organis recepti, & menti in sensorio communi oblati, diversas tales sensationum cogitationes nullam aliam ob causam excitant, quàm quia à natura ita comparati sumus" ("Moreover these various motions, which have been received by the organs and brought before the mind in the common sense, excite such diverse thoughts of sensations for no other cause than because we have been thus disposed by nature"). In the *Philosophia naturalis*, we have the following addition: "that the mind is consciously so variously affected by those various motions" (Regius 1654, 362: "ut mens ab illis variis motibus ita varie cum conscientiâ afficiatur"; Regius 1661, 429).

[65] Regius 1661, 426–427: "Haec idea cum etiam citra intellectus operationem in mente existat, hinc patet illam pro ipsa intellectus operatione non esse ponendam" ("Since this idea also exists in the mind without the operation of the intellect, it appears that this idea should not be posited as the very operation of the intellect").

[66] Regius 1661, 430: "Nulla itaque est causa, ut ullas qualitatum sensibilium ideas, ad sentiendum necessarias, menti innatas esse dicamus; praesertim cum nullae tales imagines in mente deprehendantur, nisi illae foris ab objectis adveniant; & ipsa objectorum afficiendi vis, mentisque ab illis

projected universal innate notions in the sensible particulars, since universals are constituted from the particulars.[67] It is thus very much significant, but all in all perfectly coherent, that the notion of a natural geometry is absent in Regius' account of the perception of the distance of objects through vision. He explains the perception of the distance of objects only by the process of accommodation of the eye (which was also mentioned by Descartes in addition to the natural geometry)[68] and by the more or less strong impression made on the eye by the light rays. As far as the figure of the object is concerned, Regius implicitly rejects the projection of the innate idea of the geometrical figure by imagination through vision and sticks to his empiricism. As a consequence, the figure of an object is mainly seen as the result of the situation of its various points which is itself perceived by the eye on the basis of the direction of the light rays.[69] This process also implies that the visual impression produced by the retinal picture is interpreted in relation to previous visual sensations, so that the perspectival deformations of the retinal picture, due to its flatness, can be corrected by an *aestimatio*. Of course, this approach of our sensible ideas cannot avoid raising the question that is often addressed to empiricist theories, namely: What about our first sensible idea? How can the mind operate on the data received from the senses and impressed on the retina without having at its disposal previous ideas of sensation? How is it possible to perceive an object with its three dimensions from its first visual instance? Regius does not elude this radically genetical question, as he could have by saying that the question of the origin is a metaphysical one that cannot be elucidated. Rather, he accepts that the first sensible idea be identical to the retinal image.[70] In sense perception, sensations are thus made more complex from a first mental perception which, being original, can only be identical to the two-dimensional retinal picture.[71] And even more radically, Regius considers vision to be a synthetic reconstruction of discontinuous and intermittent impressions of light and colors

---

patiendi facultas, ad illarum excitationem, & perceptionem, sufficiant" ("Therefore there is no cause to say that any of the ideas of the sensible qualities, necessary with respect to sense, are innate to the mind; especially since we do not discover any such images in the mind, unless they are brought forth by external objects; and the force that the objects have to affect the mind, and the faculty of the mind to be affected by them, are sufficient to produce the excitation and perception of those ideas of the sensible qualities"). This is an addition of the 1661 edition.

[67] Regius 1646, 285: "Universalia dico esse singularia.…Universalia itaque nihil aliud sunt, quàm tam singularia, quorum similia in multis aliis inveniuntur, vel saltem inveniri possunt" ("I say that universals are particulars.…This is why universals are nothing else than particulars, of which similar things are found, or at least can be found, in many other things"); Regius 1654, 401, 1661, 473: "Vniversalia dico esse singularia: alioqui enim de singularibus affirmari non possent, quod tamen rectè de iis fit…" ("I say that universals are particulars: for otherwise we could not attribute them to particulars, which we nevertheless rightly do…").

[68] See *Dioptrique*, AT VI 137, CSM I 170.

[69] See Regius 1646, 271, 273, 1654, 379–380, 384, 1661, 448, 454.

[70] This does not mean that Regius believes that the retinal picture is directly seen, as if we had internal eyes to look at it. Like Descartes in the *Dioptrique* (AT VI 130), Regius rejects this explanation: see Regius 1646, 274, 1654, 385, 1661, 455.

[71] See Regius 1654, 384, 1661, 454. This genetical explanation is lacking in the *Fundamenta physices*.

on the eye. This approach is mainly intended to explain how it is possible for various rays of light producing perceptions of different colors to enter into the narrow pupil. Regius considers that these rays of light enter successively in the eye, but impress their effect on the retina during a short period of time, and therefore become concomitant with that of the other successive rays of light.[72] This gives rise to a quasi-kaleidoscopic recomposition of the external world on the retina. This explanation of vision which is peculiar to Regius is convergent with his concern for appearances, and with his empiricist psychology in which the mind operates on sense impressions but without any innate ideas.

Far from being a slavish disciple, Regius therefore proposes, instead of Descartes' theory of intellect and innate ideas, a whole and radical empiricist psychology which leads to a new theory of knowledge and a new epistemology. These cannot but have a direct impact on the formulation of his natural philosophy, and in particular on what Descartes presented, on the basis of his metaphysics, as the foundations for physics.

7.2 Experience in the *Physica Generalis*

7.2.1 Descartes' Criticism of Regius' Fundamenta physices*: A Question of Method*

In 1645, Descartes had already explained to Regius why he disagreed with the text the Utrecht professor intended to publish as the *Fundamenta physices*. According to Descartes, Regius only gives probable reasons, and not certain explanations.[73] Moreover, Regius modifies the order of reasons and abbreviates the argumentation, therefore rending it only probable:

> …you do not add anything from you there, apart from the order and the brevity, two things which, if I am not mistaken, will be blamed by all people of good sense; indeed I have seen no one thus far who would disapprove the order I adopted and who would not blame me for an excessive brevity rather than for prolixity.[74]

[72] See Regius 1646, 277, 1654, 388–389, 1661, 459.

[73] Letter from Descartes to Regius. July 1645, AT IV 239: "Cumque meminerim me multa legisse in tuo compendio Physico, à vulgari opinione planè aliena, quae nudè ibi proponuntur, nullis additis rationibus, quibus lectori probabiles reddi possint…" ("And since I remembered having read a lot of things in your compendium of physics which are completely contrary to the common opinion and which are proposed there alone and without any additional reasons by means of which they could be rendered probable to the reader…"). AT IV 248–249, CSMK 254: "I admit that [my opinions] can be correctly presented through definitions and divisions, proceeding from the general to the particular, but I deny that proofs ought in that case to be omitted."

[74] Letter from Descartes to Regius. July 1645, AT IV 257: "ibi nihil de tuo addis, praeter ordinem & breuitatem, quae duo, ni fallor, ab omnibus benè sentientibus culpabuntur; neminem enim adhuc vidi, qui meum ordinem improbaret, quique non potiùs me nimiae breuitatis quam prolixitatis accusaret."

Indeed, in the *Fundamenta physices*, Regius' theory of knowledge comes only after physics and Regius therefore deprives natural philosophy from its most solid foundation which was metaphysics in Descartes' works. In the lettre-préface to the *Principes de la philosophie*, Descartes seeks to distance himself from his former disciple mainly for epistemological reasons.[75] In a letter to Elizabeth (1618–1680) of March 1647, Descartes criticizes Regius' *Fundamenta* on the ground that it contains nothing, as far as physics is concerned, apart from his assertions put in a wrong order and without their true proofs.[76] In the *Conversation with Burman*, Descartes expresses the strong divergence between himself and his former disciple as follows:

> But as far as Regius is concerned, his demonstration does not have any value; and what is surprising is that in physics he has always strived to follow and guess the author's opinions, even when he did not know them; but in metaphysics, he has contradicted the author as far as he could and as far as he knew his opinions.[77]

This comment refers to article 46 of the third part of the *Principia*, in which the fluidity of the skies and their organization in vortexes is supposed, so that a cosmogony can be derived from this. According to the above statement from Descartes' *Conversation with Burman*, Regius does not think it necessary to provide a demonstration of the way the organization of the cosmos can be deduced from the first principles of physics (that is essentially extension and movement), contrary to what Descartes attempts to do in the third part of his *Principia philosophiae*. For Regius, this is enough to say that the principles of physics *could* allow someone to deduce the cosmological organization of the material world from an originary chaos,[78] but the deduction itself is not required. Motion is the proximate cause of the world and of the situation of planets and stars. One need not develop a cosmogony to justify the present state of the world as far as this current state can be accounted for from extension and movement. Moreover, Descartes himself considers that this theoretical attitude is linked to Regius' rejection of any metaphysical commitment. Descartes wants to show how causes can produce their effects. Regius gives explanations of various phenomena, following the classical order

---

[75] CSM I 189 (my emphasis): "Last year he published a book entitled *The Foundations of Physics* in which, as far as physics and medicine are concerned, it appears that everything he wrote was taken from my writings—both from those I have published and also from a still imperfect work on the nature of animals which fell into his hands. But because he copied down the material inaccurately and *changed the order* and *denied certain truths of metaphysics* on which the whole of physics must be based, I am obliged to disavow his work entirely" (AT IXb 19).

[76] CSMK 314: "It contains nothing on physics except for my assertions in a jumbled order and without their true proofs. As a consequence they appear paradoxical, and what comes at the beginning can be proved only by what comes towards the end" (AT IV 625).

[77] AT V 170: "Regium autem quod attinet, ejusdem demonstratio nulla est; et quod mirum, in Physicis ille semper auctoris opiniones, etiam ubi eas nesciebat, sequi et conjicere studuit; in Metaphysicis autem auctori quantum potuit et ejus opiniones novit, contradixit."

[78] See Regius 1646, 76.

of subject matter,[79] going from the principles of natural things in general to cosmology, the four terrestrial elements, meteorological and terrestrial phenomena, plants, animals and man. But this topical order cannot guarantee the same kind of unity as a metaphysical foundation. The connection between each explanation and a cosmological system based on the fluidity of the skies and the vortex theory is no longer obvious.

I would like now to suggest that this difference in the attitude of both philosophers originates not only in the value given to metaphysics for the elaboration of natural philosophy, but also from a different understanding of the role of experience in their epistemology and in particular in the creation of an *order* in natural philosophy. For Regius, nature is a set of facts which can be considered independently and accounted for from mechanical principles. In the *Fundamenta physices*, we find a significant number of occurrences like "ut experientia docet," "teste experientia," etc. Experience is therefore not meant to be integrated in a holistic conception of nature. On the contrary, according to Descartes, experience has no value independently from the possibility to be linked to all the phenomenal aspects of the world through a demonstration or series of demonstrations.[80] Whereas the French philosopher defends a holistic account of natural philosophy in which experience plays a circumscribed role, Regius considers experience as a source of factual information on nature. But Regius' understanding of experience is also linked to a specific empiricist psychology which is incompatible with Descartes' conception of mind and of innate ideas. For Descartes, physics never starts with particular experiences only (not even experiments), but is based on metaphysical principles and then rendered more concrete through imagination.

7.2.2 *Experience as Constituting the Object of Physics Itself*

As Descartes notices, Regius transforms the relation between metaphysics and physics in such a way that what should stand for metaphysics (in fact a theory of knowledge) comes only after a whole treatise of natural philosophy in his works. This order is not merely anecdotal, as Descartes himself suggests. Of course,

[79] As the title of his 1646 *Fundamenta physices* indicates, Regius intends to provide the foundations or principles of nature, as Aristotle had done in the first three books of his *Physics* and as they are to be found in Scholastic textbooks of natural philosophy. These principles traditionally include the notion of nature itself, matter and form, the four causes, and natural change (including motion). In the *Fundamenta physices*, the chapter on the principles of nature which apply to all natural beings (Chap. I) is followed by an explanation of the various natural things according to the following traditional order: simple bodies which are either incorruptible (i.e., heavenly bodies: Chap. II) or corruptible (elements: Chaps. III–V), composite bodies, either inanimate (i.e., meteors and fossils: Chaps. VI–VII) or animate (plants, animals, beasts, and man: Chaps. VIII–XII). On late Aristotelian textbooks in natural philosophy, see Des Chene 1996 (in particular 9–10).

[80] See Descartes' letter to Regius from July 1645 in which Descartes insists on the need for "*probationes*" (AT IV 245).

in his *Essais* of 1637, Descartes hadn't given a full exposition of his metaphysics but only some "essais" of his method and a short metaphysical summary in the *Discours de la méthode*, and he had based his *Météores* and *Dioptrique* on a few "suppositions." But this is not to say that metaphysics played no role in his physics and, moreover, it is not the same thing to withdraw a part of metaphysics as Descartes did and to place a substitute for metaphysics *after* physics as Regius did. In the case of Regius, this corresponds to a specific epistemology in which metaphysics does not have a central function. But it also has consequences for the principles of physics he gives at the beginning of the *Fundamenta physices*. The first chapter of the *Fundamenta physices* is therefore the equivalent of the second part of the *Principia philosophiae*,[81] but it cannot rest on a metaphysical part which is obviously lacking at the beginning of the *Fundamenta*. As a consequence, the principles of natural philosophy are going to be constituted in a non-metaphysical way. Now, this non-metaphysical way implies, for Regius, that experience has a central role, as constituting the object of physics itself. For example, for the nature of body, Regius reverses the order of proof: instead of making a metaphysical reduction of body to extended substance[82] and then only selecting in experience the data linked to extension (shape, size, motion), he bases this reduction on the ground of experience:[83]

> Its essence consists only in the extension in length, width, and depth, which only differs from body by reason; and it does not consist in hardness, softness, color, taste, smell, or any other similar qualities. Indeed, as everyday experience teaches us, these can all be easily taken away from the body, without harming its essence. For hard things can become soft; soft things can become hard; colored things can lose their color; savory things their taste; smelling things their smell; and this is the same for the others.[84]

In a similar way, Regius relies on experimental medical facts to support his claim for an "organic" *mens*.[85] And he even goes as far as to conceive the essence of matter

---

[81] On this text, its structure, and its relation with the first part of the *Principia philosophiae*, see de Buzon and Carraud 1994.

[82] See *Principia philosophiae* II 4. AT VIIIa 42, CSM I, 224 where Descartes distinguishes "the nature of matter, or body considered in general" (namely the extension in length, breadth and depth) from that "which affects the senses in any way."

[83] This was already noticed by Theo Verbeek: see Verbeek 2000, 154.

[84] Regius 1646, 2: "Hujus essentia in solâ in longum, latum, & profondum extensione, quae ratione tantùm à corpore differt; non autem in duritie, mollitie, colore, sapore, odore, vel aliis similibus qualitatibus consistit. Hae enim omnes à corpore, salvâ ejus essentiâ, facilè tolluntur, ut quotidiana docet experientia. Nam res dura emolliri; mollis indirari; colorata colorem perdere; sapida sapore privari; & odorata omni odore destitui potest: atque ita de caeteris." The 1654 and 1661 editions add continuity and contiguity to the qualities which do not constitute the essence of body because they can be removed from it: see Regius 1654, 3, 1661, 3.

[85] Regius 1646, 246: "quod docet experientia in apoplexia, epilepsia et similibus aliis capitis gravibus affectibus, in quibus rerum phantasmata et imagines, seu debiti motus, menti, ob sensorii communis laesionem, a corpore offerri non possunt" ("That is what experience teaches in apoplexy, epilepsy, and similar other severe affections of the head, in which the phantasms and images of things or the appropriate motions cannot be presented to the mind by the body because of an injury of the common sense").

through an experiment, what we would nowadays call a "thought experiment": "But if extension was taken away from it, a body would soon cease to be a body, because it would no longer be an extended substance."[86] Regius considers that the modes of extension (*situs*, *figura*, *quantitas*, *motus*, *quies*) are not only perfectly intelligible, but also manifest in observation:

> …in addition to these things [the location, shape, quantity, motion, or rest of the insensible parts] which can indeed be very clearly understood, nothing can exist or be *observed* in nature that constitutes things; and these are altogether sufficient for the constitution of things; for through these the nature and effects of things are correctly explained.[87]

Because these "qualities" constitute visible bodies, we can assume them to constitute the nature of bodies in general, even in their minute insensible parts.[88]

Another aspect of the constitution of the object of physics by experience can be seen in Regius' understanding of form. Regius defines nature as twofold: "This one is twofold: the matter of natural things and their form."[89] But departing from the Aristotelian understanding of form, Regius considers this notion as immediately constituting the realm of experience since form, with matter, constitutes a *perceptible* thing, i.e., a thing that one can perceive in experience. Regius distinguishes between special form (i.e., the human mind)[90] and general or material form. The latter includes motion or rest, position, shape and size of the parts.[91]

It is noteworthy that motion is not presented, through a metaphysical reasoning, as one mode or dependence of extension, and as one way extension can be diversified.[92] But its importance at the physical level is here again justified by a fact of experience: how machines work. Regius originally relies on the way mechanical devices such as levers or pulleys work to construct his arguments on motion. The argument is the following: the form of a thing is effective only in conjunction with its movement.

---

[86] Regius 1646, 2: "Si verò extensio ab illo tolleretur, mox corpus cessaret esse corpus; quia non esset ampliùs substantia extensa."

[87] Regius 1646, 95 (my emphasis): "praeter haec [situs, figura, quantitas, motus, quies partium insensibilium] enim clarissimè intelligibilia, nil ad res constituendas in rerum naturâ dari vel *observari* potest; eaque ad earum constitutionem omnino sufficiunt; cùm per ea natura & effecta earum rectè explicentur."

[88] Gariepy claims that, "Regius, like other materialists, offered no experimental observations to warrant the existence of his insensible particles" (Gariepy 1990, 125). But Regius does rely on experience to infer by analogy that insensible parts have the same properties as sensible parts.

[89] Regius 1646, 2: "Haec duplex est: Materia rerum naturalium, earumque Forma."

[90] See Regius 1646, 29.

[91] Regius 1646, 4: "Forma generalis, (quae vulgò materialis nuncupatur, & omnibus rebus naturalibus competit,) est comprehensio motus vel quietis, item sitûs, figurae & magnitudinis partium, rebus naturalibus constituendis conveniens" ("The general form (which is commonly called material and which is applicable to all natural things) includes motion or rest, situation, the shape and magnitude of the parts, all things that are appropriate to the constitution of natural things"). See Regius 1654, 8, 1661, 10.

[92] See Descartes, *Principia philosophiae* I 56, 61; II 23. In these texts, Descartes conceives of motion not as a property of bodies which is apparent in experience, but as a modal dependence of extension and as a way through which matter can receive various "affections" or "forms."

That is precisely what we can see in machines and mechanical instruments.[93] We do not even find the kind of analysis Descartes presents in article 203 of the fourth part of the *Principia*, where the machine serves, on an epistemological level, to justify that one can conceive the minute invisible parts of bodies according to the sensible ones, just as the clockmaker can guess a clock's mechanisms just by looking at its outward appearance. In Regius, the comparison of bodies with machines works at a purely sensible level, between visible machines and visible bodies, to justify that no body could have properties if there were no motion in nature.[94]

As Theo Verbeek has shown, the distinction between perceptible parts (*partes sensibiles*) and imperceptible parts (*partes insensibiles*) in Regius' natural philosophy has its roots in his physiology.[95] In the *Physiologia*, it follows the definition of physiology as the theory of the healthy body.[96] It can be explained because health (*sanitas*) relies on a proper disposition of the parts of the body. Therefore, insensible parts are principally required to account for diseases or health. And they are reintroduced in a broader realm when applied to physics. In the *Fundamenta physices*, Regius also insists on their explanatory power.[97] They enable Regius to get rid of

[93] Regius 1646, 5: "Cùm autem haec accidentia dicimus esse efficacia & sufficientia naturae principia, non de singulis, sed conjunctis; neque de quibuslibet, sed justis & convenientibus loquimur: ut apparet in vecte, trochleâ, & aliis machinis, in quibus nec quaevis figura, nec situs omnis, nec quaelibet magnitudo, per se sunt efficaces; sed si ea sint justa, & sufficiens motus iis addatur, & illa omnia vel pleraque simul sumantur. Qui itaque his accidentibus omnem energian [sic] denegant, manifestissimae experientiae adversantur" ("Now, when we say that these accidents are effective and sufficient principles of nature, we do not speak about them taken separately but conjointly, nor do we speak about them indiscriminately, but only about the right and appropriate ones; as it appears with the lever, the pulley, and other machines, in which neither just any shape, nor just any situation or any magnitude are effective by themselves, but only if they are in the right measure and if enough motion is imparted to them, and if all those things, or a certain number of them, are taken simultaneously. That is why those who deny all efficacy to these accidents are against the most manifest experience"). See Regius 1654, 9, 1661, 10–11. Regius 1646, 11, 1654, 25, 1661, 28: "ut videmus fieri in vecte, plano inclinato, trochleâ, & aliis machinis" ("as we see it happen with the lever, the inclined plane, the pulley and other machines"). On the relation between natural and artificial bodies, see also Regius 1646, 45–47, 1654, 68–70, 1661, 76–78.

[94] This question is linked to but remains distinct from that of the individuation of bodies by motion, since it is dealt with by Regius only at the macroscopic level of the properties of bodies. On the problem of individuation of bodies which arises in Cartesian physics, see Dobre 2010.

[95] See Verbeek 2000, 157.

[96] Regius 1641, 199: "*Sanitas* est dipositio partium humani corporis actionibus recte perficiendis apta….4. *Pars humani corporis* est quaelibet corporea substantia illud complens, actionibusque perficiendis comparata. 5. Estque vel insensibilis vel sensibilis" ("Health is the disposition of the parts of the human body such that it is able to perform its actions in the right way….4. A part of the human body is a certain corporeal substance that completes that body and is disposed so as to perform its actions. 5. And it is either insensible or sensible").

[97] Regius 1646, 3–4 (my emphasis): "Insensibiles sunt, quae, propter exiguitatem aut parvitatem sensus fugientes, solo intellectu in omnibus rebus naturalibus observantur. tales sunt ramosae, ex. gr. particulae, oleum; oblongae & flexiles, aquam constituentes, de quibus postea acturi sumus./ Hae ex subtilitate, crassitie, acrimoniâ, lenitate, fluiditate, oleaginositate, aquositate, salsedine, aliisque innumeris corporum qualitatibus, postea explicandis, manifestè colliguntur. Nam his *positis*, clara & distincta illarum est *explicatio*; quae iis negatis est obscura, vel confusa"

substantial forms and to propose a fully mechanical account of phenomena in connection with experience. But, contrary to Descartes, this distinction does not come from the metaphysical definition of extended substance as indefinitely divisible. Insensible parts are first of all operative divisions of matter for medical or physical purposes.[98]

In the same way, instead of basing the principle of conservation on the sole nature of God as did Descartes,[99] he prefers to show its likelihood in its agreement with experience, in particular when applied to the conservation of the quantity of movement which is transferred from one moving body to another:

> And a movable body, once put in motion, moves perpetually, until it has communicated its motion to another body. And this *appears clearly* enough with balls which push one another; for, provided that one ball is pushed against another, if the former drives the latter forward, the former stops or moves more slowly; but if it passes by the other without touching it, it continues to move quickly.[100]

The communication of the spirits' motion to the parts of bodies in animals is an additional proof of the communicability of motion.[101]

Therefore, it clearly appears that, in Regius' natural philosophy, experience first plays a role at the level of the principles of physics themselves to constitute the

("The insensible [parts] are those that, escaping the senses because of their smallness or minuteness, are observed only by the intellect in all natural things. Such are, for example, the branch-like particles which constitute oil and the oblong and flexible particles which constitute water, which we will address in what follows./These [parts] are manifestly gathered from the subtlety, thickness, sharpness, smoothness, fluidity, oily or watery quality, saltiness and other innumerable qualities of bodies, which are to be explained in what follows. For once we have *assumed* these parts, we can give a clear and distinct *explanation* of those qualities, whereas the explanation is obscure or confused if we deny them").

[98] Regius displays a certain realism in his corpuscular theory which can recall Bacon's preoccupation to consider the real particles actually present in bodies and not fictitious particles. The parallel is established by Paolo Farina: see Farina 1977, 140.

[99] Regius nevertheless refers to matter as immutable in itself because it was created by God. Regius 1646, 7: "Ut materia universi, à Deo creata, in eo statu, in quo est, ex lege immutatibilitatis naturae, perpetuò manet; ita motus, in creatione variis materiae universae partibus certâ quantitate inditus, perseverat, ex eadem lege, in eodem quantitatis gradu" ("As the matter of the universe that was created by God remains perpetually in the state in which it is, according to the law of the immutability of nature, so the motion that was introduced in a certain quantity in the various parts of the matter of the universe at the time of Creation persists, according to the same law, in the same degree of quantity"). Regius then applies this general principle to the particular case of motion. Regius 1646, 7–8: "Et ut nullum corpus, nisi per accessum vel decessum materiae antea existentis, augetur vel imminuitur; ita nullum mobile, nisi per accessum vel decessum motûs antea existentis, magis vel minùs moveri incipit vel desinit" ("And just as no body increases or diminishes, except through the addition or substraction of a pre-existing matter, no movable body begins or ceases to move, more or less, except through the addition or substraction of a pre-existing motion").

[100] Regius 1646, 8 (my emphasis): "Et mobile semel motum, perpetuò moveri, donec motum suum alii corpori communicaverit. Atque hoc in globulis se mutuò prodrudentibus satis *clarè apparet*: dum enim unus in alterum impellitur, si ipsum propellat, iste sistitur, vel tardiùs movetur; sin ipsum intactum praetereat, pergit celeriter moveri."

[101] See Regius 1646, 8.

object of this physics (i.e., extended matter) and the more general conceptual tools such as corpuscularianism and the laws of motion. Experience works as a justificatory tool meant to replace a rigorous metaphysical deduction.

7.3 Experience in the *Physica Particularis*

Now one could ask: how such a philosopher as Regius could have been interested in Descartes' natural philosophy if his theory of knowledge and the role of experience therein were so different from Descartes'? Actually, if Descartes himself had not been so concerned with experience in the constitution of physics, maybe the *Essais* would not have been so attractive to Regius. In the *Discours de la méthode* and in the *Principia philosophiae*, Descartes himself had stressed the importance of performing experiments to complete the whole of natural philosophy. In the Lettre-préface to the 1647 *Principes de la philosophie*, he writes:

> I am also very well aware that many centuries may pass before all the truths that can be deduced from these principles are actually so deduced. For the majority of truths remaining to be discovered depend on various particular observations *which we never happen on by chance* but which must be sought out with care and expense by very intelligent people.[102]

Descartes therefore encourages those intelligent people to "continue in the search for these truths."[103] In the *Principia philosophiae*, Descartes had only published a part of his natural philosophy; a fifth part on plants and animals and a sixth part on man were missing, because of the need for more experiments to give a complete overview of these domains.[104] But he had certainly discussed several of these topics with Regius, and even maybe communicated to him some unpublished explanations on certain phenomena. Certainly, before 1647, Descartes must have considered Regius as one of those intelligent people who could enrich his philosophy by methodically inquiring into nature by experiment. But Descartes does not consider experiments as what one can find by chance when "looking around"; rather, experiments are supposed to serve specific purposes in the search for truth in natural philosophy. They must therefore be designed according to already conceived theories and hypotheses.

Experience and experiments play a central role in Descartes' natural philosophy, in particular in connection with his corpuscular physics. Indeed it is required that the mechanical properties can explain phenomena (i.e., be able to produce them), in agreement with the experience we have thereof, for these properties to be first considered as acceptable or possible, even if they are not accessible to us by the senses. The first step goes from the observable phenomena to the corpuscles'

---

[102] AT IXb 20, CSM I 189.

[103] AT IXb 20, CSM I 190.

[104] On these missing fifth and sixth parts of the *Principles*, see Gaukroger 2002, 180–246.

properties because, otherwise, we would have no reason to attribute one property rather than any other to those minute particles of matter. In the *Discours de la méthode*, Descartes sets out the general framework of such a procedure (which applies to all the mechanical explanations of phenomena):

> Then when I sought to descend to more particular things, I encountered such a variety that I did not think the human mind could possibly distinguish the forms or species of bodies that are on the earth from an infinity of others that might be there if it had been God's will to put them there. Consequently I thought the only way of making these bodies useful to us was to progress to the causes by way of the effects and to make use of many special observations. And now, reviewing in my mind all the objects that have ever been present to my senses, I venture to say that I have never noticed anything in them which I could not explain quite easily by the principles I had discovered. But I must also admit that the power of nature is so ample and so vast, and these principles so simple and so general, that I notice hardly any particular effect of which I do not know at once that it can be deduced from the principles in many different ways; and my greatest difficult is usually to discover in which of these ways it depends on them. I know no other means to discover this than by seeking further observations whose outcomes vary according to which of these ways provide the correct explanation.[105]

In a letter to Jean-Baptiste Morin, Descartes gives more detail about the nature of the relations he conceives to be between the causes of phenomena to be discovered and the effects these causes produce: "I should add also that there is nothing circular in proving a cause by several effects which are independently known, and then proving certain *other* effects from this cause."[106] We understand then that the function of experience in Cartesian natural philosophy is twofold: it is first required to collect natural facts which are what natural philosophy is meant to explain through mechanical causes, and then experience is appealed to in order to conceive other effects which can be produced by this very set of mechanical properties. But at the second level, the phenomena are not the same as those at the first level, since they are different phenomena from the ones from which the possible mechanical structure to be tested was elaborated. This mechanical structure, which was first conceived as a possible cause for the phenomena from which it was designed, will reach a higher level of certainty insofar as it enables the natural philosopher to explain phenomena different from the first ones to be explained.

Can we find anything similar in Regius? In his 1646 *Fundamenta physices*, Regius does not only confine himself to physiology, but publishes for the first time in one and the same book explanations ranging from the principles of natural things to meteors, fossils, animals and men (as living and thinking beings). De Vrijer thinks that Regius' *Fundamenta physices* heavily rely on Descartes' *Essais* and *Principia philosophiae*.[107] But it does not mean that Regius' natural philosophy, as far as the role of experience in its constitution is concerned, is only a pale copy of Descartes' physics.

---

[105] *Discours de la méthode* VI, AT VI 64–65, CSM I 144.

[106] Letter to Morin [July 13, 1638], AT II 198, CSMK 106 (my emphasis).

[107] De Vrijer 1917, 168.

Indeed, at the end of his *Philosophia naturalis*, Regius makes explicit his own methodology on which he had relied since the *Fundamenta physices* and in which experience is granted a specific role. Regius considers that at first the mind should give us an idea of the cause of a given phenomenon. But then we must "look around" (i.e., search by experience) for another or other possible causes more suitable to replace the one found by the mind. But in case experience can give us no cause as powerful and as suitable as the one established by the mind, we must stick to the first cause.[108] It can therefore appear that Regius is not as empiricist as he might have appeared at first sight. But first, the natural philosopher must stick to the cause found by the mind only by default, that is when he can find no other satisfactory one by experience. And, second, we should not forget that, for Regius, all the knowledge we can access through the mind comes, ultimately, from the senses.[109] So, in any case, the mind can rely on nothing else but experience to suggest what the cause of a phenomenon can be. Moreover, this conception of the role of experience and of the mind to find a natural cause implies that every identified cause is provisional since it can always be replaced by a better one that experience will enable us to discover.[110]

But at the same time, the role of experience in the constitution of his natural philosophy is not unproblematic for Regius. Indeed, everything that is sensible, for Regius, ultimately relies, as is the case in Descartes' natural philosophy, on invisible parts of matter.[111] As a consequence, Regius has to make a specific place for

[108] Regius 1654, 441, 1661, 522 (my emphasis): "Cum enim problema aliquod in Physicis proponitur solvendum, primo excogitanda est causa intelligibilis, qua effectum, in problemate proposito observatum, commode & intelligibiliter peragi possit. Deinde *circumspiciendum*, an non alia commodior vel aeque commoda queat inveniri. Quae si inveniatur, commodior priori est praeferenda; aequalis vero ipsi aequiparanda. Sin alia commodior vel aeque commoda excogitari nequeat, solutioni inventae tamdiu acquiescendum, donec melior vel aequalis alia fuerit inventa" ("Indeed when one proposes to solve a problem in physics, one first has to find an intelligible cause by which the effect observed in the proposed problem could be suitably and intelligibly produced. Then one has to look around to see whether it is not possible to find another more suitable or equally suitable cause. If such a cause is to be found, the more suitable has to be given preference over the first one; but an equally suitable [cause] has to be given equal standing with this first cause. If, on the contrary, it is not possible to find another more suitable or equally suitable cause, one has to be satisfied with the discovered solution until another better or equal one is discovered"). This is not to be found in the *Fundamenta physices*.

[109] Regius 1646, 251 summed up as follows in the table of contents: "Notiones nobis insculptae, ex rerum observationibus sunt ortae" ("The notions which are inscribed in us originate from the observations of things").

[110] Regius 1646, 287, 1654, 403, 1661, 476: "An autem satis clare & distincte rem perceperimus & examinaverimus, mens secundum apparentiam tantum dijudicat. Illique tamdiu acquiescendum, donec contrarium vel aliud, per experientiam, vel alia ratione, fuerit probatum" ("The mind judges only according to the appearance whether we have perceived and examined the thing clearly and distinctly enough. And we have to be satisfied with that [judgment] until the contrary or something else is proven, either through experience or by another reason").

[111] See Regius 1641, 199: "*Pars sensibilis* est, quae ex multis insensibilibus composita sub sensum cadit" ("A sensible part is one composed of numerous insensible [parts] and which falls under the sense").

imagination in his theory of knowledge.[112] As in Descartes, it is thanks to imagination that we can for example conceive the spaces between the parts of the most subtle bodies[113] or represent the insensible parts of matter.[114] Heat is also conceived as a motion of the parts of matter thanks to the observation of boiling water which is then transposed by imagination, at the subvisible level, to warm water.[115] But, more specific to Regius is the fact that the epistemic function of imagination is guaranteed by its continuity with the senses, which adequately repeats the ontological continuity between the sensible and the insensible in bodies. This continuity is first a representational one: imagination and the senses represent objects in a way that does not make them distinguishable.[116] Even if Regius seems to consider that there is an

---

[112] This is not to say that Descartes does not also have a place for imagination in his theory of knowledge and in his physics. See Lüthy 2006; Zittel 2009; Bellis 2010, 588–668. But I want to suggest that Regius' empiricism forces him to make explicit the relation between imagination and sensation in a way that was not so prominent in Descartes. In Regius, there is a specific insistence on the continuity between both faculties of the mind.

[113] Regius 1654, 4, 1661, 4: "Et quamvis in tenuissimis & subtilissimis corporibus, intervalla partium, corporaque subtilissima intercurrentia & egredientia sensibus non percipiantur, illa tamen, mentis imaginantis & judicantis aciem minimè effugiunt" ("And although, in very thin and subtle bodies, one does not perceive by the senses the space between the parts and the very subtle bodies that travel through them and go out of them, they do not escape at all the keenness of mind which imagines and judges"). The *Fundamenta physices* do not mention imagination but only, in a more general way, "the keenness of mind" (Regius 1646, 3: "mentis aciem").

[114] Regius 1654, 6, 1661, 6: "Insensibiles partes sunt, quae, propter exiguitatem aut parvitatem, sensus fugenties, solo imaginationis & judicii intellectu in omnibus rebus naturalibus observantur" ("The insensible parts are those that, escaping the senses because of their smallness or minuteness, are observed in all natural things only by the perception of imagination and judgment"). Here again, the *Fundamenta physices* are less precise and only mention "solo intellectu" (Regius 1646, 3). But this intellect already enables one to "observe" the insensible parts; it is a proto-imagination. From 1654 onwards, Regius replaces the soul or the intellect by imagination and judgment. Thus, he makes his theory of mental faculties coherent with his epistemology which had already deprived the intellect of its constituent role in knowledge (because of his rejection of innate ideas).

[115] Regius 1646, 30, 1654, 46, 1661, 51: "calor accidentarius aquae (qui est varia & vehemens ejus particularum agitatio, ut in aquâ fervidâ ad oculum apparet, & in tepidâ facillimè imaginatione intelligitur) motum suum, seu calorem vehementiorem, vicino aëri aliisque propinquis corporibus communicet…" ("Heat being accidental to water (which is a varied and violent agitation of its particles, as it appears to the eye in boiling water and as it is understood very easily by the imagination in tepid water) communicates its motion, or the more violent heat, to the neighboring air and to the other proximate bodies…").

[116] Regius 1646, 249: "hinc sequitur per naturam dubium esse, vera an falsa, seu imaginaria, mente percipiamus & dijudicemus" ("From this it follows that, by nature, it is doubtful whether we perceive true, false or imaginary things and distinguish between them with the mind"); Regius 1654, 347: "Hinc videtur manifestum, quod mens nostra aequè evidenter ab imaginariis, atque à veris, in perceptione affici possit; quodque ideo, non moralem sive probabilem verisimilemque, sed exquisitam, accuratam, & indubitabilem veritatis cognitionem quaerenti, per naturam mentis jam propositam, dubium & incertum sit, an ulla vera corpora, an verò imaginaria tantum phantasmata à nobis percipiantur…" ("From this it seems manifest that our mind can be affected with equal evidence by the perception of both imaginary and true things; and that for that reason, for the one who seeks well-scrutinized, accurate and indubitable knowledge of truth, and not moral,

appropriate way ("*rectè*") to perceive things that ensures us that these things are real and not imaginary,[117] Regius gives no more indication as to the nature of this appropriateness. The representational continuity is then ultimately rooted in a similar physiological process.[118]

Within a very different methodological framework from that of Descartes, experience is conceived as the ultimate source of information on nature, with which to find the cause of a phenomenon. A review of the French translation of the *Philosophia naturalis* (1686) indicates: "There are a lot of experiments [in this book] which have not been dealt with by Mr. Rohault or clarified by figures as they are here."[119] But what are those experiments and what does Regius manage to establish with them in his natural philosophy? We can distinguish several categories of experiences in Regius.

The first type of reference to experience in Regius would be experiments, which confirm a theory. Some observations serve, for example, to illustrate the circularity of motion by the permutation of one body to another in a circular pipe containing balls[120] or, on the contrary, the impossibility of motion when such a permutation is not possible (with the examples of a liquid which cannot fall out of a pump or of a stone stuck to a piece of leather by saliva).[121] Such observations accessible to the senses indicate the probability of similar mechanisms at the subvisible level. But they also make these subvisible processes easier to represent in imagination. These observational reports are therefore coherent with Regius' theory of knowledge and methodology: on one hand they testify to the exclusivity of experience as source of knowledge; on the other hand, they provide the sensory content which imagination and judgment can elaborate upon, in order to develop a representation of the natural causes which remain hidden to the senses. Therefore the continuity between observation and imagination is guaranteed in natural philosophy.

Conversely, experience can play an argumentative role by providing facts apparently entering in contradiction with a theory. As a possible objection, it needs to be explained according to the principles of natural philosophy. This is for example the case of the motion of the Earth.[122] Regius has to refute the arguments opposed to

probable or likely knowledge, it is doubtful and uncertain, from the nature of the mind that we have already proposed, whether we perceive any true bodies or only imaginary phantasms…"); Regius 1661, 412. On the contrary, after a first skeptical step, Descartes will strive to establish a demarcation, albeit not a clear-cut one, between the two: see *Meditationes*, AT VII 75, CSM II 52.

[117] Regius 1646, 249: "Unde patet, ea quae rectè percipimus, esse res veras, & non imaginarias…" ("From this it is obvious that the things that we perceive correctly are real things, and not imaginary ones…"); Regius 1654, 348, 1661, 413.

[118] See Regius 1654, 347, 1661, 412.

[119] *Nouvelles de la République des Lettres*, October 1686, quoted in Mouy 1934, 95 ("Il y a beaucoup d'expériences [dans ce livre] qui n'ont pas été touchées par M. Rohault, ou éclaircies par des figures comme elles le sont ici").

[120] Regius 1646, 25, 1654, 20, 1661, 23.

[121] Regius 1646, 25–26, 1654, 20–21, 1661, 23–24.

[122] Regius 1646, 67–68, 1654, 112–113, 1661, 125–126.

the motion of the Earth: if the Earth were to move, a mass projected in air or falling along a mast would fall at the West in comparison with its point of departure, since the Earth went toward the East during its fall. But, this is not what happens in experience.[123] Regius then formulates a Galilean type of answer by claiming that projectiles or falling bodies take part in the same motion as the Earth.[124] Regius also provides several examples of experiments with liquids rising from vases through pipes (including the Hero fountain) and, for each case, intends to explain the mechanical causes of the device without relying on the traditional argument of the *horror vacui*. The inexistence of the vacuum is thus confirmed by showing how circular motion is possible without any void in various determinate experimental devices.[125]

In a more constructive fashion, experience is also present, in Regius' works, as a suggestive source with which to explain analogical phenomena. For example, he considers generation to be a kind of crystallization of two mixed substances. He compares it to the formation of salt, to the concretions appearing at the end of the glassmakers' stick or formed by water springs.[126] These inorganic phenomena serve as a structural model for organic phenomena. We know that this is allowed by his conception of imagination which can gather similar sense impressions.

Finally, another category of experiences which is quite prominent in Regius' works are facts—neutral from a theoretical point of view, but which are puzzling in themselves and therefore require an explanation. Regius collects such facts and tries to give a physical explanation thereof: for example, he notices that the snow put in a bottle surrounded by salted snow remains frozen in every season and explains this by the action of subtle matter.[127] In the 1661 edition of the *Philosophia naturalis*, Regius reflects on the phenomenon of capillarity and gives a new example thereof: the water of a basin going up in a capillary tube. His explanation does not rely on subtle matter but on the viscosity of the liquid because of the presence of hooked particles.[128] Faced with a new fact, Regius gives a new explanation but without exiting the Cartesian framework. But the most striking feature can be seen at the end of the 1661 *Philosophia naturalis*, Book V, chapter XIII: "De Paralipomenis quibusdam," that is, some supplements to the previous Books, which could not be

[123] These arguments were commonly discussed, for example by Giordano Bruno, Tycho Brahe or Clavius.

[124] Regius does not seem to have performed the experiments himself and tackles the problem from a theoretical point of view, as Galileo had done, when he presented these arguments against and in favor of the movement of the Earth in the Second Day of his *Dialogo sopra i due massimi sistemi del mondo tolemaico e copernicano* (Galileo 1897, 255, 283–284, 288). Gassendi was the first to perform the experiment in 1641 in Provence. He reported it in *De motu impresso a motore translato epistolae duae* (Gassendi 1642). While he could have relied on this experimental support, Regius does not mention this text.

[125] See Regius 1646, 35, 1654, 53, 1661, 59.

[126] See Regius 1646, 209–210, 1654, 295–296, 1661, 349.

[127] Regius 1654, 147, 1661, 168–169. This experience is not mentioned in the *Fundamenta physices*.

[128] Regius 1661, 513–514.

included in the body of the work. There, Regius gives a long list of various, heterogeneous facts which are explained one after the other: glass pans being broken in different ways, contagion, the effects of water in vases, rotten apples making surrounding apples rotten as well, spiders managing to make their web between distant trees, etc. This clearly reveals that these experiences played no role in the elaboration or the exposition of the theory but are only collected by chance.

But this attitude indicates that Regius' attention is directed to the specificity of determinate phenomena, which means he does not just rely on commonsense experience. In fact, he is also aware of the puzzling phenomena that could challenge his theories. In that sense, he may have a fact-gathering attitude but experimental data are never treated themselves as the alpha and the omega of the knowledge of nature, but always call for a causal material explanation. Regius remains open-minded to the novelty of experimental data simply because he does not only turn towards experience when he is in need of a confirmation-refutation procedure of an explanatory theory. For Regius, it is not only theory that questions experience, but experience also questions theory, at least insofar as it forces to expand it in a way that it can integrate the experimental fact.

Regius' philosophy offers an original version of what Cartesian philosophy can be turned into once it is deprived of some of its most important metaphysical components. It remains Cartesian because it adopts most of Descartes' explanations and principles in natural philosophy. But the Dutch philosopher departs from Descartes by formulating a very strong form of empiricism: first, man does not need any preconceived innate ideas to have sensible ideas and second, the human mind would be empty without sense impressions (which means that all our knowledge ultimately comes from sensation). As a consequence, the entirety of the mind's activity reduces to its operations on sensations. Once the intellect is emptied from any content, it becomes more or less useless in Regius' psychology and is eventually replaced by imagination and judgment, which have an increasingly important role in his epistemology.

Regius also represents an interesting case for the relation between materialism and empiricism in the early modern period. Indeed, he can be viewed as a materialist insofar as the mind depends on the body for all its operations and because all our ideas are caused by corporeal motions. Therefore, our ideas, including our sensible ideas, are not reduced to pure appearances, pure *phantasmata*. However, Regius certainly weakens the link between materialism and empiricism that should have enabled him to dissociate himself from skepticism. Indeed, even if all our ideas are produced by corporeal mechanisms, these mechanisms can take place in our body independently from external bodies. Hence the difficulty to discriminate between sensations caused by external bodies and imaginary representations produced by my own body. Experience cannot prove the existence of the external world or, in other words, Regius' empiricism cannot provide a foundation for his materialism. Only the appeal to Biblical Revelation can guarantee the existence of the external world and prevent man from being imprisoned in his mental representations without any access to the world. Even if the very sketchy mention of the notion of resemblance seems to maintain a limited continuity between our ideas and the world,

Regius, by getting rid of the Scholastic theory of sense perception and by refusing to endorse Descartes' innate ideas, has kept a narrow gap open between materialism and empiricism. From that point of view, he could be added to the historical picture drawn by Gianni Paganini, of the fate of skepticism in the early modern period, lying in between Hobbes' phenomenalism and materialism and Hume's empiricism.[129]

Acknowledgments Research for this article was made possible by a Rubicon grant (446-10-031) funded by NWO (the Netherlands Organisation for Scientific Research) and the Marie Curie Cofund Action. I would like to thank Raphaël Chappé, Mihnea Dobre, Delphine Kolesnik-Antoine, and Tammy Nyden for their useful suggestions on earlier versions of this article. I would also like to thank Gideon Manning and Lucian Petrescu for their help with the translation into English of the Latin quotations, and Charles Wolfe for his help with the translation into English of the French quotations, as well as for his emendations to my English text. All translations are the author's except where otherwise noted.

References

Bellis, Delphine. 2010. *Le visible et l'invisible dans la pensée cartésienne: figuration, imagination et vision dans la philosophie naturelle de René Descartes*. PhD dissertation, Université Paris-Sorbonne, Paris; Radboud Universiteit Nijmegen, Nijmegen.

Bitbol-Hespériès, Annie. 1993. Descartes et Regius: leur pensée médicale. In *Descartes et Regius. Autour de l'Explication de l'esprit humain*, ed. Theo Verbeek, 47–68. Amsterdam/Atlanta: Rodopi.

Bos, Erik-Jan (ed.). 2002. *The correspondence between Descartes and Henricus Regius*. PhD dissertation, Quaestiones Infinitae, Utrecht.

Bos, Erik-Jan. 2013. Henricus Regius et les limites de la philosophie cartésienne. In *Qu'est-ce qu'être cartésien?* ed. Delphine Kolesnik-Antoine, 53–68. Lyon: ENS Editions.

Bouillier, Francisque. 1868. *Histoire de la philosophie cartésienne*, 3rd ed. Paris: Delagrave.

Caps, Géraldine. 2010. *Les médecins cartésiens: héritage et diffusion de la représentation mécaniste du corps humain (1646–1696)*. Hildesheim: Georg Olms.

Clarke, Desmond. 2010. The physics and metaphysics of the mind: Descartes and Regius. In *Mind, method, and morality. Essays in honour of Anthony Kenny*, ed. J. Cottingham and P. Hacker, 187–207. Oxford: Oxford University Press.

Damiron, Jean-Philibert. 1846. *Essai sur l'histoire de la philosophie en France, au XVIIe siècle*. Paris: Hachette.

de Buzon, Frédéric, and Vincent Carraud. 1994. *Descartes et les Principia II*. Paris: Presses Universitaires de France.

Dechange, Klaus. 1966. *Die frühe Naturphilosophie des Henricus Regius (Utrecht 1641)*. PhD dissertation, Universität Münster.

Des Chene, Dennis. 1996. *Physiologia. Natural philosophy in late Aristotelian and Cartesian thought*. Ithaca/London: Cornell University Press.

Descartes, René. 1996. *Œuvres de Descartes*, 2nd ed, ed. Ch. Adam and P. Tannery. Paris: Vrin.

Descartes, René. 1997. *Discours de la méthode*, suivi de *La Dioptrique*, 2nd ed, ed. Frédéric de Buzon. Paris: Gallimard.

Descartes, René. 1984–1991. *The philosophical writings*, 3 vols. Trans. J. Cottingham, R. Stoothoff, D. Murdoch, and A. Kenny (for the third volume). Cambridge: Cambridge University Press.

[129] See Paganini 2008 (in particular Chap. IV, 171–227).

Dibon, Paul. 1990. Sur la réception de l'œuvre de F. Bacon en Hollande dans la première moitié du XVIIe siècle. In *Regards sur la Hollande du Siècle d'Or*, 191–220. Naples: Vivarium

Dobre, Mihnea. 2010. The vanishing nature of body in Descartes's natural philosophy. In *Vanishing matter and the laws of motion. Descartes and beyond*, ed. Dana Jalobeanu and Peter Anstey, 11–30. London: Routledge.

Farina, Paolo. 1975. Sulla formazione scientifica di Henricus Regius: Santorio Santorio e il *De statica medicina*. *Rivista critica di storia della filosofia* XXX(IV): 363–399.

Farina, Paolo. 1977. Il corpuscularismo di Henricus Regius: Materialismo e medicina in un Cartesiano olandese del Seicento. In *Ricerche sull'atomismo del Seicento*, ed. Ugo Baldini, Giancarlo Zanier, Paolo Farina, and Francesco Trevisani, 119–178. Florence: La Nuova Italia Editrice.

Fichant, Michel. 1998. La géométrisation du regard. Réflexions sur la *Dioptrique* de Descartes. In *Science et métaphysique dans Descartes et Leibniz*, 29–57. Paris: Presses Universitaires de France.

Galileo. 1897. Dialogo sopra i due massimi sistemi del mondo tolemaico e copernicano. In *Le opere di Galileo Galilei. Edizione nazionale*, VII, Florence: G. Barbèra.

Gariepy, Thomas Peter. 1990. *Mechanism without metaphysics. Henricus Regius and the establishment of Cartesian medicine*. PhD dissertation, Yale University, New Haven.

Gassendi, Pierre. 1642. *De motu impresso a motore translato epistolae duae*. Paris: Louis de Heuqueville.

Gaukroger, Stephen. 2002. *Descartes' system of natural philosophy*. Cambridge: Cambridge University Press.

Kolesnik-Antoine, Delphine. 2010. *La physique de l'homme chez Regius; suivi de En quoi le traité de l'homme de Descartes peut-il être lu comme un texte matérialiste?* Québec: Presses de l'Université Laval.

Kolesnik-Antoine, Delphine. 2012. La structure passionnelle de l'âme malebranchiste: entre Descartes et Regius? In *Emotional minds. The passions and the limits of pure inquiry in early modern philosophy*, ed. Sabrina Ebbersmeyer, 51–68. Berlin: De Gruyter Verlag.

Kolesnik-Antoine, Delphine. forthcoming a. L'âme et le corps dans la philosophie naturelle de Regius: repenser en médecin l'héritage cartésien. In *Macchina e vita. Modelli epistemologici e implicazioni morali (sc. XVII–XIX)*, ed. Nunzio Alloca. Münster: Nodus Publikationen.

Kolesnik-Antoine, Delphine. forthcoming b. Peut-on à la fois être cartésien et sceptique? Le cas de Regius. In *Polémiques autour du scepticisme*, ed. Elodie Argaud and Nawal El Yadari. Paris: Champion.

Kolesnik-Antoine, Delphine. Reason and experiment in Regius and La Forge (unpublished manuscript 2013).

Lüthy, Christoph. 2006. Where logical necessity becomes visual persuasion: Descartes's clear and distinct illustrations. In *Transmitting knowledge: Words, images, and instruments in early modern Europe*, ed. Sachiko Kusukawa and Ian Maclean, 97–133. Oxford: Oxford University Press.

Mouy, Paul. 1934. *Le développement de la physique cartésienne. 1646–1712*. Paris: Vrin.

Paganini, Gianni. 2008. *Skepsis. Le débat des modernes sur le scepticisme*. Paris: Vrin.

Regius, Henricus. 1641. *Henrici Regii Medicinae Doctoris et Professoris Physiologia, sive Cognitio Sanitatis; Tribus disputationibus in Academiâ Ultrajectinâ publicè proposita*. Utrecht: Ex Officinâ AEgidii Roman, Academiae Typographi. (In *The Correspondence between Descartes and Henricus Regius*, ed. Erik-Jan Bos, 2002, 195–248. Utrecht: Quaestiones Infinitae.)

Regius, Henricus. 1646. *Fundamenta physices*. Amsterdam: Elsevier.

Regius, Henricus. 1654. *Philosophia naturalis*. Amsterdam: Elsevier.

Regius, Henricus. 1661. *Philosophia naturalis; in qua tota rerum universitas, per clara et facilia Principia, explanantur*. Amsterdam: Elsevier.

Rodis-Lewis, Geneviève. 1993. Problèmes discutés entre Descartes et Regius: l'âme et le corps. In *Descartes et Regius. Autour de l'Explication de l'esprit humain*, ed. Theo Verbeek, 35–46. Amsterdam/Atlanta: Rodopi.

Rothschuh, Karl. 1968. Henricus Regius und Descartes. Neue Einblicke in die frühe Physiologie (1640–1641) des Regius. *Archives internationales d'histoire des sciences* 21(82–83): 39–66.
Spruit, Leen. 1994–1995. *Species intelligibilis: From perception to knowledge*, 2 vols. Leiden: Brill.
Tachau, Katherine. 1988. *Vision and certitude in the age of Ockham*. Leiden: Brill.
Verbeek, Theo (ed.). 1988. *La Querelle d'Utrecht*. Paris: Les Impressions nouvelles.
Verbeek, Theo. 1992. *Descartes and the Dutch. Early reactions to Cartesian philosophy 1637–1650*. Carbondale/Edwardsville: Southern Illinois University Press.
Verbeek, Theo (ed.). 1993a. *Descartes et Regius. Autour de l'Explication de l'esprit humain*. Amsterdam/Atlanta: Rodopi.
Verbeek, Theo. 1993b. Le contexte historique des *Notae in programma quoddam*. In *Descartes et Regius. Autour de l'Explication de l'esprit humain*, ed. Theo Verbeek, 1–33. Amsterdam/Atlanta: Rodopi.
Verbeek, Theo. 1994. Regius's *Fundamenta Physices*. *Journal of the History of Ideas* 55(4): 533–551.
Verbeek, Theo. 2000. The invention of nature. Descartes and Regius. In *Descartes' natural philosophy*, ed. Stephen Gaukroger, John Schuster, and John Sutton, 149–167. London/New York: Routledge.
Vrijer, Marinus Johannes Antonie de. 1917. *Henricus Regius. Een "cartesiaansch" hoogleraar aan de Utrechtsche Hoogeschool*. PhD dissertation, University of Groningen, Groningen. The Hague: Martinus Nijhoff.
Wilson, Catherine. 2000. Descartes and the corporeal mind. Some implications of the Regius affair. In *Descartes' natural philosophy*, ed. Stephen Gaukroger, John Schuster, and John Sutton, 659–679. London/New York: Routledge.
Zittel, Claus. 2009. *Theatrum Philosophicum. Descartes und die Rolle ästhetischer Formen in der Wissenschaft*. Berlin: Akademie Verlag.

Chapter 8
Robert Desgabets on the Physics and Metaphysics of Blood Transfusion

Patricia Easton

Abstract Robert Desgabets (1610–1678), an early defender and teacher of the Cartesian philosophy, gives *expérience* a central role in metaphysics, physics, and theology. Desgabets' *Discourse on the Communication or Transfusion of Blood* (1668) delivered 10 years earlier at a Montmor Conference describes his scientific work on blood transfusion. I show how Desgabets' Cartesian conception of the body and its mechanistic functioning had important consequences for the Cartesian metaphysics and physics as well as scientific method. For Desgabets, experiments on blood transfusion were as much demonstrations of the truth of the Cartesian metaphysics as they were demonstrations of empirically known effects. The role of experiment was to provide demonstrations of the first principles or truths of physics and theology by connecting them to the way God actually made the world. I conclude that Desgabets' treatment of *expérience* plays an essential role in the content and form of physical knowledge.

Some 40 years or so after the publication of William Harvey's work on blood circulation, *De Motu Cordis*,[1] came the first allegedly successful transfusion of blood into a human subject by the French physician Jean Denis,[2] followed shortly thereafter in 1669 by a prohibition of the procedure by order of the French Parliament.[3] In England, physician Christopher Wren experimented with infusion

[1] Harvey 1628.

[2] Denis 1668, 69–72.

[3] A decision by French Justice Defita in April 1668 against the practice without the approval of the Parisian Faculty of Medicine was later upheld by the Paris Parliament in late 1669. See "Extrait des registres du greffe criminal du Chastelet de Paris, du mardy 17 avril 1668," cited in Tucker 2011, 259.

P. Easton (✉)
Department of Philosophy, Claremont Graduate University, Claremont, CA, USA
e-mail: Patricia.Easton@cgu.edu

M. Dobre and T. Nyden (eds.), *Cartesian Empiricisms*, Studies in History and Philosophy of Science 31, DOI 10.1007/978-94-007-7690-6_8,
© Springer Science+Business Media Dordrecht 2013

or injection from animal to animal, and his colleague Richard Lower, using a method of transfusion, reportedly had success transfusing from dog to dog, sometime around 1665.[4] Debates over the priority of invention notwithstanding,[5] the focus of this examination is the role of an unlikely contributor to the developments of experimentation on blood in France, an early defender and teacher of the Cartesian philosophy, Robert Desgabets (1610–1678).

Desgabets' *Discourse on the communication or transfusion of blood* contains a defence and description of the procedure of blood transfusion.[6] This short work was based on a lecture Desgabets delivered at one of the regular meetings held at the Paris residence of M. de Montmor in July 1658.[7] His interest in blood transfusion can be traced back to 1650 at Saint Arnold de Metz, where he taught philosophy.[8] While there is a lack of evidence concerning whether Desgabets himself performed experiments on blood transfusion, it is evident that he provided a framework to conceptualize and justify the procedure, and he designed an apparatus that he reports to have had made by a M. Picot in 1660.[9] In this paper I provide a brief background for Desgabets' approach to scientific questions (Sect. 8.1) and I argue that Desgabets' development and revision of the Cartesian metaphysics of matter provided the tools for reform that Desgabets envisioned in science (Sect. 8.2). In particular, Desgabets argues that duration belongs to body *per se* and only through the union of the mind with body can thought become differentiated, successive, and discursive. Thus, the method of science itself depends on body and the senses. I then turn to how Desgabets develops his views on the role of *expérience* based on Descartes' remarks in Part VI of the *Discourse on Method* and Article 46 of Part III of the *Principles of Philosophy*, that the role of experiment was to provide demonstrations of the first principles or truths of physics by connecting them to the way God actually made the world (Sect. 8.3). I conclude my examination of Desgabets on blood transfusion, by explaining how Desgabets saw the essential role of *expérience*[10] in the discovery

[4] *Philosophical Transactions* 1665–1666, 357.

[5] Géneviève Rodis-Lewis focuses on the question of priority of invention between the French and English and argues that the work done was developed independently and concurrently. See her 1974, 41–64. Harcourt Brown recounts the English-French controversy and history in his 1948, 15–29.

[6] Desgabets 1668, 3–6.

[7] Victor Cousin dates the Montmor conferences from July 1658 through July 1662. The weekly conferences were assembled to "clarify and defend the philosophy of Descartes." See Cousin 1845, 107.

[8] Desgabets MS. Epinal 64 (143), 409. A collection of Desgabets' unpublished philosophical writings. MS Epinal 43 (142) is a collection of Desgabets' unpublished theological writings. Both manuscripts are now available on-line through the Bibliothèque Municipal d'Epinal website. Epinal 64 (409) is cited in Rodis-Lewis 1974, 54n34; also cited in Victor Cousin 1845, 227. Cousin quotes the letter, "j'ai entirèment abandoné cette opération; qui est tombée en de meilleures mains que les miennes" ("I have entirely abandoned this procedure which has fallen into better hands").

[9] I am sympathetic with the view proposed by Streiff that Desgabets' main bibliographer, Dom Catelinot, embellished Desgabets' role in the experimentation and subsequent biographers repeated the mistake. See Streiff undated.

[10] I use the French term, *expérience*, hereafter to reflect the French usage which connotes both the broader notion of sensory experience and the narrower notion of experiment.

of the truth of things, both in their particular existence and in their essence. This rather unique brand of "Cartesian Empiricism" provided what Desgabets' called the "reunion of geometry and mechanics" which fostered some daring, if not successful, experiments on the human body.

8.1 The Background to Cartesian Science

Desgabets was an early defender and teacher of the Cartesian philosophy at St. Maur, in the region of Lorraine, France. He belonged to the Benedictine congregations of Saint Vanne and Saint Hydulphe, known for their spirit of reform and love for science.[11] Although Desgabets is little known today,[12] he played a significant part in the Cartesian world from 1660 to 1678, especially in the Eucharist Affair, which began shortly after Descartes' death.[13] According to Mouy, Desgabets was converted to Cartesianism while reading Descartes' *Discours de la Méthode*, and he believed that Descartes' principles could provide a solid base for physics and theology.[14] Desgabets was closely associated with the Cartesians at Paris and Toulouse, through Clerselier,[15] and his student, Pierre-Sylvain Régis.[16] Although he is perhaps the most original of the Cartesian thinkers, even lauded by Régis, as "one of the greatest metaphysicians of our century,"[17] only one book[18] and two small works[19] were published during his lifetime.[20] His correspondence indicates that he was interested

---

[11] See Levêque 1947, 62.

[12] The scholarly literature on Desgabets has continued to grow over the past 40 years, in particular since the publication of his unpublished philosophical works in 1983: Desgabets 1983. See Allen-Hermanson 2009; Beaude 1974, 1979; Cook 2002, 2005, 2008; Easton 2005; Faye 2005; Lennon 1998; Miller 2008; Rodis-Lewis 1974; Nicolas 2012; Schmaltz 2002b; Watson 1982, 1987.

[13] See Armogathe 1977, 85.

[14] Mouy 1934, 100.

[15] Claude Clerselier (1614–1684) was a lawyer of the *Parlement* of Paris, close friend of Descartes, devoted supporter of the Cartesian philosophy, and primary editor and translator of Descartes' works.

[16] Desgabets MS. Epinal 64 (143), Lettre de dom Paquin. Cited in Armogathe 1977, 85n2.

[17] From a marginal note in Régis 1704, 328.

[18] Desgabets 1675.

[19] Desgabets 1668, 1671.

[20] In addition, the following writings of Desgabets have been published posthumously in French: "Une lettre de D. Robert Desgabets à D. Jean Mabillon sur la question des azymes," in *Oeuvres des Mabillon et de D. Thierri-Ruinart* (1724); selected texts from Desgabets' unpublished manuscripts (MS Epinal 43 and 64) at Epinal, reproduced by Victor Cousin, in his 1845, 99–228; selected texts and some correspondence from manuscripts at Epinal contained in Paul Lemaire 1901, 320–413; Géneviève Rodis-Lewis and Joseph Beaude have published a valuable set of Desgabets' unpublished philosophical works in Desgabets 1983; This annotated collection is compiled from many diverse manuscripts found at Epinal, Chartres and elsewhere in France. It includes his unpublished *Traité de l'indéfectibilité des créatures, Supplément à la philosophie de M. Descartes, Guide de la*

in mechanics before 1644, well before being acquainted with Descartes.[21] In Desgabets' estimation, the only legitimate rival system to Descartes' was the one developed by Pierre Gassendi, but he concluded that the *nouvelles découvertes* weighed decisively in Descartes' favor.[22]

It was in 1654, upon appeal from Clerselier, that Desgabets entered into the Cartesian debate concerning the Eucharist. This debate was initiated by an Augustinian, Viogué,[23] who wrote to Clerselier shortly after Descartes' death concerning the Cartesian explanation of transubstantiation. Given the Cartesian view that the essence of matter is extension, Viogué wanted to know how the real presence of Christ in the Holy Sacrament could be explicated without the aid of accidents. Clerselier's response was counter-attacked by the French physician Pastel, at which point Clerselier related the correspondence to "a very learned Benedictine," Dom Robert Desgabets.[24]

The short story of Desgabets' role in the Eucharist Affair is that he argued that the Scholastic doctrines of real presence and transubstantiation were erroneous and could be cleared up by the reformation that had taken place in philosophy by "its reunion with mathematics."[25] The Scholastic doctrine, as found in Aquinas, has it that the soul of Christ unites with the matter of the bread, annihilating the substance of the bread and creating the matter of the body of Christ in its place. The appearance of the bread and the body of Christ was explained by appeal to accidents. Desgabets urged his fellow clergymen to consider Descartes' account of the nature of material substance against that of the Scholastics. The advantage of Descartes' theory, he argued, is that the mystery of the Eucharist could be explained without appealing to the creation and annihilation *ex nihilo* of substances or the phantom of accidents. Desgabets reasoned as follows: (1) matter *is* extension and is not really distinct from quantity; (2) there are no substantial material forms, just the local dispositions of insensible parts of matter; (3) therefore, after the consecration of the Holy Sacrament we have the *same* quantity of matter we had before it, what has changed is that Christ's soul is now united to the specific local dispositions once belonging to the bread. What has transpired is not the annihilation of substances,

raison naturelle, and *Traité de l'union de l'âme et du corps*. All further references to this collection will be by abbreviated title, abbreviated collection name [OPD], fascicle number, page number (which are continuous from fascicle to fascicle). For example, *Supplément*, OPD 5, 152, refers to *Supplément à la philosophie de Monsieur Descartes*, in: *Oeuvres philosophiques inédites de Dom Robert Desgabets*, fascicle 5, page 152.

[21] See Géneviève Rodis-Lewis, in her introduction, OPD 1, xvi.

[22] *Supplément*, OPD 5, 153–155. J.-R. Armogathe claims that Desgabets' Gassendism has been underestimated, "En particulier, Catelinot sous-estime le gassendisme de Desgabets, passé sous silence, pour le replacer tout entier sous le signe d'un cartésianisme amélioré" (In particular, Catelinot underestimates the Gassendism of Desgabets, which he does not comment on, placing him entirely under the label of a revised Cartesianism); Armogathe 1977, 90.

[23] According to J.-R. Armogathe, P. Viogué was with Descartes in Sweden, to aid him in his last hours.

[24] See Armogathe 1977, 88.

[25] Desgabets 1671, 5.

just the destruction of one modal being (the bread) for another (the body of Christ). In this way, Desgabets proposed to explain the real presence of Christ in the host in keeping with the Cartesian conception of matter. Desgabets took the reformation Descartes effected in physics to be one that could be brought about in theology—a reform that promised to unite the Lutherans, Calvinists, and Pelagians.[26] It was a solution that ran counter to official church doctrine however. It angered many, including Antoine Arnauld, who along with Pierre Nicole, was responsible for exposing Desgabets as the author of the *Considerations on the present state of the controversy.*[27] The culmination of the Eucharist Affair for Desgabets was an interrogation by his superiors and the suppression of his future writings.[28]

It was again Clerselier who guided Desgabets in 1658[29] to the Cartesian conferences held at M. de Montmor's, where he reportedly participated in discussions with Rohault, Clerselier, and Cordemoy.[30] Here, he revealed his scientific bent of mind, repeating numerous experiments because of his conviction that questions of physics must not be a slave to authority. For example, he viewed the Scholastic absolute belief in the axiom that *nature abhors the void* as an "inveterate prejudice."[31] After making some of his own observations, he demonstrated that the effects attributed to this so-called loathing of the void by nature were actually caused by air pressure in the atmosphere and that the "abhorrence" could be explained in terms of the mechanics of air pressure.[32] He also participated in discussions with engineers con-

---

[26] Desgabets 1671, 13.

[27] Desgabets 1671. Arnauld's role in this controversy is significant. In the "Fourth Set of Objections," of Rene Descartes', *Meditations on First Philosophy,* 1641, Arnauld objects to Descartes that the Scholastic explanation of the role of accidents in the mystery is a matter of faith that is excluded by Descartes' theory of sensible qualities. AT VII 217–218. Moreover, a letter Arnauld wrote, likely to Clerselier, speaks of the anger Desgabets' views have elicited from him. See Arnauld 1727, II, 527–29.

[28] Desgabets' interrogation is included in Ms. Epinal 64; "Extrait d'un interrogatoire fait à Dom Robert Desgabets par ses supérieurs" (Extract of an interrogation of Dom Robert Desgabets conducted by his superiors), 25–26.

[29] According to Paul Lemaire 1901, this event took place in 1648 while Desgabets was Procurer General to Paris, but J. Beaude 1974, 10, argues that many of the details of Desgabets' life cited by Lemaire are nothing more than reconstructions based on circumstantial evidence. A letter from Desgabets to Denis, July 28 1667, weighs in favor of Beaude's authority on this point.

[30] Lettre de Desgabets à J. B. Denis, July 28, 1667. Cited by G. Rodis-Lewis in Introduction, OPD 1, xvii. Mihnea Dobre has suggested to me in correspondence that Desgabets' interest in blood transfusion may relate to Rohault's interest in the phenomena of capillarity. Rohault gave public lectures at Montmor's conferences in the late 1650s while he was working on the phenomena of capillarity, and in 1664 Rohault published his treatise, *Discours de la fiévre* in Descartes' *Le Monde*, a medical work that was read at Montmor. Rohault's and Desgabets' concern with medical topics suggests that Cartesians had interests beyond physics and mechanics that has not been fully appreciated.

[31] Lemaire 1901, 47.

[32] Lemaire 1901, 47. Desgabets comments on this topic again in his critique of Cordemoy's atomism, "Raisonnement touchant les atoms et le vuide contenus dans le livre du *Discernement*" ("Arguments concerning atoms and the void contained in the book *Discernement*"), Ms Epinal 64 (685–698).

cerning various means of changing the course of the Seine River after a sudden rise in water level had flooded Pont Marie.[33] Among the topics covered in his manuscripts are a defence of Galileo's explanation of the earth revolving around the sun; a treatise on practical mechanics, which was a study of mechanics applied to windmills, chariots, buggies, chaises, wool making, pumps, fans, fountains, sound receptors, clocks, and musical instruments, and a small work on the foundations of mathematics and the laws of physics.[34] In July 1658 Desgabets delivered a lecture that outlined his invention of an apparatus and procedure for blood transfusion.[35] But, it was not until 1667, after a controversy erupted between the English and the French over who first invented the procedure, that Denis was spurred to publish a written version of Desgabets' lecture. Rodis-Lewis has sorted out many of the details of this history, and shows that the two procedures created by Lower and Desgabets are so different as to confirm the independence of their inventions.[36] But whether and when Desgabets experimented with his procedure has not been firmly established and is dubitable. What is evident, however, is that Desgabets, like Wren, was inspired by Harvey's discovery of the circulation of blood and the possibility that the motion of blood could be understood in mechanistic terms. Desgabets unlike Wren draws on Descartes' metaphysical and physical conceptions of body to create a framework for the theory and practice of blood transfusion.

8.2 Cartesian Science: The Metaphysics and Physics of Body

There are two key features of Descartes' physics that captivated Desgabets' imagination: universal mechanism and the geometrical conception of body. Cartesian mechanism was the idea that the behavior of physical bodies, including human ones, could be completely understood in terms of the shapes and configurations of their parts and their functions depended upon their local movements alone.[37] If nature could be studied as a kind of machine then the principles of mechanics should explain all physical phenomena—from the movements of the planets around

---

[33] See Beaude 1974, 10.

[34] Ms Epinal 64 (143). See "Reflexions du C de Retz sur la question si c'est la terre qui tourne contre Gaillilée" (212–213); and "Le fondemens de la philosophie et de la mathematique Christiennes avec preface" (223–278); and "Mechanique practique" (519–598).

[35] Desgabets 1668.

[36] See Rodis-Lewis 1974, 36.

[37] We see this mechanization of nature in Descartes' *Treatise on Man* (1662, written between 1629 and 1633, published in Latin in 1662 and French in 1664). Descartes compared the workings of a particular natural phenomenon, the human body, to the workings of an artificial fountain. He compared nerves to pipes, muscles to springs, animal spirits to water, and the rational soul to a fountain keeper; he compared the function of breathing to the flow of water and that of sensory response to tiles within the grottos which if stepped on cause Diana who is bathing to hide in the reeds (AT XI 119–132). In *Principles of Philosophy* IV 188, Descartes tells us that he set out to describe the world as a machine, and again compares nature to a clock (AT VIIIa 315).

the sun, the course of the Seine River through Paris, to the movements of blood through the veins of the human body.

The second feature of significance for Desgabets' adoption of the Cartesian physics is Descartes' geometrical conception of body. According to the Cartesian conception of matter, a body is a thing extended in three dimensions. The essence of body, according to Descartes is extension, which is just to say that body can be understood solely and completely in terms of its extension in three dimensions, and its dependent properties, which include size, shape, and motion.

What Desgabets saw as particularly significant about Descartes' conception of matter was that the physical magnitude that physicists study *is* the geometrical magnitude that geometers study. This meant that the non-geometrical, sensible qualities, such as tastes, colors, sounds, heat, cold, were qualities that did not belong to body. Rather, sensible qualities, according to Descartes' metaphysics, belong to mind. Thus, Descartes claimed that body just is extension and the properties that follow from it, namely, size, shape, position, rest, local motion. These properties are not only the essential properties of body but also *the only* properties of body and hence, of Cartesian physics.

Descartes' universal mechanism and geometrical conception of body provided a powerful framework for the scientific study and analysis of physical phenomena that was to have many consequences and applications. We've seen how Desgabets applied these to the theological problem of real presence, let's turn to his application of these ideas in the scientific realm, specifically blood transfusion.

8.2.1 The Science and History of Blood Transfusion

Desgabets credits Harvey's discovery of the circulation of blood—that blood is pumped by the heart through the arteries and back again through the veins—as that which permitted him to view the flow of blood as a species of the communication of movement. Harvey himself did not view blood flow quite in this mechanistic way, so it was Descartes' conception of mechanism that Desgabets drew upon.

In *Discourse* V, Descartes compares the movement of blood with the movement found in a clock, and concludes that the movement found in each follows necessarily from the disposition, situation and shape of its parts.[38] In another passage, Descartes claims that we need only suppose the laws of mechanics in order to explain how the various particles of blood move.[39] Desgabets like Descartes compared animals and human bodies to highly delicate and complex machines. According to Desgabets, the operation of the laws of nature in the form of rules for the communication of movement was the true and unique foundation of the new physics.[40] This was the

[38] AT VI 50.

[39] AT VI 54–55.

[40] *Supplément*, Part I, Chap. 9, Section vv, OPD 6, 211.

route by which Desgabets arrived at the idea of blood transfusion. Simply put, if blood circulation is a species of the communication of movement, it should be possible, and potentially beneficial as a remedy to diseases of the blood, to move, that is transfuse, blood from one body into another.

In 1658, Desgabets presented his research on blood transfusion to a Cartesian audience in Paris, later published as "Discourse on the Communication or Transfusion of Blood." The work opens as follows:

> By communication of blood I mean the effective passage of blood from a healthy man or animal into the veins of a weak or sick man. This seems possible to do without any bad consequences and with considerable benefits for the future provided it is practiced with care and precaution.[41]

From there, he sets out to accomplish three things in his discourse, which we will take in turn:

1. To describe how such a thing as the communication or transfusion of blood is possible.
2. To describe how it can be done without peril to the subjects of the procedure.
3. To prove that it can be practised with great rewards.

With regard to the first point, that is, regarding the possibility of such an operation, Desgabets says that from the discovery of the constant circulation of blood, we know that if a vein were to be opened beneath a ligature and a small tube were inserted there, any foreign blood that is forced into the tube would enter the vein and return to the heart by an extraordinary path and from there would pass into the arteries and be distributed to the rest of the body. Thus the communication of blood is made possible by the fact that blood moves in a circulatory and mechanical manner.

Still on the first point, he says that with respect to *how* this communication could be effected, he describes the apparatus to be used:

> The machine that I imagine for this operation is simple and consists only of two small silver tubes, one whose end is open like a trumpet for its gentle application against the vessels that must give the blood; and the other is of a thickness proper to the insertion into the opening of the vein. The other two ends of the tubes communicate together by means of a small leather pouch of the thickness of a walnut, which serves three functions.[42]

[41] Recueil B.N. Ms Thoisy 326: "Par la communication du sang j'entends un passage effectif du sang d'un homme sain ou de quelque autre animal, dans les veins d'un homme faibles ou malade: ce qui semble se pouvoir faire par art sans aucune mauvaise suitte, & avec des avantages qui pourraient devenir très considerable avec le temps, si la chose se pratiquait avec le soin & avec la précaution convenable" (Desgabets 1668). All translations of Desgabets' unpublished writings are the author's except where otherwise noted.

[42] Desgabets 1668, 4: "La machine que j'ai imaginée pour cette operation n'est pas fort composée, & ne consiste qu'en deux petits tuyaux d'argent, l'un desquels a l'un des bouts ouvert comme une trompette, pour estre apliqué doucement contre le vaisseau qui doit donner le sang afin de le recevoir; & l'autre est d'une grosseur convenable pour estre inseré commodement dans l'ouverture de la veine. Les deux autre bouts des tuyaux communiquent ensemble par le moyen d'une petite bourse de cuir de la grosseur d'une noix ou environ, laquelle sert 1…2…3." Desgabets reports in the

Desgabets then proceeds to his second point, which is to give a defence of the procedure, as one that can be done without peril to the subjects of the operation. Here, Desgabets addresses the specifics of the procedure relative to the requirements of the patients involved:

1. The new blood transfused into the patient all at once may cause a significant alteration. This can be avoided if only a small amount of new blood is transfused at one time, and the procedure is repeated over several days.
2. The blood must be communicated without air and without effusions (bloodshed), and the veins and arteries should be treated as vessels for carrying blood which are no more endowed with faculties than the silver tubes themselves.
3. Because blood does not mix or "communicate" with the external air, it should not be cooled; and if it should become cooled somehow, it could be prevented easily with warm linens.
4. It is often objected that the diversity of complexions should preclude this operation from being performed, but there is a way to remedy this by carefully choosing the donors; by analogy, just as we must be careful of what we ingest by the mouth—there being an infinite number of elements of different qualities and quantities which are potentially dangerous, we must be careful in our choice of quality and quantity with regard to what we take in by transfusion.[43]

To summarize, Desgabets recognized that blood should be transfused gradually, that air must not be allowed to mix with the transfused blood and must be kept at a warm temperature; and finally, that too much new blood or not enough could be toxic or poisonous to the recipient.

This brings us to Desgabets' third point and final discussion of the *Discourse*, which concerns the purpose of blood transfusion. In this section, Desgabets argues that blood transfusion has many potential fruits to bear for the betterment of human life:

> Everyone agrees that nearly all that is good and bad in the body depends on the blood; which, when balanced and tempered it is impossible that one not be in perfect health; and to the contrary, when altered, one loses one's health and sometimes life. Thus, there are many cases in which this operation should be employed.
>
> The first such case is when the vital forces leave a man whether by the loss of blood, sickness or old age; if one judges it properly the body can be restored to good temper, so long as a good choice of a donor is made, since this serves to re-establish nature, retard aging and make life longer and happier. The other case is to cure sicknesses that are caused by the intemperance of the blood itself or its insufficient quantity.[44]

---

extracted letter to Denis preceding this discourse that Abbé Picot constructed the mechanism sometime around 1660. The apparatus is nowhere to be found today nor is there evidence that it was this apparatus used by Jean Denis in his experiments.

[43] Desgabets 1668, 4–5.

[44] Desgabets 1668, 5–6: "Tout le monde est d'accord que presque tout ce qu'il y a de bien & de mal dans nos corps, depend du sang, lequel estant louable & bien temperé, il est impossible qu'on ne jouisse d'une parfaite santé, & au contraire estant notablement alteré, on ne peut manquer de

Desgabets then gives a specific example to illustrate the application of the procedure in such a case: "…if before the viscera become blocked, we were to give a Hydropic [one suffering from extensive bodily swelling], new blood in place of the old blood from the veins, which have hardly any vital heat, there would be a great recovery of his strength and his health."[45] This is the note upon which Desgabets ends his *Discourse*, with a specific application of blood transfusion as a remedy for certain diseases and conditions.

Thus, Desgabets viewed the procedure of blood transfusion as a specific application of Cartesian physics aimed at the betterment of human life. He provided a scientific framework, a procedure, the design of an apparatus, and a rationale for blood transfusion. He directly inspired the experiments and work of Jean Denis whose failures do not diminish the significance of their place in the history of science or of Cartesian science. Rather, it highlights the extent to which the Cartesians of the period were influenced by Descartes to model the human body after a machine and to perform experiments to test the nature of its operations.

This picture of Cartesian science as one that placed experiment and observation at the heart of the theory and practice of science admittedly departs from the standard story of a hopelessly speculative and a priori science.[46] By standard story I mean one in which *expérience* leads us to nothing but error, whereas reason leads to truth, and wherein science is a deductive system constructed on the model of Euclidean geometry.[47] However, there is textual as well as historical evidence that experiment and observation played an important and essential role in the discovery of truth in science. The exact nature and extent of the role of *expérience*, however, is controversial in Descartes, and varies among the Cartesians.[48] In the next section, I begin to assess the extent and nature of Desgabets' empiricism.

---

perdre la santé, & quelque fois la vie. Or il y a plusieurs cas auxquels cette operation semble principalement devoir estre employée.

La premier, c'est lors que les forces manquent à un homme, soit par la perte de son sang, soit par la maladie, ou par la vieillesse, auquel cas il ne faut pas douter que luy ostant [sic] si on juge à propos, une grande partie de son sang mauvais ou inutile, & lui en donnant du nouveau qui soit bien temperé selon le choix qu'on feroit d'un bon sujet pour le donner, cela serviroit à restablir la nature, à retarder la vieillesse, & à rendre la vie beaucoup plus longue & heureuse.

L'autre cas est pour guérir plusieurs maladies qui sont causées par l'intemperie du sang où par sa petite quantité."

[45] Desgabets 1668, 6: "Par exemple si auparavant que les visceres soient gastez, on donnoit à un Hydropique un sang loüable au lieu de celuy qu'il a dans les veines, lequel n'a presque plus de chaleur vitale; il y a grande apparence qu'il recouvreroit ses forces & sa santé."

[46] There are several challenges to the "standard" picture of Cartesian science as purely a priori. For example, see Liard 1882; Milhaud 1918, 221–240; reprinted as Chap. 9 in his: *Descartes savant* of 1921; Gewirtz 1941, 183–210; Laporte 1945; Buchdahl 1963, 399–417; Olscamp 1965; Denissoff 1961, 31–75; Garber 1978, 114–151; Clarke 1989; Gaukroger 2002.

[47] For good examples of the standard story of Cartesian rationalism in science, see Schouls 1972, 220–234; and Blackwell 1966, 220–234.

[48] See Chap. 1 by Dobre and Nyden.

8.3 From Metaphysics to Physics to *expérience*: Cartesian Empiricism

Descartes more than once remarks that the role of experiment is to provide demonstrations of the first principles or truths of physics by connecting them to the way God actually made the world. In Part IV of his *Discourse on Method* he says:

> But I must also admit that the power of nature is so ample and so vast, and the principles so simple and so general, that I notice hardly any particular effect of which I do not know at once that it can be deduced from the principles in many different ways; and my greatest difficulty is usually to discover in which of these ways it depends upon them. I know no other means to discover this than by seeking further observations whose outcomes vary according to which of these ways provides the correct explanation.[49]

In Part III, Article 46 of Descartes' *Principles of Philosophy* he says:

> For, seeing that these parts could have been regulated by God in an infinity of diverse ways; experience alone should teach us which of all these ways He chose. That is why we are now at liberty to assume anything we please, provided that everything we shall deduce from it is {entirely} in conformity with experience.[50]

The idea is that human reason alone cannot discover how God chose to create the world—for the possible ways that reason can conceive exceed the one actual world that the senses come into contact with. The appeal to experience and observation is what delimits the merely conceivable, possible ways to the actual one, and so plays a necessary role in scientific knowledge. We must appeal to experience in order to find our way back from effects to their causes.

A similar idea runs throughout Desgabets' *Supplement*:

> He [Descartes] founds the laws of nature for physics only upon the simple supposition that God, in creating the world, put as much movement in the totality of matter as is found there at present, which *we know from experience*: this is sufficient for Descartes to deduce the formation and nature of all things that make up the visible world, in reasoning always from the cause to effects with consequences similar to those of mathematics. [author's emphasis][51]

Thus, the simple proposition that the total matter and motion is conserved provides the principle from which the particular consequences can be derived; yet, these actual quantities of matter and motion found in nature produce the sensible effects that can only be known by experience. The particular configurations of matter and movements in nature, once known, can be "deduced" from their cause much

[49] AT VI 64–65, CSM I 144.

[50] AT VIIIa 101, Descartes 1983–1984, 106.

[51] "Il ne fonde les lois de la nature pour la Physique que sur la supposition très simple que Dieu en créant le monde a mis autant de mouvement dans le total de la matière qu'il y en a présentement et que l'on connaît par expérience: ce qui suffit à M. Descartes pour en déduire la formation et la nature de toutes choses qui composent ce monde visible, en raisonnant toujours de la cause aux effets par des conséquences toutes semblables à celles des Mathématiques." In "Nouvelle ou autre Préface," *Supplément,* OPD 5, 154.

in the way a triangle with its specific measurements can be "deduced" from its geometrical principles.

Expérience, then, must discover which of the many possible worlds God actually created, and reason connects that actuality to the laws. *Expérience* is therefore necessary in delimiting what is metaphysically possible to what is actual. According to Desgabets, Descartes' great contribution is to have established the true metaphysics, a science that deals with actual things not purely possible things: "True metaphysics can no longer amuse itself by considering being as indifferent to being or non-being, i.e., as purely possible, or as actual nothing. Now metaphysics considers being as actually possessing all the perfections discovered in it."[52]

It remains to be shown whether Desgabets' actualism in metaphysics provides the framework for a peculiar brand of Cartesian empiricism that can be compared to that of John Locke. So far what can be said is that *expérience* delimits reason to reasoning about the actual, but it is not clear it contributes any sensible content to ideas.[53]

Monte Cooke notes the long tradition that regards Desgabets as an empiricist "… from Cousin and Bouillier to Rodis-Lewis, Easton, and Lennon."[54] Cook raises an important challenge to any suggestion that the senses provide content to our ideas, one that Allen-Hermanson has responded to at length.[55] The crux of Cook's challenge is that although Desgabets repeatedly asserts and argues for the view that all thoughts depend on body, this sense of dependence is purely causal and metaphysical whereby the body acts as the occasion for our imaginations and intellections, but does not provide the content or representation of those thoughts or ideas. Cook argues that Desgabets is clearly a *knowledge* empiricist, but not a *content* empiricist. Cook rightly and pointedly challenges any more robust content reading that attributes to the senses more than an extrinsic role in the generation of thoughts.

Sean Allen-Hermanson proposes that Desgabets' conception of simple conception is one rooted in perception and conscious experience.[56] Our simple conceptions proceed from our senses and precede judgment and hence error, and as such they provide the basis for all our ideas and knowledge. He observes that "Desgabets appears to be a content empiricist who offers an early sense-data theory, motivated by anti-skepticism."[57] This is an evaluation of which I am sympathetic but Desgabets' argument in the *Supplement* is admittedly difficult to decipher. Perhaps an illustration or two may help to clarify Desgabets' view. An architect forms the plan for a house by mentally combining, uniting, and separating matter in thought into extrinsic

---

[52] "La vraie métaphysique ne s'amuse plus à considérer l'être comme indifférent à être ou n'être pas, c'est-à-dire comme purement possible, ou comme un néant actuel. Mais elle le considère comme possédant actuellement toutes les perfections qu'elle y découvre." *Supplément*, Part II, Chap. 8, Section ii. OPD 6, 245.

[53] Monte Cook distinguishes between *content* and *knowledge* empiricism. I thank him for pressing this point with me in correspondence. Cook 2008, 501–516.

[54] Cook 2008, 501–516.

[55] Allen-Hermanson 2009, 57–85.

[56] Allen-Hermanson 2009, 57–85.

[57] Allen-Hermanson 2009, 75.

denominations. The house built according to this plan is composed of actual stone and wood, an object that is a real division and configuration of material parts. As Desgabets puts it, "…a house that an architect composes in his mind, which he actually builds, or which he destroys, is the same real house, although it exists differently."[58] The house exists extrinsically by thought and intrinsically by the composition of its material parts. The leader of an army divides and combines its men into regiments and companies, and these units are real, i.e., the regiments extrinsically contain a specific number of soldiers. As Desgabets puts it:

> Men divided into regiments, companies etc., really form these bodies. 20, 30, 40, pistols are actually such a number, and it cannot be said that this has an existence only in thought, because regiment, foot, 20 pistols are real and corporeal things, and not of thoughts, although their being such as they come by thought, gives them this form and this extrinsic denomination.[59]

Thus, according to Desgabets, matter provides the actual being and existence of the house and the regiment, which are the sensible objects of our thought, whereas mental combination, and separation provides the extrinsic denominations of those objects. Moreover, Desgabets explains how the senses give us knowledge of the nature of things:

> …by exciting the thoughts and the ideas that we have of things, is indubitable and in conformity with their objects; thus it is that the soul knows itself by the senses, in the manner that we said, finding nothing clearer, nor better known than its own nature; thus matter being known by extension and according to its properties which are figure, movement and rest, which we know by the senses, is the thing of the world best exposed to the eyes of the soul.[60]

Metaphysicians, geometers, and mathematicians give extrinsic being to things by thought; matter itself and its division by motion found their intrinsic being. Recall Desgabets' appeal to Descartes' reformation in philosophy as rooted in the idea that mathematics and physics are at long last reunited. Although the mathematicians and physicists study the same "object," i.e., physical magnitude, physicists consider it not as it is in itself, but as it relates to its modes of being. In other words, physicists study particular extended bodies whereas mathematicians study extension itself. Rather than assimilate physics to mathematics in order to make it more certain, Desgabets offers a materialist foundation for mathematics.

---

[58] *Supplément*, OPD 6, 242: "Une maison qu'un architecte compose dans son esprit, qu'il bâtit effectivement, ou qu'il détruit, est la même maison réelle, quoiqu'elle existe différemment."

[59] *Supplément*, OPD 6, 240: "Les hommes partagés en régiments, en compagnies etc. forment réellement ces corps. 20, 30, 40, pistoles sont effectivement un tel nombre, et on ne peut pas dire que cela n'a d'existence que dans la pensée, parce que régiment, pied, 20 pistoles sont des choses réelles et corporelles, et non pas des pensées, quoique leur être tel leur vienne par la pensée, qui leur donne cette forme et cette dénomination extrinsèque."

[60] *Supplément*, OPD 7, 265: "…en excitant les pensées et les idées que nous en avons, sont indubitables et conformes à leurs objets; c'est ainsi que l'ame se connaissant par les sens, en la manière que nous avons dit, ne trouve rien de plus clair, ni de mieux connu que sa propre nature; c'est ainsi que la matière étant connue étendue et selon ses appartenances qui sont la figure, le mouvement et le repos, que nous connaissons par les sens, est la chose du monde la mieux exposée aux yeux de l'âme."

It is the senses not reason that present us with matter as it exists at a particular place and moment. A purely a priori physics would not only be incomplete but impossible because reasoning itself *cannot provide the actual artifices and compositions of matter*. In order to know the actual artifices and compositions of matter, we must have recourse to sensible qualities, which are nothing but the local dispositions of matter as comprehended by the mind. These are the mind's tie to the movements of body, and thus, the actual compositions of matter, and so are equally a part of the object of physics and mathematics:

> …these modes of matter, or these sensible qualities, are obviously part of the proper object of mathematics and mechanics because there is no question here but of the sizes, movements, rest, figures and arrangements of the parts of matter, and what can result from their assemblage and combination, and in all of this there is nothing that is not governed by the laws of mechanics. It is upon this truth that the happy reunion of physics and mathematics is made in our day.[61]

Physics or mechanics must "look to *expérience* which operates by the senses, rather than by reasoning."[62] It is *expérience* that puts us in contact with the modes of matter, which would otherwise be unknowable. Matter itself in its essence would be unknowable without its particular configurations and our senses putting us in contact with them.

Desgabets meant more than that the senses merely put us in contact with the modes of matter, he saw the senses as necessary to the discovery of truth in physics. Without the senses, we would be left to reason about mere possibilities, having no means to discover the actual quantities, shapes, sizes, and position of bodies. In his writing against the skeptic Simon Foucher, Desgabets distances himself from Malebranche on the question of the role of the senses in the search after truth:

> …you have perhaps followed the author of the *Search*, who has absolutely declared himself against the use of the senses to know the truth. Nevertheless, as often as the senses make us perceive what is in fact in exterior things, we are right to say then that those things are the objects of the senses. Thus when they [exterior things] act on us and we think of body, of movement, of rest, of figure, of arrangement of parts, and of all that can result from their assemblage, all of that is the object of a distinct knowledge and of a very pure intellection *that the senses gives us,* and we have an idea of it as clear as any other thing whatever it should be.[63] [author's emphasis]

---

[61] *Guide*, OPD 4, 110: "…ces modes de la matière, ou ces qualités sensibles, se rapportent visiblement à mathematique et à la mécanique comme une partie de leur propre objet, parce qu'il n'est ici question | que de grandeurs, mouvements, repos, figures, et arrangements des parties de la matière, et de ce qui peut résulter de leur assemblage et combinaison, et en fondée la réunion qu'on a faite heureusement en nos jours de la physique et de la mathématique." Note that while Desgabets refers to "these sensible qualities," he means the qualities of extension, such as size and shape that are sensed, he does not mean sensible qualities such as color and taste, which belong to mind.

[62] *Supplément*, OPD 6, 245.

[63] Desgabets 1675, 102–103: "Vous avez peut-estre suivy en cela l'Auteur de la Recherche qui s'est absolument déclaré contre l'usage des sens pour connoistre la verité. Cepéndant autant de fois que les sens nous font appercevoir ce qui est effectivement dans les choses exterieures, on a raison de dire, que pour lors elles en sont les objets. Ainsi, lors qu'elles agissent sur nous, & que nous pensons à corps, à mouvement, à repos, à figure, à arengement [sic] des parties, & à tout ce qui

Beyond knowledge of material extension and its modes, the senses give us the object of a distinct and intellectual knowledge. Desgabets recognizes that he parts from Descartes on this point, "…to feel and to know by intellection are the same thing when the feeling is related to its true object."[64] In the case of knowing the objects of physics—bodies in their nature and actual existence—we must come to know the physical magnitude through its sensible signs and mental denominations.

8.4 Conclusion

I have argued that Desgabets' Cartesian conception of mechanism and body had metaphysical as well as epistemological underpinnings and consequences. Matter and its motion make possible the presence to the mind of an actual physical object in the form of a sensible object. Moreover, motions in bodies via the senses provide the means by which thoughts are differentiated and individuated, given endurance and succession. Thus, matter and motion make sensible to the soul the particularity of the sensible object, which is necessary to connect essences (possibles) to actuality.

For Desgabets, experiments on blood transfusion were as much demonstrations of the truth of the Cartesian metaphysics as they were demonstrations of the way the world actually is. The role of experiment was to provide demonstrations of the first principles or truths of physics by connecting them to the way God actually made the world. The upshot of my position is that Desgabets was an empiricist in all the important senses of that term: his engagement with empirical research, most notably his work on blood transfusion, his rejection of hyperbolic doubt and pure intellection, and his insistence on the necessary role of sensible signs in the formation of our ideas of bodies in their form and content. In fact, I have claimed that the essential role Desgabets assigns to *expérience* is not merely consequential but integral to his system.[65]

In placing *expérience* at the very foundation of true knowledge, Desgabets revises and reconstructs Cartesianism on an empiricist foundation. This unique brand of "Cartesian Empiricism" provided what Desgabets' called the "reunion of

---

peut resulter de leur assemblage, tout cela est-l'objet d'une connoissance distincte & d'une intellection tres-pure que nous donne le sens, & nous en avons une idée aussi claire que de quelque autre chose que ce soit."

[64] *Supplément*, OPD 5, 180: "…sentir et connaître par l'intellection pure sont une même chose, quand on rapporte le sentir à son vrai objet."

[65] Tad Schmaltz uses the term "empiricism" to describe Desgabets' position, but he thinks that his radical Cartesianism is largely consequential to metaphysical positions Desgabets holds, not to epistemological considerations. However, given how Desgabets sees the connection between the metaphysical and epistemological theses, I don't think they can be separated. See Schmaltz 2002b, 149, 181. See also his 2002a, 513–540.

geometry and mechanics" which fostered some daring, if not successful, experiments on the human body.

Acknowledgments I gave an early version of this paper, "The Father of Cartesian Empiricism: Robert Desgabets on the physics and metaphysics of blood transfusion" at the International Society for the History of Philosophy of Science (HOPOS), Panel on Cartesian Empiricism, Budapest, Hungary, June 24–27, 2010. The discussion in the session by the participants and audience, in particular Gideon Manning, was very helpful. I especially want to thank my fellow participants and editors of this volume, Tammy Nyden and Mihnea Dobre for our many discussions and their careful comments and suggestions. I also am grateful to Monte Cook for his written comments, which were enormously helpful. Finally, I thank Tom Lennon for transcribing for me the *Extrait d'une Lettre de M. Denis,* 1667 when he came upon it in the Bibliothèque Nationale many years ago. All errors remain my own, but I have benefitted much from the ideas and discussions of others. All translations are the author's except where otherwise noted.

References

Allen-Hermanson, Sean. 2009. Desgabets: Rationalist or Cartesian empiricist? In *Topics in early modern philosophy of mind. Studies in the history of philosophy of mind*, vol. 9, ed. Jon Miller, 57–85. Dordrecht: Springer.

Armogathe, Jean-Robert. 1977. *Theologia Cartesiana: L'Explication physique de l'Eucharistie chez Descartes et dom Desgabets*. The Hague: Martinus Nijhoff.

Arnauld, Antoine. 1727. *Lettres*, 5 vols. Nancy: Nocolai.

Beaude, Joseph. 1974. Desgabets et son oeuvre. *Revue de sythèse* 95: 7–17.

Beaude, Joseph. 1979. Cartésianisme et anticartésianisme de Desgabets. *Studia Cartesiana* 1: 1–24.

Blackwell, Richard J. 1966. Descartes' laws of motion. *Isis* 57: 220–234.

Brown, Harcourt. 1948. Jean Denis and transfusion of blood, Paris, 1667–1668. *Isis* 39(1–2): 15–29.

Buchdahl, Gerd. 1963. Descartes' anticipation of a "logic of discovery.". In *Scientific change: Symposium on the history of science*, ed. A.C. Crombie, 399–417. London: Heinemann.

Clarke, Desmond M. 1989. *Occult powers and hypotheses: Cartesian natural philosophy under Louis XIV*. Oxford: Oxford University Press.

Cook, Monte. 2002. Robert Desgabets's representation principle. *Journal of the History of Philosophy* 40(2): 189–200.

Cook, Monte. 2005. Desgabets on the creation of eternal truths. *Journal of the History of Philosophy* 43(1): 21–36.

Cook, Monte. 2008. Desgabets as a Cartesian empiricist. *Journal of the History of Philosophy* 46(4): 501–516.

Cousin, Victor. 1845. *Fragments de la philosophie cartesienne*. Paris: Charpentier.

Denis, Jean. 1668. *Lettre de Denis à Monsieur Sorbière Docteur en Médicine par Jean Denis aussi Docteur en Médicine. Touchant l'origine de la Transfusion du sang, et la manière de pratiquer sur les hommes*, 12p. Paris: J. Cusson.

Denissoff, E. 1961. L'énigme de la science cartésienne: La physique de Descartes est-elle positive ou déductive? *Revue Philosophique de Louvain* 59: 31–75.

Descartes, René. 1970. *Oeuvres de Descartes*, ed. Charles Adam and Paul Tannery. Paris: J. Vrin.

Descartes, Rene. 1983–1984. *Principles of philosophy*. Trans. V. R. Miller and R. P. Miller. Dordrecht: Kluwer Academic. (Reprinted in 1984 with corrections by D. Reidel Publishing Company, Dordrecht, Holland.)

Descartes, Rene. 1984–1991. *The philosophical writings of Descartes*, 3 vols. Trans. John Cottingham, Robert Stoothoff, and Dugald Murdoch (Volume 3 including Anthony Kenny). Cambridge: Cambridge University Press.

Desgabets, Robert. 1668. *Discourse de la communication ou transfusion du sang*. Published with Denis, Jean. 1668. Paris.

Desgabets, Dom Robert. 1671. *Considérations sur l'état présent de la controverse touchant le T. S. Sacrement de l'autel*. Published anonymously, Holland.

Desgabets, Robert. 1675. *Critique de la critique de la Recherche de la vérité*. Paris.

Desgabets, Dom Robert, 1983. *Oeuvres philosophiques inedités de Desgabets*. Introduction by Généviève Rodis-Lewis, text and annotations by Joseph Beaude. 7 opuscules. Paris: CNRS.

Desgabets, Robert. MS. Epinal 43 (142) (Théologie). Bibliothèque municipale d'Epinal. (Ms. is available on-line at the Epinal Municipal Library website.)

Desgabets, Robert. MS. Epinal 64 (143) (Philosophique). Biblothèque municipale d'Epinal. (Ms. is available on-line at the Epinal Municipal Library website.)

Easton, Patricia. 2005. Desgabets's indefectibility thesis—A step too far? In *Receptions of Descartes*, ed. Tad M. Schmaltz, 27–41. London: Routledge.

Faye, Emmanuel. 2005. The Cartesianism of Desgabets and Arnauld and the problem of the eternal truths. In *Oxford studies in early modern philosophy*, 2 vols., ed. Daniel Garber, vol. 2, 193–209. Oxford: Clarendon Press.

Garber, Daniel. 1978. Science and certainty in Descartes. In *Descartes: Critical and interpretive essays*, ed. Michael Hooker, 114–151. Baltimore: John Hopkins University Press.

Gaukroger, Stephen. 2002. *Descartes' system of natural philosophy*. Cambridge: Cambridge University Press.

Gewirtz, Alan. 1941. Experience and the non-mathematical in the Cartesian method. *Journal of the History of Ideas* 2: 183–210.

Harvey, William. 1628. *De Motu Cordis*. London.

Laporte, Jean. 1945. *Le rationalisme de Descartes*. Paris: Presses Universitaires de France.

Lemaire, Paul. 1901. *Le Cartésianisme chez les Bénédictins: Dom Robert Desgabets son système, son influence et son école*. Paris: Alcan.

Lennon, Thomas M. 1998. The Cartesian dialectic of creation. In *The Cambridge history of seventeenth century philosophy*, 2 vols., ed. Michael Ayers and Daniel Garber, vol. 1, 331–362. Cambridge: Cambridge University Press.

Levêque, Chanoine. 1947. *Petite histoire religieuse de nos Vosges*. Mirecourt: Imprimerie Géhin.

Liard, Louis. 1882. Du rôle de l'expérience dans la physique cartésienne. In his *Descartes*. Paris.

Milhaud, Gaston. 1918. Descartes expérimentateur. *Revue philosophique de France et d'etranger* 89: 221–240.

Milhaud, Gaston. 1921. *Descartes savant*. Paris: F. Alcan.

Miller, Timothy. 2008. Desgabets on Cartesian minds. *British Journal for the History of Philosophy* 16(4): 723–745.

Mouy, Paul. 1934. *Le Développement de la physique Cartésienne, 1646–1712*. Paris: J. Vrin.

Nicolas, Paola. 2012. Les instances de Desgabets: Supplément aux objections de Gassendi et Arnauld. *Revue Philosophique de Louvain* 110(3): 471–493.

Olscamp, Paul J. 1965. Translator's introduction. In Rene Descartes. 2001. *Discourse on method, optics, geometry, and meteorology*. Indianapolis: Hackett.

Philosophical Transactions. 1665–1666. Tryals Proposed by Mr. Boyle to Dr. Lower, to be made by him, for the Improvement of Transfusing Blood out of One Live Animal into another. Promised Numb. 20. *Philosophical Transactions*. London: Sold by Lockyer Davis, and Peter Elmsly, printers to the Royal Society.

Régis, Pierre-Sylvain. 1704. *L'usage de la raison et de la foi*. Paris.

Rodis-Lewis, Généviève. 1974. L'écrit de Desgabets sur la transfusion du sang et sa place dans les polémiques contemporaines. *Revue de synthèse* 95: 31–64.

Schmaltz, Tad. 2002a. The Cartesian refutation of idealism. *British Journal for the History of Philosophy* 10: 513–540.

Schmaltz, Tad. 2002b. *Radical Cartesianism: The French reception of Descartes*. Cambridge: Cambridge University Press.
Schouls, Peter. 1972. Reason, method, and science in the philosophy of Descartes. *Australasian Journal of Philosophy* 50: 30–39.
Streiff, François. Undated. *Dom Robert bénédictin lorrain 'inventeur' de la transfusion sanguine en 1650*, http://professeurs-medecine-nancy.fr/Desgabets.htm. Accessed 17 June 2010.
Tucker, Holly. 2011. *Blood work: A tale of medicine and murder in the scientific revolution*. New York: Norton & Company.
Watson, Richard. 1982. Transubstantiation among the Cartesians. In *Problems of Cartesianism*, ed. T.M. Lennon, J.M. Nicholas, and J.W. Davis, 127–148. Kingston/Montreal: McGill-Queens University Press.
Watson, Richard. 1987. *The breakdown of Cartesian metaphysics*. Atlantic Highlands: Humanities Press.

Chapter 9
Rohault's Cartesian Physics

Mihnea Dobre

Abstract In 1671, Jacques Rohault published his *Traité de physique*, a textbook on physics relying on his weekly conferences held in Paris. A good mathematician and at the same time a curious experimenter, Rohault was one of the main Cartesian figures of his time. Connected to Parisian philosophical circles, Rohault was deeply concerned with the reception of Descartes' philosophical views. He was associated with Claude Clerselier and he encouraged Pierre-Sylvain Régis to spread Cartesianism in Toulouse. Performing experiments and using instruments in his observations, allowed for a very good reception of Rohault's natural philosophy in the late seventeenth century. Thus, his textbook on physics was quickly translated and disseminated across Europe. Of a particular interest is the English version of this book, which was annotated by the celebrated Newtonian, Samuel Clarke. This chapter will provide a deep analysis of Rohault's system of physics, with an emphasis on his experimental approach. Equally important, the Newtonian reception of Rohault's treatise will be discussed in close connection to the structure of his philosophical system and the methodological novelties introduced by the French philosopher.

M. Dobre (✉)
The Center for the Logic, History and Philosophy of Science (CELFIS),
Faculty of Philosophy, University of Bucharest, Splaiul Independentei 204,
060024 Bucharest, Romania
e-mail: mihneadobre@yahoo.com

M. Dobre and T. Nyden (eds.), *Cartesian Empiricisms*, Studies in History and Philosophy of Science 31, DOI 10.1007/978-94-007-7690-6_9,
© Springer Science+Business Media Dordrecht 2013

9.1 Jacques Rohault: Life and Works

Jacques Rohault (1618–1672) was one of the leading Cartesians in seventeenth-century France.[1] Born in 1618, in Amiens, Rohault studied within a Jesuit college and in the 1640s he moved to Paris, where he became known as a "professeur des mathématiques."[2] With more extensive knowledge and practice of the sciences than any other Cartesian, Jacques Rohault has been described by Paul Mouy thus: "[he] has tried to build an experimental physics on Cartesian foundations. To his 'Wednesdays', all Paris, provinces, even the foreigners came to witness the wonderful experiments by which he confirmed Descartes' physics."[3]

Little is known about his early life in Paris. However, more details are available from 1650s onward. What comes as a surprise is the apparently high public esteem of Rohault's experimental work, which seems to have predated any reference to his name.

A possible reason for the relatively late recognition of his name can be discovered in a small fragment from the *Nouvelles de la République des Lettres*, where the editor was complaining against the lack of constancy in how news travelled in the Republic of Letters:

> Quite often, we do not find the name of an Author, if his name is not pronounced as it is written. Thus, only a few could have properly cited M. Rohault before he published his name. Some have cited M. *Roo* and others M. *Roho*. Who knows if one day we shall not make 2 or 3 different Authors from that, like we have done with several ancients due to the lack or misplace of a letter?[4]

Hence, one can conjecture that Rohault's name was used quite indistinctly in that period. Such a reading is compatible with various other examples, including one

---

[1] For a good discussion of Rohault's life, see Clair 1978. This is the still the best biography available and it sheds a good light on Rohault's social status and on his personal connection with Clerselier. On Clerselier and Rohault as promoters of Cartesianism in France, see Balz 1930. For the general placement of Rohault within the Cartesian movement, see Bouillier 1868 and Mouy 1934; and for a discussion of Rohault as an experimental philosopher, see McClaughlin 1996, 2000.

[2] Rohault's birth was debated in the literature. For instance, Gérard Milhaud ascribed it to May 23, 1623 (see Milhaud 1972); a famous portrait of him has 1620 inscribed as his year of birth; but more convincing is the view expressed by McClaughlin (see McClaughlin 1976), which ascribes it to 1617. In this chapter, I shall follow the chronology suggested by Clair (see Clair 1978).

[3] Mouy 1934, 108: "S'efforçait de constituer, sur les bases cartésiennes, une physique expérimentale. A ses 'mercredis', tout Paris, la province, l'étranger même se pressaient pour assister aux belles expériences par lesquelles il confirmait la physique de Descartes." For an alternative reading of Rohault's role in the Parisian academies, see Chap. 3 by Roux.

[4] See *Nouvelles de la République des Lettres* Octobre 1685, 1014: "Ce qui fait souvent que l'on ne rencontre pas le nom d'un Auteur, c'est qu'on ne le prononce pas comme on l'écrit. De là est venu que si peut de gens ont bien cité M. *Rohault*, avant qu'il eût publié lui-même son nom. Les uns le citoient M. *Roo*, & les autres M. *Roho*. Qui sçait si un jour l'on ne fera pas de tout cela 2 ou 3 Auteurs differens come l'on a fait à l'égard de plusieurs anciens à cause de l'oubli ou de la transposition d'une seule lettre?"

famous reference to his name in a brief history of the pneumatic experiments in France, where his name is spelled "Roho."[5] In a recently discovered manuscript-copy called the *Physique nouvelle par Monsieur Rho disciple de Monsieur Des-Cartes*, the modern editor of the book, Silvain Matton, notes the same difficulty.[6] But regardless of the way in which his name was spelled, Rohault was an acknowledged Cartesian and experimenter. What is difficult, however, is to find the moment when he started contributing to the philosophical debates in the mid-seventeenth-century Paris.

Take, for example, this recently discovered and published manuscript—*Physique nouvelle (1667)*—that was difficult to date and, in fact, was only conventionally ascribed to that year. Matton convincingly argues that 1667 is the year when the book came under Billy's possession, but not the one when it was written.[7] Moreover, the lack of evidence with respect to the final form of the manuscript seems to indicate a notebook compiled after Rohault's public lectures, but not his involvement in rendering the manuscript in its final form. Yet, a close examination of the *Physique nouvelle* and the *Traité de physique*, reveals only small differences. Thus, a plausible explanation would be that Rohault's philosophical views were already developed by the time such pirated-copies as the one of the *Physique nouvelle* were in circulation. At least from the point of view of experimental activity, Rohault was already a reputed performer.[8]

Rohault's name appears in various correspondences from around 1660. For example, repeating Col. Tuke's impressions from his visit to Paris, Thomas Birch presents him in close connection to the Parisian philosophical circles: "the next day the president, Monsieur de Montmor, came to my lodging [Col. Tuke's] at Palais Royal, and carried me to one Monsieur de Rohault's house, who is of their society, from whom I heard a very ingenious lecture of the nature of the load-stone, and his hypothesis made out by diverse curious experiments."[9] This report is from 1661, but other similar remarks come from 1659, when Henry Oldenburg visited Paris.[10] Combined with Clerselier's flattering words from his preface to Descartes' second volume of letters (1659)—"A great number of people…participated in the assembly

[5] See *Journal des sçavans* April 26, 1666, 208. Another variation of spelling his name is to be found in Denis, Jean Baptiste. 1668. *Lettreècrite à M. Sorbière…*, Paris, where he is called "Roh."

[6] See Matton's "Remarques sur le manuscrit de la *Physique nouvelle*" in Rohault 2009.

[7] Rohault 2009, LXXXV.

[8] This is further supported by the content of the Ms. 2225 of the Bibliothèque Sainte-Geneviève, which collects reports from Rohault's Wednesday conferences from November 1660 to April 1661 and a discourse from 1669. There seem to be no differences between the Ms. 2225 and the *Physique nouvelle* with respect to Rohault's famous experimental cases—his experiments with magnets, his demonstrations of the effects caused by the glass drops, etc. See Rohault 1660.

[9] Birch 1756, I, 28.

[10] Oldenburg addresses two letters to Saporta where he discusses about Rohault's public conferences. Thus, on June 28, 1659, Oldenburg gives a high praise to Rohault's account of vision and on August 11, 1659, he refers to the Frenchman's explanation of colours. For Oldenburg's reports, see Oldenburg 1965–1986, I.

held every Wednesday at Mr. Rohault, very learned Mathematician & extremely skilled in Mechanics"—the public image of Rohault is quite clear.[11]

We may infer from all these details that Rohault was involved in the various learned circles active in Paris, most probably first in the literary circles associated with Cyrano de Bergerac and Moliére, but more and more connected to the philosophical ones, and especially with the Cartesian movement centered upon Clerselier.[12] His experimental work from the 1650s and early 1660s seems to be completed in the *Physique nouvelle*, which was to take a more systematic exposition in the published *Traité de physique* of 1671. During the past decade of his life, Rohault does not seem to trouble much with new experimental activities. He might have still performed experiments during his famous conferences, but there are no traces of new empirical data in the passage from his early manuscript to his first published book.[13] At the same time, as Desmond Clarke and Trevor McClaughlin have noticed, the 1660s are dominated by Clerselier's efforts to make palatable the Cartesian explanation of Eucharist.[14] Rohault got engaged into the strategy developed by his father-in-law and wrote a small treatise published in the same year with the *Traité*, the *Entretiens sur la philosophie*. This seems to support further our reading that places Rohault's original work at an earlier date. The rest of his writings on mechanical and mathematical problems were published posthumously by Clerselier in 1682.

What is puzzling, however, in the reception of Rohault's natural philosophy is the long life of his *Traité* within the Newtonian context. After Rohault's death in 1672, the treatise knew a period of great glory, being translated and published in a number of places across Europe.[15] It was used as a physics textbook in various universities, including Louvain, Cambridge, and Oxford. If one looks at the book review made in the *Nouvelles de la République des Lettres*, one gains a better understanding of how Samuel Clarke's edition of Rohault came into being.[16] According to this review, Rohault's views circulated for a while in manuscript form. This would have been the text of the *Physique nouvelle (1667)*, to which we have referred

[11] See Claude Clerselier's unpaginated preface in Descartes 1659: "d'un tres-grand nombre de personnes…a fait aller à l'assemblée qui se tient tous les Mercredis chez Mr. Rohault, tres-sçavant Mathematicien, & fort experimenté dans les Mechaniques."

[12] For the connection between Rohault, Cyrano, and Moliére, see McClaughlin 1976, 179.

[13] Another important biographical detail about Rohault is the name of his first wife, Nicole Fillassier, to whom he married in 1650. After her death in 1663, Rohault remarried in 1664 with Geneviève Clerselier and became, thus, the son-in-law of Claude Clerselier. See McClaughlin 1976, 181 and Clair 1978, 27–28. Fillassier was a family of merchants and craftsmen, which would have facilitated Rohault's better access to different artisans and workshops. This can explain both Rohault's familiarity with handling instruments, but also his possibility to get less common materials for his experiments (e.g., glass-drops). I would like to thank Trevor McClaughlin for pressing this point.

[14] See McClaughlin 1979; Clarke 1989. For a general discussion of the Eucharist affair and its consequences for Cartesian philosophy, see Schmaltz 2002; Chap. 2 by Ariew; Chap. 8 by Easton.

[15] Pierre Clair makes a list of the various editions of Rohault's book. See Clair 1978, 5–8.

[16] See *Nouvelles de la République des Lettres* October 1706, 455–460.

earlier. Immediately after the publication of the official version of the text as the *Traité de physique*, Théophile Bonnet made a Latin translation. The narrative of this report further argues that only the haste of the publisher based on Geneva, who commissioned Bonnet and eventually printed the book in 1674 is the reason why the Latin was not entirely accurate. Still, Bonnet's Latin text remained the official translation that supported numerous other editions in different places in Europe.

Antoine Le Grand's annotated version (London, 1682 and Amsterdam, 1691) enjoyed some success and this might have stirred Samuel Clarke's attention.[17] Moreover, from the *Nouvelles de la République des Lettres*, we find that Clarke was unsatisfied by the Latin text available and made his own translation: "we have to admit with Mr. Clarke that this first Translator has made some crass mistakes."[18] What is interesting in this early report on Clarke's edition is the good summary of the historical context. The author comments also on the annotations, which are compared to Le Grand's earlier notes. Translated again in its entirety and briefly annotated, the edition prepared by Samuel Clarke is, from the very beginning, presented as Newtonian. In fact, the Newtonian character of Clarke's annotations is clearly stated in the book's subtitle, such as in the one for the 1702 edition: *Jacobi Rohaulti Physica. Latinè verit, recensuit, & uberioribus jam Adnotationibus, ex illustrissimi Isaaci Newtoni Philosophiâ maximam partem haustis, amplificavit & ornavit Samuel Clarke, A.M. Admodum Reyendo in Christo Patri, Joanni Episcopo Norvicensi à Sacris Domesticis*. Building upon his prior editions, Samuel Clarke added more comments to the text and in 1710 he included Charles Morgan's mathematical notes.[19] Eventually, Samuel Clarke's brother, John, made an English translation, which included all the annotations, corrected once more by his more famous brother. The text in its final form was printed in 1723 and other English editions followed in 1729 and 1735.

Michael Hoskin has convincingly argued that Clarke's additions to Rohault's text were increasingly Newtonian.[20] However, historians of both philosophy and science have been puzzled by the success of this book in the context of the intense spread of Newtonianism. In a classic essay on *The Study of Early Scientific Textbooks*, George Sarton noted the importance of this treatise on physics: "now the Rohault-Clarke treatise could be defined not as a Cartesian Newtonian textbook (that would be nonsense) but as a Cartesian textbook including, in the footnotes, a Newtonian refutation."[21] More recent, Dennis Des Chene characterized the abovementioned editions as a "battleground between Newton and Descartes" and Volkmar Schüler has argued that "in the end, it was a strong circulating textbook

[17] On Le Grand's Cartesianism, see Chap. 11 by Hatfield.

[18] See *Nouvelles de la République des Lettres* October 1706, 457: "il faut convenir avec Mr. *Clarke*, que ce premier Traducteur a commis des fautes assez grossiéres...." The book review is concerned with the Latin edition printed in 1702, which is the second edition prepared by Samuel Clarke. The first edition was printed in 1697.

[19] For the history and the evolution of these annotations, see Hoskin 1961; Schüler 2001.

[20] See Hoskin 1961.

[21] Sarton 1948, 145.

on Cartesian physics that played, circumstantially, a significant role in making Newtonian physics palatable among these latter circles [philosophers and theologians], namely the *Traité de physique* by Jacques Rohault—that is, Samuel Clarke's edition of this book."[22] Yet, a satisfactory account of what made Rohault's text so influential is missing and the intricate problems of its reception still await more in-depth exploration. On account of this, in what follows we shall focus mainly on the English version of Rohault's treatise on natural philosophy.

9.2 Rohault's Natural Philosophy

When he printed the *Traité de physique* in 1671, Rohault was the main Cartesian philosopher of his time.[23] As we have argued above, this was mainly based on the image produced by his very popular conferences. Combining good pedagogical skills with an extensive knowledge of Cartesian philosophy and an unbounded curiosity regarding wondrous phenomena, Rohault was providing mechanical explanations that pushed ahead the science of his day. In this section, we shall examine Rohault's method in physics, which, we shall argue, parallels and even predates that of his English contemporaries.

Rohault's first publication, a small treatise called *Discours de la fiévre*, was printed together with Descartes' manuscript of *Le monde* in 1664. From Clerselier's preface to the book, one learns that Rohault's small treatise was read in the Montmor academy.[24] This is important both in terms of Rohault's biography—and, as a matter of fact, it confirms Col. Tuke's testimony mentioned in the previous section—and for the type of intellectual setting in which he was formed. By reading in public their scientific contributions, attendants of the Montmor academy were invited to give brief arguments without rhetorical attempts to convince the audience and to refrain from transforming their lectures into a mere narration of authorities.[25] Rohault's text shares all these elements; it is a brief medical treatise that does not contain any reference to names of so-called authorities and focuses on the argument rather than on rhetoric and persuasion.

Rohault was to further develop these traits in his own "Wednesday" conferences. Yet, his laborious activity as a *savant* failed to earn him a place in the newly established *Académie des sciences*. Once more, Henry Oldenburg's correspondence is helpful for uncovering the context. On June 22, 1663, Oldenburg

[22] See Des Chene 2002, 185; Schüler 2001, 96.

[23] For Rohault's first edition of the *Traité*, see Rohault 1671. However, due to the role of Clarke's annotations to Rohault's text, I shall mainly refer to the *System* (Rohault 1987).

[24] See Clerselier's unpaginated preface in Descartes 1664. For a discussion of Montmor academy, see Chap. 8 by Easton. For a general discussion of the relation between the French academies and Cartesianism, see Brown 1934; Chap. 3 by Roux.

[25] For more on this, see the rules drawn by Samuel Sorbière in his letter to Thomas Hobbes from February 1, 1658, in Sorbière 1660, 663–664.

wrote to Robert Boyle that natural philosophers such as "Roberval, Fermat, Frenicle, Rohault, Ozou" have been left out from the newly established stipend of the King of France.[26] However, even without institutional support Rohault developed an empirically oriented natural philosophy, employing visual aids and experimentation in his explanations.[27] His famous account of magnetism attracted public appreciation and dazzled witnesses. Another spectacular experiment involved the use of a large-scale model of the human eye in order to support his Cartesian account of vision. But what made these cases important in the context of seventeenth-century natural philosophy?

One effect is pedagogical. Rohault made readily available some Cartesian explanations and he forced the audience to reason from the most common properties of a phenomenon, going through a series of observations, up to causal explanation. In 1659, Clerselier remarked that Rohault lectured in such a way that "Witnesses of his experiments transform also into Judges and Referees of his explanations."[28] Moreover, in his preface to Rohault's posthumous writings, Clerselier highlights this mélange between the heuristic use of experimentation and a new methodological approach in physics:

> Those sort of proofs, so clear and convincing, & so evident to the senses, and so much different from those virtues and occult qualities that other Philosophers are accustomed to use in order to give reason to things they do not know, seem to me to explain clearly the truth of principles on which they depend; because it is not possible to derive such a great number of correct consequences & confirm their effects by such a small number of Principles, if these Principles were not true.[29]

The second, and most important outcome is methodological and it reflects a trend of transformation within Cartesian natural philosophy. By carefully designing mechanical explanations for all available phenomena, Rohault combined reasoning with experimentation in a new way. In an age when Cartesianism became accused of too much speculative thinking, Rohault is on a par with his contemporary so-called "experimental philosophers." But just as our volume rejects a clear-cut divide such as the traditional opposition between British Empiricists and Continental Rationalists, the case discussed in this chapter escapes other more recent divisions, such as the experimental philosophers versus speculative thinkers.[30]

---

[26] See Oldenburg 1965–1986, II, 73.

[27] For a list of the instruments and devices found in Rohault's possession at his death in 1672, see McClaughlin and Picolet 1976.

[28] Descartes 1659, unpaginated preface: "les Spectateurs des experiences que l'on y fait, se render aussi les Juges & les Arbitres des explications qu'on leur donne."

[29] Rohault 1682, unpaginated preface: "Ces sortes de preuves si claires, si convaincantes, & si sensibles, fort differentes de ces vertus & qualitez occultes dont les autres Philosophes ont coûtume de se servir pour rendre raison de ce qu'ils ignorent, justifient ce me semble bien clairement la verité des principes dont elles dépendent; car moyen de pouvoir tirer un si grand nombre de consequences justes, & que les effets verifient, d'un si petit nombre de Principes, si ces Principes n'estoient veritables."

[30] See Chap. 1 by Dobre and Nyden.

Rohault and other Cartesian philosophers, like Pierre-Sylvain Régis or François Bayle, later expanded a more integrated use of reason and observation in natural philosophy.[31]

Commenting the newly published *Physique nouvelle*, Michel Blay construed this transformation as "a kind of mechanical-experimental view replacing Descartes' metaphysical physics and, thus, taking part in the foundation of a new type of scientific practice which slowly escapes to the question of senses."[32] And Rohault is the prime representative of this change.

After many years of lecturing in public and worried that his ideas were being misinterpreted, Rohault produced an official version of his physics.[33] He introduced it with a long preface aimed at giving a systematic exposition of the impediments he found in the evolution of natural philosophy. If this looks at first as a mere historically oriented task, it soon turns into a programmatic account of his method.

Rohault confesses his unsettling worries for the state of physics—"when I came to consider Philosophy, particularly Natural Philosophy, I was very much surprised to see it so barren as not to have produced any Fruit"—which are further commented on five points.[34] He begins with a widespread objection against the old philosophy, namely, the blind reliance on authority. Always turning back to Aristotle (or other ancients) impedes one's use of reason. Hence, for him, old views should be examined critically and accepted only if the light of reason confirms them.

Rohault's second objection is that natural philosophy has been traditionally discussed in ways that are too metaphysical and this is accompanied by an interesting example. In order to argue that "every useful Science ought to descend immediately to Particulars," Rohault reacts against disputes concerning the divisibility of matter.[35] Obviously pointing to his contemporaries' debates regarding the ontological commitments expressed in their matter theories, the Frenchman tries to take a neutral attitude: "For though it could not be accurately determined, whether it be infinitely divisible or no; it would be sufficient to know, that it can be divided into Parts small enough to serve for all Purposes that can be."[36] In his *Le Développment de la physique cartésienne: 1646–1712*, Paul Mouy attributed this claim to Rohault's attempt to differentiate himself from Cordemoy's atomistic views.[37] But comparable

---

[31] For some examples of Cartesian philosophers who put emphasis on experiment and observation, see Ariew 2006 and other chapters in this volume.

[32] Blay, introduction to Rohault 2009, xxix: "Une sorte de conception mécanisto-expérimentale se substitue à la physique métaphysique de Descartes et participe ainsi à la construction d'une nouvelle manière de faire de la science échappant progressivement à la question du sens."

[33] The unofficial version was the *Physique nouvelle*, as we have seen above.

[34] Rohault 1987, unpaginated preface.

[35] Rohault 1987, unpaginated preface.

[36] Rohault 1987, unpaginated preface.

[37] See Mouy 1934, 114–115: "Ce n'est pas contre Descartes que Rohault se prononce ici, évidement. C'est plutôt contre Cordemoy, inventeur d'un atomisme fondé sur des principes métaphysiques soi-disant cartésiens, et aussi contre les scolastiques qui s'embarrassent de définir l'idée métaphysique du mouvement au lieu d'en étudier les lois et les effets" (It is, of course, not against

remarks can be found in other places. In 1661, Robert Boyle made a similar claim in his "Some Specimens of an Attempt to Make Chymical Experiments Useful to Illustrate the Notions of the Corpuscular Philosophy," where he associated Atomists and Cartesians under the label of corpuscular philosophers.[38] However, within the Cartesian milieu, Clerselier expressed the same neutral view "that every body can be divided in very tiny parts; I do not wish to determine whether their number is *infinite* or *not*, but, at least to our knowledge, it is certain that their number is *indefinite*."[39] Interestingly, while the text is presented as a reply to Roberval and was published in 1667 in the third volume of Descartes' *Lettres*, it belongs to a public lecture delivered in the Montmor Academy in July 1658.[40] Rohault turns this neutrality concerning matter's divisibility into an argument for exploring the properties of bodies. He denies the existence of occult qualities, which he considers mere inventions to cover one's vanity. But he still tries to discover the properties and causes that are hidden in nature by turning to experimentation. Again, Rohault makes a similar claim to Boyle, and, we shall immediately see, this parallelism will cover some other points.

These two impediments in the progress of natural philosophy are followed by a third observation regarding "the Method of Philosophers." Rohault denounces the two methodological "extremes" of natural philosophy, which put too much emphasis either on experiment or on reason, although neither of them—taken separately—can give good explanations. For him, the conclusion is "wherefore it cannot but be very advantagious [sic] to mix Experiments and Arguments together."[41] This is a very important remark, because it introduces Rohault's methodological novelty and it is connected to the previous objection in its emphasis on experimental practice. Reason, on the one hand, is considered to stay at general levels and unable to descend by itself to particulars. On the other hand, "by Experiments we can come to the Knowledge of gross and sensible Things only," which hardly makes it an option from the philosophical point of view.[42] Instead, Rohault claims that a joint use of them will produce the best results in natural philosophy. This is expressed in his

Descartes that Rohault speaks about here. It is more against Cordemoy, supporter of an atomism founded upon so-called Cartesian metaphysical principles and also against the Scholastics…) For Cordemoy's natural philosophy, see Ablondi 2005.

[38] For Boyle's so-called nescience regarding the divisibility of matter, see Boyle 1999, I, 355–356. For a general discussion of this problem in Boyle, see Anstey 2000, 41–45.

[39] Descartes 1667, 543: "que chaque corps peut estre divisé en des parties extremement petites; Je ne veux pas determiner si leur nombre est *infiny*, ou *non*, mais à tout le moins il est certain qu'au regard de nostre connoissance il est *indefiny*." This third volume of correspondence contains several letters in defence of Cartesian views, written by Clerselier and Rohault.

[40] See the title for this letter [97] "Lettre de M. Clerselier, (qui fut luë dans l'assemblée de M. de Montmor le treiziéme Juillet 1658, sous le nom de Monsieur Descartes, & comme si c'eust esté luy qui l'eust autrefois écrite à quelqu'un de ses Amis) servant de réponse aux difficultez que Monsieur de Roberval y avoit proposées en son absence, touchant le mouvement dans le plein" in Descartes 1667, 538.

[41] Rohault 1987, unpaginated preface.

[42] See Rohault 1987, unpaginated preface.

tripartite classification of experiment. First, there is simple observation, or "the mere simple using [of] our Senses." But this is not a rigorous task as sometime fruitful observation is accidental. The second type of experiment seems to surpass this difficulty, because it is employed "when we deliberately and designedly make Tryal of any Thing, without knowing or foreseeing what will come to pass."[43] For Rohault, this is the observation of nature, which is sometimes aimed, but quite often represents a random variation in the experimental setting. Yet, it does not represent the proper use of experiment and for this reason, Rohault gives a third category. Thus, he claims: "the third Sort of Experiments are those which are made in Consequence of some *Reasoning* in order to discover whether *it* was just or not."[44] Only this third type of experiment properly connects reason and experience, making them continuously communicate with each other, such that reason can guide observation and, in turn, by revealing some particulars through empirical investigation, experience can force new theoretical conclusions.[45] As we shall argue below, Rohault will use this method of enquiry in his physics.

Rohault identifies a fourth historical impediment in the evolution of natural philosophy. He denounces the Scholastics' rejection of mathematics. Reminiscent of some of Descartes' own claims, Rohault ascribes to the study of mathematics the formation of a style of thought, which by far exceeds the simple logic of the Scholastics. For him, natural philosophy can greatly benefit from a prior training in mathematics, which acts both as a logic of discovery, but, at the same time, gives a better knowledge of the particulars through geometry. Rohault shares with Descartes and other early modern thinkers the view that geometry will help one understand the figures and properties of bodies.

The fifth and final objection expressed by Rohault might seem odd at first. He says "I observed a fifth Defect, not in the *Method* of those who study Philosophy, but in *that* of a great many who read their Works."[46] One way to interpret this objection is to take Rohault's later considerations about Aristotle as an attempt to

---

[43] Rohault 1987, unpaginated preface. The second type of experiment listed by Rohault is remarkable for several reasons, including his similarity with Baconian methodology of variation within experiment and its reference to the method of the chymists. For a discussion of seventeenth-century chymists' reaction to Cartesianism, see Chap. 6 by Joly.

[44] Rohault 1987, unpaginated preface. This claim encompasses in a nutshell the double meaning of experiment in Rohault: a negative one, where experiment is used to confirm a particular hypothesis, and a positive one, where experiment is employed in the expansion of a theoretical model to new phenomena. For a discussion of the meaning of Cartesian *expérience*, see Clarke 1989, 209.

[45] The mutual benefit of experiment and speculation for each other is a recurring theme in Robert Boyle's natural philosophy. For example, commenting Boyle's method from the *Designe about Natural History*, Peter Anstey and Michael Hunter argue: "Boyle envisages a reciprocal relation between histories and hypotheses, the one informing the other and vice versa.... Boyle's methodology is better described as a two-stage reciprocal enterprise in which theory informs experiment with a view to constructing a natural history, which in turn informs theory" (see Anstey and Hunter 2008, 107).

[46] Rohault 1987, unpaginated preface. On the same page, his words looks even more confusing, "for scarce can a Philosopher present the Publick with any Fruits of his Studies, but some unknown Person who has a Mind to signalize himself, attacks them before he understands them."

accuse the Scholastics of misinterpreting the Stagirite. He supports such a reading through other claims—"I have taken all the general Notions from *Aristotle*"—or his later announcement that the principles of all things are matter, form, and privation. But there is another component of this objection. We have argued above that Rohault's popular lectures started to circulate in the form of unofficial manuscript copies. This biographical detail provides a hint concerning Rohault's project in the *Traité*. His aim is not only to explain various phenomena or to produce an encyclopedia or a natural history, but also to create a textbook on physics. Carefully ordered, the topics of his book share much of the empirical elements of the unofficial *Physique nouvelle*, yet, the theoretical core of his natural philosophy is much more carefully designed.

We shall now see how Rohault's physics is built. As with any textbook on natural philosophy, Rohault opens by defining the discipline, which is simply presented as "the Knowledge of natural Things, that is, that Knowledge which leads us to the Reasons and Causes of every Effect which Nature produces."[47] He further gives an abridged version of Descartes' metaphysics, but without making the claim that his physics is metaphysically grounded. Recalling Descartes' doubt, he denounces prejudice, which can only impede a natural philosopher in his examination of nature. And just before stating his Cartesian guiding principles, Rohault introduces "the Notions which precede the Study of Natural Philosophy, [which] may be reduced to two general Heads. For first, we know that there are *Things* really *existing* in the World; and from hence we think we know, at least in part *what* they are."[48] Ontology and epistemology join at the very core of Rohault's physics, creating the possibility to change the method of natural philosophy. Both of the principles listed here leave room for further interpretation and clarification. On the one hand, the existence of things is not subject to a metaphysically impregnated definition of matter, something that was severely problematic in Descartes.[49] On the other hand, how a physicist knows nature is reevaluated from the traditional Cartesian picture. If Descartes ascribed "more than moral certainty" to his physics, for Rohault certainty ceases to be an issue.[50] Instead, he operates with approximate knowledge, which will deliver more or less probability for his natural philosophy.

Rohault's two principles are thus loosening up criteria that were problematic in Descartes' natural philosophy and eventually produced the "breakdown" of

---

[47] Rohault 1987, I, 1.

[48] Rohault 1987, I, 3.

[49] Descartes' move from metaphysics to physics is achieved through his theory of matter. The existence of bodies and the identification between body, matter, and extension are mainly done on the basis of his prior metaphysics. However, there is a gap in the passage from one *res extensa* to a multitude of *res extensae*, which Descartes' philosophy fails to accommodate. For a detailed discussion of the individuation problems in Descartes, see Dobre 2011. For an overview of problems caused by Descartes' general physics in the *Principia philosophiae*, see de Buzon and Carraud 1994.

[50] For some of the problems raised by Descartes' classification of types of certainty, see Ariew 2011; Dobre 2013a; Chap. 2 by Ariew; Chap. 10 by Nyden; Chap. 11 by Hatfield.

Cartesian system.[51] Even if traces of Cartesian metaphysics can be found in this section of Rohault's treatise—see for example the shortened form of the *cogito*, which is given under the form of a syllogism—he builds up an explanation that takes greater account of the joint use of *perception*, *judgment*, *reason*, and *sensation*.[52] Rohault further introduces his method "of applying Philosophy to particular Subjects," which will later be used in various experimental cases, such as the experiments for the rejection of vacuum.[53] Everyone reading this section of the text will notice the combination of the positive statement expressing the possibility of grasping some properties related to particular phenomena and the apparently negative claim that when uncovering hidden causes, one can only hope to get merely probable explanations. Regarding the first, Rohault claims "that in order to find out what the Nature of any Thing is, we are to search for some one Particular in it, that will account for all the Effects which Experience shows us it is capable of producing."[54] His experimental search for common traits and properties is subject to confirmation, which is the second part of his strategy:

> If that which we fix upon, to explain the particular Nature of any Thing, do not account clearly and plainly for every Property of that thing, or if it be evidently contradicted by any one Experiment; then we are to look upon our Conjecture as false; but if it perfectly agrees with all the Properties of the Thing, then we may esteem it well grounded, and it may pass for very probable.[55]

We shall return shortly to Rohault's meaning of "conjectures," but before that, we should take notice of the similarity between the views expressed in this passage and Descartes' own solution from the *Principia philosophiae*.

In the final paragraphs of his book, Descartes discusses the type of certainty allowed in his physics. He refers to the code-cracker, for whom, attempting to decipher a letter is to find a coherent pattern by replacing the original words such that the new text will become meaningful. His findings become more certain if more correlations are to be found. However, despite increasing the number of instances, it is not clear whether this will allow for a higher certainty, which forces Descartes to introduce an intermediary stage of "more than morally certain."[56] His words denote such an ambiguity that severely departs from his earlier metaphysically grounded physics. Thus, in the *Principia* IV 205, Descartes confesses his hopes:

---

[51] For the "breakdown" of Cartesianism, see Watson 1966. For the general problems posed by the Cartesian tensioned relation between metaphysics and physics, with a detailed discussion of four cases of early French Cartesians, including Jacques Rohault, see Dobre 2010.

[52] For the corresponding section in the *Physique nouvelle*, see Rohault 2009, 6: "les facultés de *comprendre*, ou apprehender, de *iuger*, de *raisonner* ou conclure, et enfin de *sentir*."

[53] See Rohault's title for Chap. III of the first part of his book: "The Manner of Applying Philosophy to Particular Subjects." Rohault 1987, I, 13.

[54] Rohault 1987, I, 13.

[55] Rohault 1987, I, 13–14.

[56] Besides Ariew 2011, where Descartes' quest for certainty is discussed in close connection to the scholastic tradition, for an overview of how the problem evolves within Cartesian philosophy, see Dobre 2013a. See also Chap. 2 by Ariew; Chap. 11 by Hatfield.

> Now if people look at all the many properties relating to magnetism, fire and the fabric of the entire world, which I have deduced in this book from just a few principles, then, even if they think that my assumption of these principles was arbitrary and groundless, they will still perhaps acknowledge that it would hardly have been possible for so many items to fit into a coherent pattern if the original principles had been false.[57]

Rohault picks this idea and turns it into a method, focusing on the coherence and simplicity of his conjectures, which are subsequently subjected to the empirical test. Echoing Descartes, he states "And indeed there may be so many, and so very different Properties in the same Thing, that we shall find it very difficult to believe…In which Case, our Conjecture is not only to be looked upon as highly probable, but we have Reason to believe it to be the very Truth."[58] And in the next paragraph, he adds his reliance on experimentation: "we must consider, that, if our Conjecture be otherwise well grounded, it does not lose its Probability, because we cannot upon the Spot explain by it a Property, which appears from some new Experiment, or which we did not before think of."[59] Rohault specifically refers to the telescopic observations, "which were not in use till our Days, [and] have confirmed the Hypothesis of *Copernicus*."[60] With this example of instrumentally aided observation he ends the third chapter of his book, which was also the most theoretical exposition of his method.

Rohault continues by defining some of the terms he will use further. He lists classical metaphysical concepts, such as: being, substance, mode, accident, quality, essence, corruption, etc.[61] But this task is not a return to Scholastic classificatory schemes; instead, Rohault's aim is twofold. On the one hand, it is a preparatory step differentiating him from the Scholastics while trying to recover Aristotle. On the other hand, it coheres with the method of research presented earlier, because Rohault draws here a distinction between primary and secondary qualities. While such a distinction received more in-depth treatment in the philosophical works of English philosophers, most famously in John Locke, seeds of this idea can be found in various other places.[62] Rohault makes it clear in his subsequent explanations of hot and cold or taste and smell. But for the moment, he wraps up Aristotelian vocabulary in a Cartesian framework centered upon the renewed meaning of "modes."[63]

---

[57] AT IXb 328, CSM I 290.

[58] Rohault 1987, I, 14.

[59] Rohault 1987, I, 14. For the similar passage in the *Physique nouvelle*, see Rohault 2009, 15–16. We should notice that in his earlier text Rohault does not provide any examples, as it will later be the case with Copernicus.

[60] Rohault 1987, I, 14.

[61] See Rohault 1987, I, Part 1, Chap. IV, 15–18.

[62] See for example, Galileo's *Assayer* (1623), where the difference is between qualities pertaining to senses (i.e., touch), which are supposed to be in the object and qualities that are in the mind of the observer (e.g., colours). Another important place to find this distinction is Boyle's *Origin of Forms and Qualities* (1666). For the current chapter, it is not important to discuss the shift of nuances and the evolution of this distinction in the context of seventeenth-century thought, suffice is to say that Rohault draws a similar distinction. For a survey of the problem in the historical context, see Nolan 2011.

[63] Rohault will use the same Cartesian view regarding "modes" in his so-called Aristotelian acceptation of the three principles of material things, both in the *Traité* and the *Entretiens*. For a discussion of this use in both writings, see Dobre 2010.

Next, the French natural philosopher lists "The principal Axioms of Natural Philosophy."[64] Yet, they should not be taken as replacement for Descartes' laws of nature; despite being presented as "the Foundation of all Philosophical Truths," the list is merely provisional.[65] After he lists eight of them, Rohault acknowledges the great number of "axioms," but he considers that only the ones he presented are general enough to reside in the core part of his system of natural philosophy:

> The first is; that *Nothing, or that which has no Existence, has no Properties....*
> *Secondly; It is impossible that Something should be made of absolute Nothing; or that mere Nothing can become any Thing....*
> *Thirdly; No Thing or Substance can be wholly annihilated; that is, so cease to be, that there shall remain nothing at all of it....*
> *Fourthly; Every Effect presupposes some Cause....*
> *Fifthly,* Which is a Consequence of the foregoing Axiom; *If we our selves are not the Cause of any Effect, it must necessarily depend upon some other Cause....*
> *Sixthly; Every Thing, as much as it can, endeavours to continue in that State in which it is....*
> *Seventhly; That every Alteration is made by some external Cause....*
> *Eighthly; Every Alteration is always proportionable to the Force of the Agent which causes it:* So that the Thing which is altered continues, as much as it can, in its first State....[66]

But if Rohault wanted to give here only the most general principles of his natural philosophy, why did he list so many axioms? Take, for example, axioms four, five, and seven, which can generally be grouped under the claim made in the fourth. Moreover, axiom eight seems to be derived from axiom six. As for the rest of Rohault's axioms, they are customary metaphysical claims. By limiting their number to only these eight propositions and announcing that he will employ other axioms when needed, Rohault takes an inflationist approach to natural philosophy, taking its core subject to change; something that Descartes would not have allowed for his laws of nature.

Yet, there is something more in the central part of his physics, the theory of matter. Again, he restricts his views to the Cartesian identification of body with matter and extension. The argument presented by Rohault is similar to what one can find in Descartes' physics: there is only one property that is not accidental to matter and this is extension. He takes the same eliminative step as his master, building his argument on the observation that "though we do not perfectly understand what *Hardness, Liquidity, Heat, Cold, Heaviness, Lightness, Taste, Smell, Sound, Light, Colour, Transparency, Opacity,* and the like, are," they should not be taken as more than accidents of matter because "they are none of them inseparable from Matter, that is, it may exist without any of them...wherefore we say, that the *Essence* of Matter does not consist in any of these Things, but that these are accidental only."[67] Rohault finds four properties inherent in matter that cannot

[64] See Rohault 1987, I, Part 1, Chap. V.

[65] Rohault 1987, I, 20.

[66] See Rohault 1987, I, 18–20. For his claim that "there are yet more *Axioms* which I shall afterwards draw many Conclusions from; but because they are not so general as these, I shall content my self with mentioning them, when I have occasion to make use of them," see p. 20.

[67] Rohault 1987, I, 23.

be eliminated: extension, divisibility, figure, and impenetrability. Although they seem to be on a par with each other, Rohault does not devote much space to his attempt to reinforce Cartesian position and simply restates it: "because we conceive *Extension* before the other Three, and because we cannot conceive the other Three, without first supposing *Extension*, we ought to think that *Extension* is that in which the Essence of Matter consists."[68] Samuel Clarke reacts quickly to this conclusion, making two observations in his annotation. First, he argues that if Rohault would have extended his argument to its final consequences, he should have listed "existence" as the basic property, because "*Existence* is conceived before all other Properties of Matter."[69] However, even if allowed to be less strict, Clarke claims that Rohault should have ascribed the essence of matter to impenetrability, because only this belongs to material things. His argument is very Newtonian, although this time there is no direct reference to Newton. However, both Clarke's example of the pendulum and his reference to gravity and the difference between space and matter are deeply rooted in Newton. As will happen in numerous other places, Clarke's background opposition rests on his ontological commitment to the absolute space. In this respect, Rohault is not Newton and if one closely examines Clarke's objections, one finds that most of them can be reduced to this difference between space and matter.

However, Clarke does give Rohault credit for deriving some of his conclusions correctly, but he blames the starting premises, which are identified with his theory of matter. Such is the case of Rohault's rejection of vacuum. From the Cartesian perspective, the case of void seems rather clear. It is a contradictory notion that cannot have a place in natural philosophy. Yet, Rohault does not take the easy route of simply denouncing it as an impossible concept. Instead, he discusses the possibility of the void on both theoretical and experimental bases.

9.3 The Rejection of Vacuum: A Case Study

In Descartes' natural philosophy, the world is described as a plenum filled with matter. Hence, there is no room for something that is not matter. One of Descartes' arguments for the rejection of vacuum is presented under the form of a thought experiment:

> If someone asks what would happen if God were to take away every single body contained in a vessel, without allowing any other body to take the place of what had been removed, the answer must be that the sides of the vessel would, in that case, have to be in contact. For when there is nothing between two bodies they must necessarily touch each other. And it is a manifest contradiction for them to be apart, or to have a distance between them, when the distance in question is nothing; for every distance is a mode of extension, and therefore cannot exist without an extended substance.[70]

[68] Rohault 1987, I, 24.

[69] See Clarke's footnote in Rohault 1987, I, 24.

[70] AT IXb 73, CSM I 231.

When Rohault deals with this topic in the core part of his physics, he almost repeats Descartes. He asks the same question as in the above, "what the Consequence would be, if God should annihilate the Air in a Room?"[71] By building his theory of matter in the same way as Descartes and defining vacuum as "a Space void of all Matter," he cannot allow the existence of a completely "empty space." Instead, matter—even if it is not visible—should always be present in spaces that fill in distances between bodies.[72] Thus, he quickly answers the question about God's annihilation of matter by describing the natural consequence "the Walls would approach one another so near, that there would remain no Space betwixt them."[73]

Sharing different ontological commitments, Clarke and Rohault make a good case so far for the intricate relation between two competing systems of physics. A historian of science supporting the traditional view of the Scientific Revolution will probably point here a tension between the metaphysically-minded Cartesian natural philosophers and open-minded experimentalists. Yet, this is a superficial view grounded on the other historiographical misconception of a clear-cut divide between Rationalism and Empiricism.[74] This obvious tension between Samuel Clarke and Jacques Rohault comes in many degrees that will allow a more historically minded philosopher to get a better grasp of what is at stake in this change that marked the history of both philosophy and science. Such a case is provided by Rohault's experimental rejection of vacuum.

We mentioned above that Rohault was a very keen observer of nature. Connected to a tradition of experimenting with various instruments, Rohault used such devices for his Cartesian conferences.[75] With respect to vacuums, one should take notice of the correspondence between the Huygens brothers—Christiaan and Ludewijk—where, in the early 1660s, the first was contemplating the idea of commissioning the production of an air-pump to Rohault and asked his brother to communicate some of his own experimental findings to the Frenchman.[76] Above, we have referred to a passage from the *Journal des Sçavans*, where Rohault was placed in the context of French barometric research. Now, we see once again that his pneumatic experiments were famous in the learned community of his time. This notoriety seems paradoxical if we take into account that Rohault has already rejected vacuum as contradictory notion. Thus, let us see how Rohault builds his explanation.

First, he differentiates his account from the common view that nature has a fear of vacuum. His explanation covers both philosophical and empirical levels. From the

[71] Rohault 1987, I, 28.

[72] For the Newtonian reaction against Rohault's explanation, see Rohault 1987, I, 27. But just before listing several (Newtonian) reasons for rejecting it, Clarke agrees: "This is consistently enough said by him, who affirms the Essence of Matter to be Extension."

[73] Rohault 1987, I, 28.

[74] See Chap. 1 by Dobre and Nyden.

[75] For Rohault's connection to craftsmen and artisans, see McClaughlin 1996, 475–476.

[76] See letters no. 823 (December 18, 1660), 924 (December 7, 1661), and especially 952 (January 4, 1662) in Huygens 1890–1891, III–IV. Note that in some of these letters, Rohault's name is spelled "Rohaut."

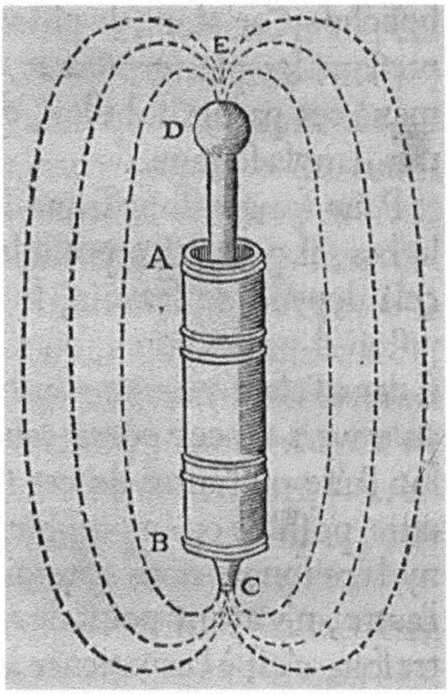

Fig. 9.1 An image of the syringe that was used by Rohault in his experiments. AB represents the tube of the syringe, C is the open end, D is the piston, and E represents the maximum height at which the piston can be drawn from the tube. The current illustration is from Rohault's *Traité*, see Rohault 1671, 73. (Courtesy of ETH-Bibliothek Zürich, Alte und Seltene Drucke) A similar image is depicted in Rohault 2009, 34

philosophical point of view, the fear of vacuum is rejected due to its emphasis on the final cause, when the efficient was needed. Rohault's analogy with the wood that is transported for winter from the provinces to Paris is very successful if one examines the book reviews of his *Traité* in both the *Journal des Sçavans* and the *Philosophical Transactions*.[77] Both reviewers pick up the example where Rohault claims that "if any one should ask, how Wood came from very remote Parts to *Paris*, and it should be answered, it came *out of the Fear of Cold*; this is no Answer to the Question; because the *final* Cause is alledged [sic] instead of the *efficient* Cause, which was the Thing demanded."[78]

From the perspective of experimental practice, Rohault simply draws the conclusion that if nature has a fear of vacuum, a syringe (see Fig. 9.1) that is immersed into water will raise the liquid no matter how long the tube is. However, experimental practice—both Rohault's own results, but also of his contemporaries, such as Pascal—show that water can ascend only to a given level, which makes him conclude that such view is untenable.[79]

In order to make an explanation convincing for his contemporaries, Rohault announces: "I shall offer some Particulars…in order to draw some certain and

[77] For the reception of Rohault's *Traité* in the two scientific journals, see *Journal des Sçavans* June 22, 1671, 25–30 and the *Philosophical Transactions* 1671, 2138–2141.

[78] See Rohault 1987, I, 56. The same example is given in Rohault 2009, 32.

[79] For a comparison between the barometric experiments performed by Pascal, Roberval, and Rohault, see Mouy 1934, 126–132. The aim of the current chapter is not to make a similar comparison, but rather to focus on a different aspect of Rohault's natural philosophy.

undoubted Consequences from a Foundation which cannot be contested."[80] Consistent with the method presented above, Rohault passes here from the theoretical part of his natural philosophy to experimental practice. He begins by initiating a series of so-called "conjectures," which are scenarios built upon the use of his instrument.

First, he claims that when the syringe is held in the air and has the hole on C opened, the motion of the piston, D, will be unimpeded and will push the air out of the tube AB, such that "the Air was moved by a real Impulse."[81] For Rohault, this experimental case is helpful in establishing the type of motion acceptable in his natural philosophy: local motion, described by the rules of mechanics and transmitted through impulse.

Now that he has argued for the search of the efficient cause and the mechanical explanation of phenomena, he moves to a claim that is much more connected to his Cartesian views. His second "conjecture" reveals a new possible experimental scenario: if C is blocked and there are no pores on the tube AB and neither in the piston D, then—based on his theory of the plenum—D will not move. Quite surprisingly, Clarke comments on this passage and agrees with the consequence of Rohault's reasoning: "this would indeed be true, if the World were full." However, just like in other cases, he quickly dismisses it on the basis of his Newtonian views, ending his comment with the following claim: "Nor need we here trouble our selves with any occult Pores or subtile [sic] Matter."[82]

Rohault's next set-up is presented as an alternative to the one just mentioned: C is (still) blocked, but there are pores in the tube AB. This time he adds a comparison between the pores and the parts of air, claiming that the latter are much more subtle than any existing pore in the tube. His conclusion—not commented this time by Samuel Clarke—is that D can move.

Now, Rohault's reasoning seems to have reached an impasse. His last two "conjectures" deal with the pores from the tube of his instrument. He has to answer now to the problem of their existence: "we must first know, whether the Syringe or the Sucker have any Pores in them or no; and after that, whether there be any Particles in the Air subtil [sic] enough to enter in at these Pores."[83] Leaving aside the ironic goal for someone who, at the theoretical level, denied the existence of void space as a contradictory notion and now tries to find out whether pores exist or not, Rohault's main issue in deciding between alternatives is that "neither of them can be determined by our Senses or by Reason, and there being no Contradiction in either, it must be decided by Experience."[84] If, at this point of his argument, we turn back to Rohault's preface, we can get a better grasp of his method of "conjectures." If the existence of pores in bodies is obscured to any of the two faculties, reason and sensation—in other words,

---

[80] Rohault 1987, I, 57.

[81] Rohault 1987, I, 57.

[82] Rohault 1987, I, 57n1.

[83] Rohault 1987, I, 58.

[84] Rohault 1987, I, 58.

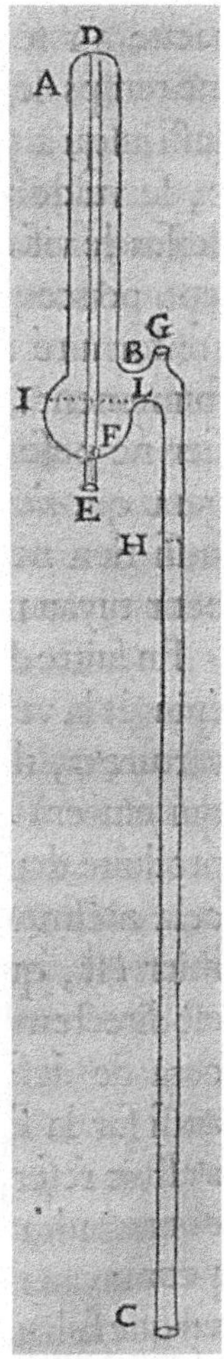

Fig. 9.2 Image of an instrument invented by Rohault and known as "la chambre de Rohault." BC is a tube of 27 inches, with C open. It communicates with a larger glass object (AIFLB) through BL. Within the AIFLB object, a very narrow tube (DE) is placed, such that D is closed and E is opened through F. G is another opened part of the apparatus. Rohault 1671, 93 (Courtesy of ETH-Bibliothek Zürich, Alte und Seltene Drucke). The same device is depicted in Rohault 2009, 40

their existence is not a theoretical consequence of the matter theory, nor subject to direct observation—experiment is called to settle the dispute.

At the same time, instruments play an important role, because they can influence the future "conjectures." And Rohault's argument unfolds by modifying the experimental setting and using a new device (see Fig. 9.2). This is Rohault's own variation after Pascal and Roberval and for a while it was known as the "chamber de Rohault." Rohault dresses all his argument in the form of an elaborate explanation about the motion and constituents of the air. By delving into the hidden parts of nature he only has the option to rely on the matter theory presented in the beginning of his treatise, which is none other than that of Descartes. Air and liquids are extended matter and can be divided indefinitely and variously distributed in experimental devices. Rohault—and Clarke in his footnote—refers to the experiments concerning the equilibrium of liquids in order to argue that invisible particles can have a very minute extension. For Rohault, this is a sufficient proof for accepting the other part of his problem: the pores in the bodies exist and they can allow the subtle particles of air to pass through them.[85]

Having been assured by his experimental procedure joined with the consequences deduced from his matter theory—"for having made it appear, that both the Syringe and the Sucker are full of Pores, and that the Air is full of Matter, subtle enough to pass through them"—Rohault explores another empirical consequence.[86] This time, he takes the case when (in the same syringe as in Fig. 9.1) D moves toward C all the way from A down to B, while C is immersed into water. With the help of the prior conclusion about the existence of pores into the tube and his explanation of the subtlety of matter, Rohault completes his explanation of this phenomenon. It can be best understood by looking at the image of the syringe, where the whirling air makes a full turn from E to C and back from C to E (just like Descartes' circles of matter). The theoretical boundaries supporting this conclusion are those of a strict mechanism, which are further revealed in his conclusion: "the Moving of the Sucker, is the general Cause of the Entrance of some Matter into the Place which it leaves."[87]

There is not enough space here to explore other cases of Rohault's experimentalism, suffice is to say that in various other places of the treatise, his strategy is similar: starting from some of the commonly known properties of the phenomenon under investigation, Rohault conjectures about the hidden mechanical causes, which are experimentally validated.[88] In a nutshell, his explanation refers strictly to efficient causes ascribed accordingly to the mechanical model of Cartesian natural philosophy. Motion is transmitted through contact and, since everything existing in nature is extension and extension only, it will require a transition of parts of matter from one vicinity to another.

9.4 Conclusion

In several places of his natural philosophy, Rohault remarked on the need to descend from general principles to particulars. Thus, he is not concerned with a physics that only gives classifications and general explanations, but with the one that jointly offers explanatory power and empirical confirmation. While his commitment to Cartesianism is not subject to debate, his adherence to a metaphysically oriented

[85] See Rohault 1987, I, 58–61.

[86] Rohault 1987, I, 61.

[87] Rohault 1987, I, 61.

[88] See for example Rohault's explanation of the production of glass-drops in Rohault 1987, I, 136–140; for the explanation in *Physique nouvelle*, see Rohault 2009, 80–84. The same strategy is applied in his discussion of magnetism; see Rohault 1987, II, 163–187 or Rohault 2009, 370–390. For the glass-drop experiment, see Dobre 2013b, 117–120.

physics raise a number of concerns. As we have seen above, Rohault dismisses metaphysics from natural philosophy, but at the same time, he smuggles in some metaphysical suppositions—as we would call them today. Yet, theory and its consequences are experimentally presented. Of course, this could be a consequence of Rohault's public conferences, and one can interpret it rhetorically. But, just like in the case study discussed above, that would only represent a scratch on the surface. Highlighting the circularity in Rohault's explanation (i.e., his commitment to a Cartesian metaphysical rejection of vacuum) is trivial and it obstructs us from getting a better grasp of his scientific methodology.

Thus, the aim of this chapter was to uncover Rohault's method in natural philosophy. His use of experiment is quite consistent with what he has presented in the programmatic preface to the *Traité*: not the mere use of the senses (observation) or random trials, but investigations "made in Consequence of some *Reasoning* in order to discover whether *it* was just or not."[89] This constant interplay between theory and experimentation is generated also by the lack of certainty with which one can know nature. Whereas by the sole use of the theory, some of the properties of the phenomenon under investigation (e.g., the breaking of a glass drop; the production of rarefied air in vacuum pumps; the so-called magnetic attraction; etc.) can be derived, still, there are others that can only be discovered through experimentation. Thus, Rohault adds a series of "conjectures" that have an intermediary status between what has been derived from his theory and what can be empirically acquired. Only, at this moment, he turns back again to experience in order to validate some of his prior conjectures. Experiment receives a double role: on the one hand, it discriminates between conjectures pointing to the ones that are empirically confirmed; on the other hand, it is a great pedagogical tool that can convince his audience to support his explanations.

Regarding the later, Rohault's book has been praised from the very beginning. For example, an entry in the *Journal des sçavans* says "this whole Book is nothing else than a collection of reasoned experiments, ordered methodically."[90] It also influenced Burchard de Volder's teaching of experimental physics.[91] Elsewhere, Rohault's *Traité* was quickly adopted in the university teaching. For example, Geert Vanpaemel convincingly argues that during the 1670s, in Louvain, "elements of the *Traité* were incorporated in practically every section of the curriculum."[92] Thus, we find that

> Rohault's lasting contribution to the physics course was his extensive treatment of the sensible qualities, which became a focus feature of the lectures on natural philosophy. With it, he introduced a new style of natural science, in which both hypothesis and experiment had a definite function.[93]

---

[89] Rohault 1987, I, unpaginated preface.

[90] *Journal des sçavans* June 22, 1671, 26: "tout ce Livre n'est qu'une suite d'experiences raisonnées & arranges methodiquement."

[91] By following Le Clerc, Ruestow refers to de Volder's classes as relying on Rohault, among others; see Ruestow 1973. See Chap. 4 by van Bunge and Chap. 10 by Nyden.

[92] Vanpaemel 1984, 33.

[93] Vanpaemel 1984, 39.

Rohault's discussion of sensible qualities in terms of modes and substances lies halfway between an Aristotelian explanation in terms of matter, form, and privation and Descartes' mechanistic account, which reduced everything to matter in motion.

By developing his own system of natural philosophy, Rohault picks up some of Descartes' unresolved issues, such as the tensioned relation between epistemology and ontology. Yet, he avoids most of the intricate problems associated with this relation by overlooking the metaphysics and focusing, instead, on physics. His natural philosophy is founded on two principles that leave out the metaphysical infused search for certainty and replaces it with a search for probable and always perfectible knowledge. By looking at Rohault's contemporaries, one will notice a striking similarity with Robert Boyle.[94] Some decades ago, Larry Laudan explored the connection between the clock metaphor and probabilism, pointing out several similarities between Descartes and Boyle.[95] More recently, Sophie Roux convincingly argued that in the second half of the seventeenth-century France, as well as in England, hypothetical reasoning constituted one of the main methods of natural philosophy.[96] Rohault is nevertheless part of these stories, but there is a biographical detail that was not available until the publication of the *Physique nouvelle* in 2009, which should be taken into account. Time and again in this chapter, we have seen that Rohault's practice was difficult to place at a certain date, but reports in various correspondences, statements in the scientific journals or in book prefaces, and the relation between Rohault's experiments in the *Physique nouvelle (1667)* and his *Traité* of 1671 suggest that Descartes' follower had developed his method as early as the 1650s. If this reading is correct, then our available histories about the transformation of natural philosophy in the period from Descartes to Newton are missing an important aspect. Rohault with his experimental method discussed in this chapter points to a lineage between French and English developments of natural philosophy in the second half of the seventeenth century. Moreover, this sheds a new light on Samuel Clarke's editions of the book. If Rohault developed an experimental physics that overlapped in several important aspects with those of the English "experimental philosophers," then, none should be puzzled by Clarke's so-called Newtonian annotations. Revealing continuity rather than pointing to a paradigmatic change, Rohault makes a good case study for how Cartesianism can adapt and evolve even in the late seventeenth century, challenging current histories of both philosophy and science.

Acknowledgments This work was supported by the strategic grant POSDRU/89/1.5/S/62259 (University of Bucharest). I would like to thank Peter Anstey, Trevor McClaughlin, and Tammy Nyden for their comments on the previous draft of this chapter.

---

[94] On the similarity between Rohault and Boyle with respect to causation, see Kenneth Clatterbaugh's excellent chapter on "The Limits of Classical Mechanism: Boyle, Rohault, and Newton" in Clatterbaugh 1999.

[95] See Laudan 1966.

[96] See Roux 1998.

References

Ablondi, Fred. 2005. *Gerauld de Cordemoy. Atomist, occasionalist, Cartesian*. Milwaukee: Marquette University Press.

Anstey, Peter. 2000. *The philosophy of Robert Boyle*. London: Routledge.

Anstey, Peter, and Michael Hunter. 2008. Robert Boyle's 'Designe about Natural History'. *Early Science and Medicine* 13: 83–126.

Ariew, Roger. 2006. Cartesian empiricism. *Revue Roumaine de Philosophie* 50: 71–85.

Ariew, Roger. 2011. The new matter theory and its epistemology. Descartes (and late scholastics) on hypotheses and moral certainty. In *Vanishing matter and the laws of motion. Descartes and beyond*, ed. Dana Jalobeanu and Peter Anstey, 31–46. London: Routledge.

Balz, Albert. 1930. Clerselier (1614–1684) and Rohault (1620–1675). *The Philosophical Review* 40(3): 221–245.

Birch, Thomas. 1756. *History of the Royal Society*, 4 vols. London: A. Millar.

Blay, Michel. 2009. Introduction. In *Physique nouvelle (1667)*, Jacques Rohault, V–XXIX, Paris/ Milan: Séha/Archè.

Bouillier, Francisque. 1868. *Histoire de la philosophie cartésienne*. 2 vols., 3rd ed. Paris: C. Delgrave.

Boyle, Robert. 1999–2000. *The works of Robert Boyle*. London: Pickering & Chatto.

Brown, Harcourt. 1934. *Scientific organizations in seventeenth century France (1620–1680)*. Baltimore: The Williams and Wilkins Company.

Buzon, Frédéric de., and Carraud Vincent. 1994. *Descartes et les "Principia" II. Corps et mouvement*. Paris: Presses Universitaires de France.

Clair, Pierre. 1978. *Jacques Rohault (1618–1672). Bio-Bibliographie*. Paris: Centre National de la Recherche Scientifique.

Clarke, Desmond. 1989. *Occult powers and hypotheses*. Oxford: Clarendon.

Clatterbaugh, Kenneth. 1999. *The causation debate in modern philosophy, 1637–1739*. London: Routledge.

Des Chene, Dennis. 2002. Cartesian science: Régis and Rohault. In *Blackwell companions to philosophy: A companion to early modern philosophy*, ed. Steven Nadler, 183–196. Malden: Blackwell.

Descartes, René. 1659. *Lettres de M. Descartes*, vol. II. Paris: Charles Angot.

Descartes, René. 1664. *Le monde*. Paris: Jacques le Gras.

Descartes, René. 1667. *Lettres de M. Descartes*, vol. III. Paris: Charles Angot.

Descartes, René. 1970. *Oeuvres de Descartes*, ed. Charles Adam and Paul Tannery. Paris: J. Vrin.

Descartes, René. 1985–1991. *The philosophical writings of Descartes*, 3 vols., ed. John Cottingham, Robert Stoothoff, and Dugald Murdoch. Cambridge: Cambridge University Press.

Dobre, Mihnea. 2010. *Metaphysics and physics in Cartesian natural philosophy: Descartes and early French Cartesians on the foundation of natural philosophy*. Proefschrift Radboud Universiteit Nijmegen, Nijmegen.

Dobre, Mihnea. 2011. The vanishing nature of body in Descartes's natural philosophy. In *Vanishing matter and the laws of motion. Descartes and beyond*, ed. Dana Jalobeanu and Peter Anstey, 11–30. London: Routledge.

Dobre, Mihnea. 2013a. Knowledge and certainty in the foundation of Cartesian natural philosophy. *Revue Roumaine de Philosophie* 57(1): 95–110.

Dobre, Mihnea. 2013b. On glass-drops: A case study of the interplay between experimentation and explanation in seventeenth-century natural philosophy. *Journal of Early Modern Studies* 2(1): 105–124.

Hoskin, Michael. 1961. 'Mining all within': Clarke's notes to Rohault's *Traité de Physique*. *The Tomist* 24: 353–363.

Huygens, Christiaan. 1890–1891. *Oeuvres complètes*, vols. III–IV. The Hague: Martinus Nijhoff.

Journal des Sçavans. 1666, 1671. Paris.

Laudan, Laurens. 1966. The clock metaphor and probabilism: The impact of Descartes on English methodological thought, 1650–65. *Annals of Science* 22(2): 73–104.
Matton, Sylvain. 2009. Remarques sur le manuscrit de la 'Physique nouvelle'. In *Physique nouvelle (1667)*, Jacques Rohault, LXXIII–XC, Paris/Milan: Séha/Archè.
McClaughlin, Trevor. 1976. Quelques mots sur Rohault et Molière. *Europe* 54(569): 178–183.
McClaughlin, Trevor. 1979. Censorship and defenders of the Cartesian faith in mid-seventeenth century France. *Journal of the History of Ideas* 40(4): 563–581.
McClaughlin, Trevor. 1996. Was there an empirical movement in mid-seventeenth century France? Experiments in Jacques Rohault's *Traité de physique*. *Revue d'Histoire des Sciences* 49(4): 459–481.
McClaughlin, Trevor. 2000. Descartes, experiments, and a first generation Cartesian, Jacques Rohault. In *Descartes' natural philosophy*, ed. Stephen Gaukroger, John Schuster, and John Sutton, 330–346. London: Routledge.
McClaughlin, Trevor, and Guy Picolet. 1976. Un exemple d'utilisation du Minutier central de Paris: la bibliothèque et les instruments scientifiques du physicien Jacques Rohault selon son inventaire après décès. *Revue d'histoire des sciences* 29(1): 3–20.
Milhaud, Gérard. 1972. Rohault et Molière. *Europe* 50(523–524): 37–49.
Mouy, Paul. 1934. *Le développement de la physique cartésienne: 1646–1712*. Paris: J. Vrin.
Nolan, Lawrence. 2011. *Primary and secondary qualities: The historical and ongoing debate*. New York: Oxford University Press.
Nouvelles de la République des Lettres. 1685, 1706. Amsterdam.
Oldenburg, Henry. 1965–1986. *The correspondence of Henry Oldenburg*, ed. Alfred Rupert Hall and Marie Boas Hall. Madison: University of Wisconsin Press.
Philosophical Transactions. 1671. London.
Rohault, Jacques. 1660. Ms 2225 *Conférences sur la physique, faites en 1660–1661, par Jacques Rohault, et recueillies par un de ses auditeurs qui, dit-il, y a ajouté du sien*. Paris: Manuscript in the archives of the Bibliothèque Sainte-Geneviève.
Rohault, Jacques. 1671. *Traité de physique*. Paris. doi:dx.doi.org/10.3931/erara-4795.
Rohault, Jacques. 1682. *Oeuvres posthumes de M. Rohault*. Paris: Guillaume Desprez.
Rohault, Jacques. 1987. *System of natural philosophy, illustrated with Dr. Samuel Clarke's notes, taken mostly out of Sir Isaac Newton's philosophy*. New York/London: Garland Publishing.
Rohault, Jacques. 2009. *Physique nouvelle (1667)*. Paris/Milan: Séha/Archè.
Roux, Sophie. 1998. Le scepticisme et les hypothèses de la physique. *Revue de Synthèse* 4(2–3): 211–255.
Ruestow, Edward G. 1973. *Physics at seventeenth and eighteenth-century Leiden: Philosophy and the new science in the university*. The Hague: Martinus Nijhoff.
Sarton, George. 1948. Second preface to volume 38: The study of early scientific textbooks. *Isis* 38(3–4): 137–148.
Schmaltz, Tad. 2002. *Radical Cartesianism. The French Reception to Descartes*. Cambridge: Cambridge University Press.
Schüler, Volkmar. 2001. Samuel Clarke's annotations in Jacques Rohault's 'Traité de Physique', and how they contributed to popularising Newton's physics. In *Between Leibniz, Newton, and Kant: Philosophy and science in the 18th century*, ed. Wolfgang Lefèvre, 95–110. Boston: Kluwer Academic Publishers.
Sorbière, Samuel de. 1660. *Lettres et discours de M. de Sorbière: sur diverses matières curieuses*. Paris: François Clousier.
Vanpaemel, Geert. 1984. Rohault's *Traité de physique* and the teaching of the Cartesian physics. *Janus: Revue Internationale de l'Histoire des Sciences, de la Médecine et de la Technique* 71: 31–40.
Watson, Richard. 1966. *The downfall of Cartesianism*. The Hague: Martinus Nijhoff.

Chapter 10
De Volder's Cartesian Physics and Experimental Pedagogy

Tammy Nyden

Abstract In 1675, Burchard de Volder (1643–1709) was the first professor to introduce the demonstration of experiment into a university physics course and built the Leiden Physics Theatre to accommodate this new pedagogy. When he requested the funds from the university to build the facility, he claimed that the performance of experiments would demonstrate the "truth and certainty" of the postulates of theoretical physics. Such a claim is interesting given de Volder's lifelong commitment to Cartesian *scientia*. This chapter will examine de Volder's views on experiment and show that they are not Newtonian or inductivist, as is sometimes claimed. While de Volder thinks we need deductive reasoning from first principles to provide evidence of the certainty of the content of our physical theories, he also contends that we need experiment to provide evidence of the certainty of the *existence* of the particular bodies those theories discuss. This approach to experiment is based on a distinction between rational certainty and the certainty of material bodies in the actual world. While this account is deeply influenced by Descartes, it is importantly different than Descartes' distinction between absolute and moral certainty. De Volder's "Cartesian Empiricism" is best understood as a continuation and further development of a long tradition of teaching through observation at Leiden.

10.1 Introduction

Burchard de Volder (1643–1709) was a Cartesian professor of mathematics and natural philosophy at the University of Leiden from 1670 to 1705. De Volder did a great deal to raise the status of pre-Newtonian experimental physics through his

T. Nyden (✉)
Department of Philosophy, Grinnell College, Grinnell, IA, USA
e-mail: nyden@grinnell.edu

M. Dobre and T. Nyden (eds.), *Cartesian Empiricisms*, Studies in History and Philosophy of Science 31, DOI 10.1007/978-94-007-7690-6_10,

© Springer Science+Business Media Dordrecht 2013

teaching innovations and contributions to the manufacture of scientific instruments. He set up an infrastructure at the University of Leiden that, one generation later, would be essential to the spread of Newtonianism. However, de Volder's understanding of the role of experiment in natural philosophy was quite different from that of his Newtonian successors. De Volder dreamed that mathematical methods would one day unite knowledge gained by experiments with Cartesian *scientia*: that is, deduction of phenomena from clear and distinct first principles. De Volder became more critical of Cartesianism toward the end of his career (perhaps it would be more accurate to say he was frustrated with its lack of results) but, he never rejected Cartesian methods. Though he met Newton and was one of the first on the Continent to read the *Principia*,[1] he never became a Newtonian as is sometimes claimed.[2] He was not satisfied with Newton's focus on effects. For de Volder, a priori knowledge of the *causes* of phenomena is the true object of physics and we do not reason back to first principles from effects.

De Volder never sees experiment as part of *scientia*, but rather takes it to be an essential partner to *scientia* in the broader enterprise of natural philosophy. While de Volder's philosophy is deeply influenced by Descartes and his experimental practices were initially influenced by his interaction with Robert Boyle and the Royal Society,[3] his philosophy of science is interestingly different from both of these influences and is best understood within the context of a long tradition of teaching through observation at Leiden. This chapter offers a preliminary study of his experimental physics, which I take to be pre-Newtonian. De Volder offers a promising case study of one of the many perspectives on experiment, certainty, and method in the seventeenth and early eighteenth centuries.

Despite de Volder's role in the spread of experimental physics, relatively little scholarship is dedicated to him. Most of the secondary sources that mention de Volder do merely that, dedicating perhaps a paragraph or two, if not just one sentence, about him within a work on another focus, be it the history of experimental physics,[4] the history of science in The Netherlands,[5] the history of scientific instruments,[6] the history of the University of Leiden,[7] the Dutch reception of

[1] Thijssen-Schoute 1954, 55; Ruestow 1973, 110.

[2] For example, see Sassen 1970, 7.

[3] Later in his career, de Volder taught Rohault's *Traité de Physique* in his courses. Le Clerc 1709, 398; repeated in Thijssen-Schoute 1954, 654; and Ruestow 1973, 103. Initially de Volder used Boyle's text. It is not clear when he switched to Rohault's or when he came to know the work of the French philosopher. I mention this to point out that his documented early exposure to experimental physics comes by way of England, though at some point, he clearly was familiar with the French Rohault. For Rohault's Cartesian physics, see Chap. 9 by Dobre.

[4] Heilbron 1982.

[5] See van Berkel 1985; Cook 1992, 2007; van Berkel et al. 1999.

[6] See De Clercq 1988, 1989, 1997a, b; Daumas 1972; Crommelin 1926, 1935, 1951.

[7] See Otterspeer 2008, 2001; van Poelgeest 1984 is a document in the Western Special Collections Room at Leiden University library, it has a small entry on de Volder; Suringer 1865; Siegenbeek 1832.

Cartesianism[8] or Newtonianism,[9] or de Volder's famous student Herman Boerhaave (1668–1738).[10] There are, however, two works on the history of physics at Leiden that discuss de Volder in greater detail: Ruestow's *Physics at Seventeenth and Eighteenth-Century Leiden: Philosophy and the New Science in the University* and Gerhard Wiesenfeldt's *Leerer Raum in Minervas Haus: Experimentelle Naturlehre an der Universität Leiden, 1675–1715.*[11] Another body of literature that discusses de Volder focuses on his 8-year correspondence with Leibniz.[12] This is where de Volder receives the most attention of philosophers, though it is largely within the context of Leibniz studies. There are only a few sources that focus on de Volder's own philosophy. These works question the extent to which de Volder should be considered a Spinozist.[13] This chapter hopes to look at de Volder in a new way—not in terms of the reception of a particular philosopher, be it Descartes, Newton, or Spinoza, but rather, as a foot soldier in the seventeenth-century battle of *scientia*, that is, as someone in the thick of the crossfire of competing understandings and justifications of the new science/philosophy. This approach contends that what makes de Volder an interesting and valuable object of study is precisely that he does not fit neatly into the categories of Cartesian, Newtonian, and Spinozist, or perhaps we should say, his case indicates how untidy these categories actually were in the seventeenth century. Further, a study of his ideas about natural philosophy and how they relate to his views about pedagogy provide insight into the development of pre-Newtonian experimental physics. In the following, I will present a brief overview of his life and legacy, to be followed by a discussion of the pedagogical tradition at Leiden, which had a significant influence on de Volder's own teaching. The final section will discuss de Volder's pedagogy and give his account of the role of experiment in natural philosophy.

10.2 De Volder's Career and Legacy

Most of the little we know about de Volder's life comes from three sources: The eulogy given by fellow professor, Jacob Gronovio at the order of the academic senate,[14] the biography by Jean Le Clerc, which appeared in the *Bibliotheque Choisie* soon after

[8] See Thijssen-Schoute 1954; Schuurman 2004 (while this is a work on the reception of Descartes' and Locke's epistemologies, it discusses de Volder in relation to the British influence on Dutch science); Wiesenfeldt 2000.

[9] Vermij 2003.

[10] See Sassen 1970; Lindeboom 1972 (Chap. 2 is a brief biography of de Volder), 1974; Kegal-Brinkgreve and Luyendijk-Elshout 1983; Knoeff 2002; Klein 2003.

[11] Ruestow 1973 gives considerable attention to de Volder in Chaps. 5 and 6; Wiesenfeldt 2002, Chap. 11 is on de Volder.

[12] See Loemker 1946; Ross 1986; Cardoso 1996; Jesseph 1998; Lodge 1998, 2001, 2004; Rey 2009a, b.

[13] See Klever 1997, Chap. 11, 1988; Lodge 2005.

[14] Gronovio 1709.

de Volder's death,[15] and Jean Pierre Niceron's entry on de Volder in his *Mémoires pour servir á l'histoire des hommes illustres dans la république des lettres.*[16] De Volder was born in Amsterdam where he began studies in philosophy and mathematics at the Illustrious School, also called the Amsterdam Athenaeum. The Dutch illustrious schools did not offer degrees and functioned as an intermediary between grammar school and university.[17] At the Amsterdam Athenaeum de Volder studied philosophy under Arnauld Senguerd (1610–1667) and mathematics under Alexandre de Bie (1623–1690).[18] Senguerd most likely taught Franco Burgersdijk's scholastic logic,[19] however de Volder was probably exposed to some Cartesian thought through de Bie, who was, by this time, gradually opening up to the new philosophy.[20] De Volder continued his education at the University of Utrecht, where he studied under Johannes De Bruyn[21] (1620–1675), who was committed to the new science and spent his career defending Cartesianism despite Utrecht's ban on teaching Descartes' philosophy in the classroom. De Volder earned his Master of Arts on October 18, 1660 and then pursued a medical degree at the University of Leiden under the instruction of Fransiscus de la Boë Sylvius (1614–1672).[22] It is under Sylvius' influence that de Volder is thought to have abandoned Aristotelian philosophy for that of Descartes'.[23] This is significant, for Sylvius' approach involved "affirming observation and experiment and, at the same time, attempting to introduce mathematics into medicine…he insisted in his lecture that for natural knowledge to be truly certain, it must be *demonstrated mathematically….*"[24] As we will see, de Volder's own pedagogy was greatly influenced by Sylvius. After receiving his medical degree from Leiden in 1664, de Volder returned to Amsterdam to practice medicine in the Remonstrant municipality.[25]

In 1670, the curators at Leiden offered de Volder a chair in logic on the recommendation of Johannes Hudde (1628–1704), a mathematician and the mayor of Amsterdam. However, de Volder was only allowed to take the position on the condition that he gives up his association with the Mennonite faith. He agreed, moved to Leiden, and joined the Walloon church, which served a French-speaking Calvinist population hailing from the Southern Provinces and was slightly more

[15] Le Clerc 1709.

[16] Niceron 1733.

[17] Van Miert 2009, 32–34.

[18] Le Clerc 1709, 349.

[19] Wiesenfeldt 2003, 910. Arnauld Senguerd studied under Burgersdijk at Leiden and Burgersdijk's *Institutionum Logicarum* (1626) was the standard logic textbook in The Netherlands at the time.

[20] Van Miert 2003, 102.

[21] Klever 1988, 199.

[22] Le Clerc 1709, 348.

[23] Knoeff 2002, 24.

[24] Smith 2004, 223.

[25] Le Clerc 1709, 350.

liberal than its Dutch Reformed counterpart. De Volder began by teaching the logic of Franco Burgersdijk (1590–1635), an eclectic neo-Aristotelian with Ramist influences, the traditional lesson plan he would have been expected to teach at the time.[26] However, de Volder had a low opinion of syllogistic logic. Within 1 year of his appointment at Leiden, he requested to teach physics. The curators allowed him to do so alongside his logic courses. Even though he was officially serving as the chair of logic, de Volder soon replaced Aristotelian logic and the Burgersdijk curriculum with more metaphysical and mathematical material. He particularly emphasized the philosophy of Descartes, both in his university and private courses, and quickly took on the role as a representative of Cartesianism at the University.

Controversy surrounding Cartesianism had a long history at Leiden, going back to the 1640s. In 1672 such controversies came to a head. They mainly consisted of disputes between Voetian and Coccejan members of the institution, an academic theological division that ran deep and aligned with the class and political divisions of the Dutch Republic as a whole. "Voetians" take their name from their leader, Gisbertus Voetius (1589–1676), a conservative Utrecht theologian who called for a "further Reformation" of Dutch society and academia.[27] Coccejans, on the other hand, followed the liberal Leiden theologian Johannes Cocceius (1603–1669), who led a counter-Voetian movement within the Reformed Church, which maintained that philosophy had an independent status from theology. Both the Voetians and Coccejans were Orthodox Calvinist, however, they differed in how they defended the truth of the Reformed Faith. Voetians appealed to Aristotelian philosophy while Coccejans appealed to the new philosophy, particularly Cartesian rationalism, and created new approaches to the interpretation of Scripture.[28]

At stake was the power of the liberal arts faculty relative to that of the theology faculty. The new science gave philosophy a foundational status, one independent of the censorship of theologians and the Voetians were not keen to give up that power. Moreover, philosophy was beginning to be seen as more than a mere preparation for the study of medicine, law, and theology. It was now seen to have value in its own right. By the 1670s, the Cartesians had made a lot of headway. Not only did Cartesianism dominate in the second half of the seventeenth century, but the liberal arts faculty at Leiden enjoyed a great deal of status and autonomy relative to other institutions. Leiden was unique in that the prestige of its professors in the liberal arts faculty approached if not equaled the status of theology faculty.

The debates between Cartesian Coccejans and Aristotelian Voetians had class and political dimensions as well. The Voetians appealed to the lower-middle class. They were conservative and strict in practice, being influenced by the Puritanism of England and Scotland. Politically, they strongly supported the House of Orange and

[26] For more on Burgersdijk, see Bos and Krop 1993.

[27] McGahagan 1976, 109.

[28] Spaans 2007, 333. See Nyden-Bullock 1999, 2007, 13 and 46–47.

the return of Stadholder rule. The Coccejans, on the other hand, sided with the republican regime during the Stadholderless Period (1651–1672). Their sensibilities were those of the higher classes and were less strict in dress and worship and much more tolerant of non-orthodox religions. Burgersdijk's neo-Aristotelian legacy of separating the disciplines helped the success of Cartesianism at Leiden. It gave Cartesian professors the ability to argue for autonomy of the philosophy department from the theology department.

Between 1672 and 1674, there was a particularly bad flare up of tensions in Leiden between the Cartesian Coccejans and Aristotelian Voetians, resulting in disruptions during lectures, often so violent they sometimes resulted in injury. The timing was not arbitrary. France declared war on The Netherlands and Johan de Witt (1625–1672), the Grand Pensionary of Holland and the effective leader of the Dutch Republic during the First Stadholderless Era, was unable to stop France from occupying a significant part of the country. After his defeat, the people became disillusioned with de Witt's republican cause and embraced the return of Orangist rule. William III of the House of Orange was appointed Stadholder and as commander of the army successfully blocked any further advance by the French. Sentiment was so high that a mob murdered and mutilated de Witt and his brother. Still, not everyone was thrilled to see the Stadholderless Era end. Cartesians flourished during this period and enjoyed their close associations with de Witt. Philosophically, Cartesians aligned with the republican values of religious tolerance and freedom of speech and because of republican rule were able to advance the status of philosophy in great measure. The orthodox Calvinists, on the other hand, supported the House of Orange and hoped to use the new political situation to better their position in the university. From their point of view, this was the time to put philosophy back into its proper place—as subservient to theology and the synods of the Reformed Church. The Cartesians fought back—often in the form of Cartesian students ridiculing their Aristotelian Professors to the point that it was impossible to continue the lecture.[29] De Volder was on the side of the Cartesians, however, he was disgusted at the tactics they were using. He felt that there must be a better way to maintain and further the progress that the new philosophy had made within the institutional structure, even in the new political situation. Experiment would hold the key.

In 1674, he traveled to England, where he met Newton[30] and it is believed he met Boyle and attended meetings of the Royal Society.[31] Immediately after his return, de Volder requested the Leiden curators to fund the creation of a Physics Theatre in which he would demonstrate experiments during physics lectures.[32] The curators agreed and in 1675 de Volder taught the first experimental physics course offered by

[29] Wiesenfeldt 2000, 2.

[30] November 1684 letter from de Volder to Newton in Hall 1982, 11.

[31] Boas Hall 1966, 1n14; De Clercq 1997a, 134–135; There is nothing known with certainty about the scientists de Volder met in England, see Lindeboom 1974, 6.

[32] Molhuysen 1918, 293.

a European university.[33] Some secondary sources imply that de Volder's creation of the Physics Theatre was the result of this trip, that de Volder adopted an "English" approach to experiment, often characterized as Baconian, and brought it back to Leiden.[34] He certainly was impressed with the experiments and instruments he witnessed there, which made up most of the experiments he demonstrated in his early courses, but it would be a mistake to assume he *understood* experiment in the same way as those he met in England. First of all, there was not *one* view of experiment in England in the seventeenth century, but a multiplicity of views.[35] Second, as we will see in the last two sections of this paper, de Volder never took an inductivist approach to natural philosophy and did not see experiment as providing the content of our knowledge of natural laws.

Wiesenfeldt hypothesizes that de Volder went to England with the purpose of bringing experimental physics to Leiden as a means for overcoming the deadlock between the Cartesians and Scholastics. De Volder's introduction of experiment at Leiden did end up having this effect.[36] On December 3, 1674, after returning from England, he requested funds from the curators to build the Physics Theatre and to acquire equipment. Leiden Faculty-Senate archives record this request:

> Doctor Burchard de Volder, Philosophy professor at this university, represented both orally and in writing to H.C. and B. the usefulness and great advantages that can be brought to this university by following the example of foreign universities and illustrious schools so that by experiment the truth and certainty of the theories taught in theoretical physics might be demonstrated. He thinks that it would not only be useful and entertaining for the students, but would also attract other students to the study of Physics. The matter was discussed, and they would look into the costs for buying and maintaining instruments for doing the experiments.[37]

---

[33] Crommelin 1926, 1–2, 1951, 2; De Clercq 1997b, 42–43; Turner 1998, 103; Klever 1988, 199; Otterspeer 2001, 325; Knoeff 2002, 23.

[34] Lindeboom 1972, 6; De Clercq 1997a, 134–135, 1989, 5; Knoeff 2002, 24; Schuurman 2004, 64.

[35] On the diversity of approaches to experiment, see Jalobeanu 2013.

[36] See Wiesenfeldt 2000.

[37] Molhuysen 1918, 298: "D. Burchardus de Volder, Professor Philosophiae in dese Universiteyt, heeft aen de H.C. ende B. soo mondelingh als by geschrifte gerepresenteert de nuttigheyt ende de groote avantages, die dese Universiteyt soude konnen werden toegebraght, indien nae het exempel van andere uytlandsche academien en illustre scholen alhier in dese Universiteyt by experimenten moghten werden gedoceert en aengewesen de waerheyt ende seekerheyt van die stellingen ende leeren, die in Physica theoretica de studenten werden voorgehouden, met presentatie dat hy niets soo lieff soude sien dan dat C. ende B. voorn. hem tot de exercitie praefatae Physicae experimentalis soude believen te admitteren ende daer toe te subministreren soodanigh een plaets, mitsgaders alsulke instrumenta ende verdere nootsaekelycheden, als de voors. demonstratie soude komen te vereyschen; sijnde de voorn. de Volder absolutelyck gepersuadeert dat, behalve de nuttigheyt ende het vermaek van de voors. te doene demonstratie, door deselve veele studenten van andere academien ende scholen herwaarts aengeloct ende het studium Physices seer gefaciliteert soude konnen werden. Waerop gedelibereert sijnde is goedgevouden ende geresolveert dat dese saeke wat nader sal werden geexamineert, ende overwoogen welke ende hoedanige plaetsen dairtoe best soude konnen werden geappliceert en hoeverre dese Universiteyt daer mede als oock met den incoop ende onderhoud van de nodige instrumenten ende preparatien soude werden belast, om 't selve gedaen sijnde, alsdan soodanigh te werden geresolveert als bevouden sal werden te behooren."

Notice that de Volder promised the Theatre would be useful and advantageous for the University, not only for its pedagogical advantages (which we will discuss in the final section), but also for increasing student interest in physics, which had in the past been a big draw of foreign students to Leiden. At this moment, status would have been a concern for the curators because of a recent drop in foreign enrollments. De Volder reinforced this point by hitting on the curators' insecurities, the Leiden University's relatively new competition in matters of the new philosophy, the scientific societies and academies popping up throughout Europe, as well as the Dutch Illustrious Schools, which were causing competition for both professors and students. In this context, we can see that de Volder's request suggested that if the University of Leiden did not teach experimental physics, then the University would be failing to meet a need that these institutions would be pleased to offer. Further, his recent trip to England and visit to the Royal Society would have given him an air of authority in regard to foreign opinion.

The curators enthusiastically granted de Volder's request as soon as January 26, 1675. He bought a house near the main Academic Building and Botanical Gardens to house the Theatre and began to furnish it with the latest equipment, spending one eighth of the university's annual budget.[38] De Volder shared the Theatre with his Neo-Aristotelian colleague, Wolferd Senguerd (1646–1724), son of his former teacher, who was just about to be appointed to full professor of Aristotelian philosophy in response to the Cartesian disturbances. (The curators took the approach of hiring both Cartesian and Aristotelians in the hope that by balancing the philosophy faculty, they would have avoided further disputes.)[39] Notice that experiment was not considered the domain of the new science, but rather, a new tool that could be employed by both camps. The controversies continued the year after the Theatre was created, coming to a head in a resolution forbidding the teaching or discussing of 20 Cartesian propositions surrounding "the theological consequences of Cartesianism, such as issues of universal doubt, the nature of God, the role of philosophy in interpreting the Scriptures and the nature of the will."[40] De Volder responded by co-authoring a pamphlet with the Cartesian Coccejan theologians Abraham Heidanus and Christoph Wittichius.[41] "They argued that the Cartesian distinction between mind and body secured the important distinction between theology and philosophy."[42] Heidanus was the only one to put his name on the pamphlet, and the 78 year-old theologian was fired as a result. Eventually things calmed down and the Cartesians were thereafter allowed to teach in peace.

---

[38] Wiesenfeldt 2000, 3.

[39] Van Bunge 2001, 45.

[40] Molhuyusen 1918, III, 320; Knoeff 2002, 25–26. For other condemnations of Cartesian philosophy, see Chap. 2 by Ariew.

[41] *Consideratien over de Resolutie van de Ed. Achtbare Heeren Curateuren der Universiteyt binnin Leyden, en Borgermeestern de selver Stede, &c*, Aernoudt Doude, 1676.

[42] Wiesenfeldt 2002, 3.

Notice that during this year of heightened controversy, both de Volder and Senguerd centered their physics courses on the demonstration of physical experiments. It is worth pointing out here that while they represented different sides in the debates between Aristotelian Voetians and Cartesian Coccejans, they were both part of a long eclectic tradition of philosophy at Leiden. This eclecticism had pre-Cartesian roots in Burgersdijk, whose humanist interpretation of Aristotle influenced his student and predecessor, Adriaan Heereboord (1613–1671), as well as Heereboord's student and predecessor, Johannes De Raey (1602–1702). Heereboord and de Raey used this eclectic approach to introduce Cartesianism, presenting it as consistent with an accurate reading of Aristotle and mixing it with other modern philosophies, such as that of Bacon and Gassendi.[43] This mixing of new and old philosophies is partly responsible for Leiden's early introduction of Cartesianism, and later Newtonianism, in its classrooms. It allowed Cartesianism to be presented as consistent with Ancient and humanist philosophies, thus making it less threatening to the status quo.[44] It allowed experimentation and mathematics to be combined with Cartesian and Aristotelian philosophies in a way that did not raise eyebrows, preparing fertile soil for the seeds of Newtonianism one generation later.[45]

The Physics Theatre did not bring an immediate end to the controversies, but its early history does shed light on how the categories of old and new philosophies were not clearly separated at Leiden. As part of this eclectic tradition, both de Volder and Senguerd considered old interpretations (whether that of Aristotle or Descartes) to be in need of revision and for both of them experiment would play a role.[46] While it is the case that de Volder and Senguerd's positions were meant to represent competing sides, they both had a very similar training emphasizing observation and were both well disposed to some aspects of Cartesian natural philosophy and somewhat critical of others. They did have differences, but many of them were not what we might expect. For instance, Senguerd's views were closer to Gassendi's atomism while de Volder was committed to Descartes' plenum theory.[47] The truth is that their most straightforward differences were more political than philosophical. Senguerd was a pious Calvinist who tended to support the anti-Cartesian resolutions, but philosophically, he was more of an eclectic thinker than an anti-Cartesian and in fact did promote some Cartesian ideas about matter and motion.[48] He did lean towards Aristotle, but saw no reason that the new philosophy could not be seen as

---

[43] See Van Bunge 2001.

[44] Wiesenfeldt 2000, 6.

[45] On the continuities between Dutch Cartesianism and Newtonianism at Leiden see Krop 2003. On the ways professors at Leiden combined various natural philosophies see Ruestow 1973 (Despite Ruestow's recognition of eclecticism and synthesis, he does represent the movement from Cartesianism to Newtonianism at Leiden as continuous).

[46] Wiesenfeldt 2000, 6.

[47] Knoeff 2002, 23.

[48] Wiesenfelt 2000, 3; Knoeff 2002, 24.

consistent with the ancients. Nonetheless, experiment is something new de Volder and Senguerd had in common: they both initially took their lesson plan from Boyle's experiments.[49] Both professors used experiment to verify conclusions already held. They did not promote induction or try to gather new information about the world through experiments.[50]

De Volder was right, the Theatre did bring the university and himself great acclaim. But it did more than just raise the status of de Volder and his institution, it helped to maintain and build upon the recently acquired status of philosophy within the structure of the Dutch academy. Both Cartesianism and experimental physics necessitated modifications in the university's structure, both sought a role for philosophy that was independent of synods and theologians.[51] However, after the fall of de Witt in 1672, the Voetian theologians were doing all they could to take back the power Cartesians had acquired for the philosophy faculty during the Stadholderless Era. Proposed 2 years after the fall of de Witt, during a time of political uncertainty for the Cartesians, the Physics Theatre would help the philosophy faculty hold on to its status by providing its own prestigious facilities, equivalent to, but separate from the Anatomy and Chemistry Theatres of the medical school.[52] The medical faculty always enjoyed remarkable autonomy from the theologians and the philosophy faculty hoped to secure the same.

De Volder's creation of the Physics Theatre left two important legacies in the history of experimental physics: (1) a contribution to the production of scientific instruments and (2) the fame and emulation of his pedagogy. His courses were most famous for his demonstration of the air pump, but university inventories and student notes indicate that in addition to pneumatics, he performed experiments in mechanics, hydraulics, optics and magnetism. His Physics Theatre contained 64 scientific instruments, including many apparatus for his air pump, a collision machine, microscopes, fountains, and a magic lantern. Further, there were Magdeburg hemispheres hung on a platform so as to illustrate how, when evacuated of air, they could support the weight of a person.[53] Lecture notes by Carolus Vinson, one of de Volder's physics students in the 1676–1677 school year, provides evidence that de Volder's early lectures centered on the demonstration of many experiments taken from Boyle's *New Experiments Physico-Mechanicall Touching the Spring of the Air and its Effects* (1660).[54] Le Clerc's biography tells us that de Volder taught Jacques Rohault's *Traité de physique* (1671) later in his teaching career.[55]

---

[49] Wiesenfeldt 2000, 3.

[50] De Clercq 1997a, 17, 1989, 10.

[51] Wiesenfeldt 2000, 5.

[52] Otterspeer 2001, 331.

[53] De Clercq 1997b, 10.

[54] Vinson 1676–1677.

[55] Le Clerc 1709, 398, cited in Ruestow 1973, 103.

Scientific equipment was not readily available for purchase in Holland at the time, so de Volder contacted a local artisan, Samuel Musschenbroek (1660–1681), who made chandeliers, lamps and other brassware. De Volder produced his own design for an air pump and commissioned Musschenbroek to make it, along with most of his other scientific equipment. This completely changed the nature of the Musschenbroek workshop. Not only did it begin an ongoing working relationship with Leiden physics professors who wanted to further furnish the Physics Theatre and their own private collections, but it initiated an overall change in the business, which began to focus on scientific equipment and became the most important supplier in The Netherlands and indeed throughout much of Europe.[56]

De Volder's teaching became famous and within the next 30 years his pedagogy spread from Leiden to other institutions. By 1705, the other four Dutch universities were teaching physics through experiments. In Utrecht, Franeker, and Harderwijk the professors were graduates from Leiden and most likely had been students of de Volder himself. In the fourth, the University of Groningen, Johann Bernoulli (1667–1748) introduced the experimental approach. Though he did not attend Leiden, we know from his January 8, 1698 letter to Leibniz that his pedagogy was directly influenced by de Volder:

> The governors have set me a new teaching task, and to this end have appropriated a certain sum to buy experimental instruments, so that, after the example of de Volder in Leiden, I shall occupy and amuse our students with mathematic-physical experiments.[57]

While this experimental pedagogy spread most quickly within Dutch universities, it eventually made its way to other Northern European universities. Just after 1700 there were still only a few cases outside of Dutch Republic: J. C. Sturm (1635–1703) at Altdorf, G. A. Hamberger (1662–1716) at Jena and Pierre Varignon (1654–1722) at the *Collège Mazarin*.[58] Interestingly, most of these cases have links to Leiden. Sturm studied at Leiden and was a friend of de Volder's. He would have influenced his student Hamberger through his *Collegium Experimentale sive Curiosum* at the University of Altdorf, where beginning in 1679, he demonstrated experiments based on those from the *Accademia del Cimento*.[59] Pierre Varignon began teaching at the *Collège Mazarin* 13 years after De Volder's first experimental course. What he knew of de Volder's activities is unclear; after all, he trained and worked in Catholic institutions, which were less likely to take the Protestant Leiden University as a model for excellence. However, it is noteworthy that he too was a longtime correspondent of Johann Bernoulli and Leibniz. Perhaps he knew of de Volder's work through them.

While experiments had been demonstrated at Italian universities, it wasn't until 1690 that a special laboratory was built for the demonstration of experiment

[56] De Clercq 1997a, 108 and 152.

[57] Bernoulli to Leibniz January 8, 1698, translation from De Clercq 1997a, 143.

[58] Heilbron 1982, 132–133.

[59] See Ahnert 2002.

in Bologna.[60] This is 15 years after the founding of Leiden's Physics Theatre. It was even longer, over 31 years, before non-Dutch universities demonstrated experiments within their *physics* courses. The first English *public* physics course included experiment and was taught by Francis Hauksbee (1666–1713) in London in 1704, but it was not until 1706 that an English *university* taught experimental physics in a course by William Whiston and Roger Cotes at Cambridge.[61] Christian Wolff continued in the tradition began in Germany by Sturm and Hamberger by teaching a course in experimental physics at Halle in 1706 and experiment did not enter the university physics classroom in Italy or Switzerland until 1737 (at Bologna and Geneva).[62]

The spread of de Volder's teaching innovation played a little-known, but important role in setting up the infrastructure that would, one generation later, facilitate the spread of Newtonian experimental physics on the Continent.[63] First, his Physics Theatre would one generation later become the epicenter of the early spread of Newtonian physics on the Continent. Students, including Voltaire, came from all over Europe to study with Willem J. 's Gravesande to see Newtonian theory demonstrated with physical experiments, most by 's Gravesande's own design. Second, as we have seen, de Volder was the catalyst for the Musschenbroek production of scientific instruments. This should not be overlooked, because 's Gravesande used this workshop, now run by Jan van Musschenbroek, Samuel's son, to create his instruments, which were made precisely to demonstrate the truth of Newtonian physics. These instruments were illustrated in 's Gravesande's famous Newtonian textbook, *Mathematical Elements of Natural Philosophy Confirmed by Experiments* and conveniently sold with a sales list from the Musschenbroek workshop. Most instrument collections, whether by scientific societies or individuals, including that of the Tsar of Russia, contained equipment either purchased from the Musschenbroek workshop or built off the precise illustrations of their equipment in 's Gravesande's texts.[64]

While de Volder's work did much to bring about an infrastructure that would be ideal for the early transmission of Newtonianism, his own pedagogy of experiment was not Newtonian, but an extension of a long tradition at Leiden of teaching through observation. As we will see in the final section of this paper, he never held a Baconian or Newtonian understanding of experiment. Nor do his views strictly adhere to that of Descartes, though they have deep Cartesian influences. It is helpful to understand his pedagogy as a continuation of Leiden's tradition of teaching through observation and his natural philosophy within the context of Leiden's eclecticism.

---

[60] Schuurman 2004, 66.

[61] Heilbron 1982, 134 and 153.

[62] Heilbron 1982, 133–139.

[63] See Gori 1972.

[64] De Clercq 1989, 8.

10.3 Leiden's Tradition of Teaching Through Experiment

From its beginnings in 1575, the University of Leiden was committed to a balance of theory and practical experience.[65] It made a point of teaching through practice in order to prepare students to meet the needs of the country.[66] This commitment is expressed in a letter from Willem of Orange to the States of Holland and Zeeland regarding the founding of Leiden University:

> [It is necessary to establish a university] to be a firm support and sustenance of freedom and good legal administration of the country not only in matters of religion, but also with regard to the general welfare of the people.... [Students are to be trained] in both the right knowledge of God and all sorts of good, honourable, liberal arts and sciences, serving the legal administration of countries.[67]

The essence of this practical approach was providing students opportunities to observe and interact with the natural world. In 1593 Leiden's famous Botanical Gardens and Anatomy Theatre were founded for this purpose. The tradition of observation continued in 1636 with the innovation of teaching medical students through clinical rounds by Sylvius (de Volder's medical professor) and in 1669, with the addition of a Chemistry Theatre. All of these, the theatres, gardens, and clinical rounds, were integral parts of lecture. When viewed in this context, de Volder's Physics Theatre, directly modeled on the Anatomy and Chemistry Theatres of his institution, seemed the next obvious step for a professor recently exposed to the physical experiments of the Royal Society.

Perhaps the most important example of this Leiden tradition in terms of its influence on de Volder is the pedagogy of Sylvius, which centered on teaching through observation and experiment and at the same time was deeply influenced by the philosophy of Descartes. Sylvius held a Cartesian theory of sensory knowledge and his pedagogy of observation must be understood in these terms. He held that the passions were excited by external objects and therefore connected to sensory knowledge, a view influenced by Descartes' *Passions of the Soul*.[68] Sylvius appreciated the pedagogical benefits of observation in getting student's attention and creating in them a *desire to learn* more about the causes of the phenomena they observed, knowledge only gained through mathematical deduction. Experiment inspired students to appreciate the importance of and take on the task of a priori methods. While sensory knowledge had the power to inspire students, Sylvius warned that it had dangers as well. First, there are moral dangers when immersed in

---

[65] Otterspeer 2001, 333.

[66] Smith 2004, 221. The commitment to practical knowledge is particularly evident in Leiden's engineering school, *School voor Nederduytsche Mathematique*, set up by Simon Stevin, which was closely associated with the University (Van Bunge 2001, 2).

[67] Letter from Willem of Orange to the States of Holland and Zelland (December 28, 1574), quoted from Otterspeer 2001, 325.

[68] Smith 2004, 226. For a discussion of this aspect of Descartes' philosophy see the first section of Chap. 5 by Smith.

the senses. Natural philosophers must protect themselves by having moderate, if not chaste, bodily habits so as to prevent the distortion of knowledge by desire.[69] Second, sensory knowledge is unreliable, and so Sylvius' lectures combined experiment with mathematics and stressed that natural knowledge is only certain when demonstrated mathematically.[70]

10.4 De Volder, Experiment, and Certainty

De Volder's teaching continued in the tradition of Sylvius. His innovation was to bring this tradition to the philosophy faculty and the study of physics. In a 1679 funeral oration for his friend Siberti Coeman, de Volder recounts how he and Coeman both studied mathematics under Sylvius, immediately before claiming that mathematics and philosophy are one science and necessary to attain extraordinary knowledge of physical things,[71] a theme he repeatedly imparted to his students. De Volder says that it was the separation of philosophy from mathematics that kept philosophy in "impurity and darkness" for centuries, for mathematical philosophy is what makes it possible to discover and recognize truth.[72] De Volder, like Sylvius, emphasizes the importance of mathematical certainty and distinguishes it from what we gain through experiment. Throughout his writings de Volder accepts the Cartesian reformulation of scholastic *scientia* as systematic knowledge deduced from clear and distinct ideas, which are known a priori through pure reason. For de Volder, *scientia* is the method of physics and is only achieved through mathematical methods, not through sensory knowledge gained through experiment. De Volder's mathematical method is not that of Galileo or Newton, but, like that of Sylvius, pure deduction itself.[73] For de Volder, the goal of physics is to deduce phenomena from their causes, that is, from the first principles.[74] De Volder is quick to differentiate his method from the Scholastic understanding of *scientia*: while he too sees physics as systematic knowledge, it is not acquired nor does it explain through syllogisms. De Volder declared that what the Scholastic method promises with its words can only be delivered by mathematics.[75] Logic is useless speculation, while physics must investigate the very nature of things.[76] Physics and mathematics must be integrated.[77]

---

[69] Smith 2004, 228.

[70] Smith 2004, 223.

[71] De Volder 1679, E1v-E2r

[72] De Volder 1679, E2r. Note that similar themes are found in Rohault's preface to *Traité de Physique*. For a longer discussion of Rohault's natural philosophy, see Chap. 9 by Dobre.

[73] Ruestow 1973, 107.

[74] De Volder 1698, 5.

[75] De Volder 1679, E2r; see Klever 1988, 202.

[76] De Volder 1679, E1v, 1681a, section XI, 11 and section XLVIII, 42; 1682, 32–33.

[77] De Volder 1682, 4, 14, 17, 20, and 25.

De Volder claimed that it was precisely the Aristotelian division between these two disciplines that has up until the seventeenth century held physics back.[78]

De Volder's commitment to a Cartesian *scientia* that emphasizes a priori methods is clear throughout his entire career, from the first paragraph of his dissertation to his correspondence with Leibniz a few years before his death. For instance, his medical dissertation, *Disputatio Medicalinauguralis De Natura* (1664), begins with a statement that mathematics provides the only foundation of certainty available to all the arts and sciences[79] and as we saw, his 1679 *Oratio Funebris in Obitum Consultissimi Viri Siberti Coeman* claims that physics cannot be separated from mathematics.[80] The 1681 *Disputationes Philosophicae Sive Cogitations Rationales de Rerum Naturalium Principiis* is his first work to deal with these claims in a systematic way. It asserts that scientific explanations must be deduced from first principles, the primary concern of the physicist.[81] This work provides the criteria for identifying first principles: (1) they must be clearly and distinctly perceived; (2) they cannot be the effects of further natural causes; (3) they are mechanical, that is, they do not appeal to properties of mind or thought; (4) all phenomena in the world can be deduced from them.[82] He points out that it is impossible for humans to deduce all phenomena, as they are infinite, but we must demonstrate that they can be deduced and de Volder attempts to do this by showing that the principles of matter and motion are sufficient to explain our sense perceptions.[83] Perhaps nowhere does de Volder make the case as strongly as in the *Oratio de Conjugendis Philosophicis et Mathematicis Disciplinis*, his inaugural lecture as chair of mathematics in 1682. This work claims that mathematics describes the physical properties of bodies. Mathematics does not consist of mere tautologies or relations of ideas, but rather indicates physical reality insofar as it is understood in terms of its relations.[84] Further, the mathematical training of the mind is key to all the other branches of knowledge, for the mathematical arts:

> are those which, not only by their rules, but also—and this is of paramount importance and of universal application—by being frequently exercised, strengthen that most beneficial habit of mind, in respect of all the other arts or branches of knowledge, and indeed in respect of all things that occur in life, and, if the mind is equipped with it, it refuses to be deceived by any apparent truth, but drives out errors…[85]

---

[78] De Volder 1682, 15, 1681a, section X and section XI, 10–12.

[79] De Volder 1664, section I, 1.

[80] De Volder 1679, E1v.

[81] De Volder 1681a, 12; see Ruestow 1973, 93.

[82] De Volder 1681a, 12.

[83] Ruestow 1973, 95.

[84] Klever 1988, 209.

[85] De Volder 1682, 13–14: "Hae itaque illae artes sunt, quae non modo regulis, sed quod maximum est, quod que omnem rem consicit, frequenti exercitio eum menti nostrae ad omnes caeteras artes, scientiasve, ad omnia quae in vita occurrunt, fructuossimum confirment habitum, quo instructa nulla se veri specie decipi patiatur, sed errores expellat, nullas non…."

For de Volder, mathematics not only tells us about relations of the bodies in the world, but also indicates the very structure of the mind and can help form proper habits in the mind. Finally, in his 1698–1706 correspondence with Leibniz, which focuses on the living force controversy, 14 of the 18 letters from de Volder explicitly request an a priori demonstration of Leibniz's claims about the active nature of substances. For instance, in a letter dated November 12, 1699, de Volder says:

> I acknowledge that, 'for a proposition to be said to be understood, it is enough that it follows necessarily from things that are understood', if it follows evidently from things understood a priori, but not from things understood a posteriori. Consider the nature of gravity, so that what I mean may be clarified with an example. I understand this as well as possible if I have understood that the descent of bodies necessarily follows from the universal structure of the world or from the particular structure of our earth. But I do not understand it at all if I infer, from the fact that I see bodies descend, that some explanation of why this happens, which I call *gravity* is required.[86]

In other words, induction from particular observations is not de Volder's method of scientific explanation, it is not how one goes about identifying causes.

As is the case with Sylvius, a commitment to mathematical certainty does not imply that de Volder doesn't value in experiment. Like Sylvius, de Volder takes it to have a pedagogical value. However, unlike Sylvius, de Volder does not characterize experiment in terms of its affects on the students' passions. Instead, de Volder uses experiment to place in his students a certain belief in the existence of the phenomena under discussion. This belief will motivate the students to understand the causes of the phenomena, an understanding only achieved through mathematical deduction from first principles. This pedagogy is rooted in de Volder's distinction between two types of certainty: that gained through mathematical deduction and that gained through experiment. As an examination of de Volder's distinction will indicate, de Volder also differs from Sylvius in that he considers sensory knowledge to be a reliable source of knowledge, that is, of the knowledge of the existence of particulars.

De Volder's distinction is also importantly different than Descartes' distinction between moral and absolute certainty.[87] For Descartes, it is impossible to doubt ideas of which we are absolutely certain, whereas ideas that are only attended by moral certainty can be doubted, though we may never have occasion to do so.[88] Descartes' *Principles* Part IV defines morally certain ideas as "having sufficient certainty for application to ordinary life, even though they may be uncertain in relation to the absolute power of God."[89] Here we see another aspect of Descartes' distinction: moral certainty is useful to daily life, whereas metaphysical certainty is

[86] De Volder to Leibniz, November 12, 1699, translation in Leibniz 2013, 138–139.

[87] For a discussion of Descartes' concept of moral certainty in relation to other Cartesians, see Dobre 2013. For a discussion of the Scholastic roots of the concept of moral certainty, see Ariew 2011.

[88] From French edition of Descartes' *Principles*, see CSM I 289n2.

[89] AT VIIIa 327, CSM I 289–290.

required for philosophical deliberations. Descartes provides a commonly used example to illustrate his concept of moral certainty:[90] "Thus those who have never been in Rome have no doubt that it is a town in Italy, even though it could be the case that everyone who has told them this has been deceiving them."[91]

In the *Oratio de Rationis Viribus, Et Usu in Scientiis*, de Volder also makes a distinction between two kinds of certainty, "rational" certainty and a certainty that is obtained through experiment. His distinction differs from Descartes' not only in terminology, but also in its content. What de Volder terms rational certainty, is basically Descartes' absolute certainty. Rational certainty cannot be doubted, its content is necessary, consisting of innate ideas, known through reason, such as the essential nature of body being extension. For de Volder, this is the domain of *scientia*. However, de Volder does not use the phrase "moral certainty" when speaking about sensory knowledge and experiment.[92] He does not have a handy adjective for this second kind of certainty, but importantly, he does express that it *is* a type of certainty.[93] Further, the concept is quite different from Descartes' moral certainty. Unlike Descartes, de Volder never appeals to the trustworthiness of our sensations in relation to their utility. De Volder points out that God could have created the world in different ways and chose to create this one.[94] Through reason we can come to understand what worlds are possible, that is, what God could have created. However, reason alone can never tell us which world God did actually create, that is, which world exists, nor can it help us with the particulars about that world, which are of key importance for natural philosophy, including medicine. For this we need experience:

> Ultimately all considerations subsequently return to this one point, that it is only in metaphysics and mathematics, which deal simply with ideas which are clearly and distinctly understood, that calculations of certainty flourish. Among these it is reason which is the ruler and master. In Physics however, irrespective of the certainty of the conclusions we draw from our hypotheses, uncertainty nevertheless remains as to whether the physical entities, which we assumed in our reasonings, truly exist or not.[95]

---

[90] For discussion of Locke's use of this example see van Leeuwen 1963, 135. For discussion of William Chillingworth's use of this example see van Leeuwen 1963, 25; Orr 1967, 52; and Remer 1992, 30.

[91] AT VIIIa 327, CSM I 290.

[92] As Dobre explains, Descartes' use of the intermediary category of "more than morally certain" was very confusing to his followers, who often tried to come up with new explanations and distinctions, Dobre 2013.

[93] De Volder 1698, 17.

[94] Klever overstates de Volder's "Spinozism." Klever suggests that de Volder, like Spinoza, accepts an absolute necessity and determinism. However, de Volder consistently attributes free will to God. See Klever 1988, 206.

[95] De Volder 1698, 33–34: "Quae tandem omnia huc redeunt denique; In Metaphysicis, Mathematisque, quae solas ideas, easque clare & distincte perceptas tractant certa demum vigere ratiocinia. In hisce imperium obtinere & dominari Rationem. In Physicis vero, utut ex Hypothesi quam certissime concludamus, incertum tamen relinqui, eane, quae in ratiocinio assumsimus, corpora revera sint, an minus."

Earlier in the same work, de Volder tells us that all branches of learning concerned with physical entities:

> are incapable of determining without the assistance of the senses and experiment whether conclusions drawn using even the most precise reasoning are certain in respect of any physical entity in the natural world…[96]

In other words, it is only through experiment that we can come to determine that what we learn through theoretical physics actually is instantiated in the existing world, that is, that particular bodies with the characteristics we come to understand through theoretical physical actually exist. *Scientia* tells us about universal properties, which are necessary, whereas experience tells us about the existing particular things that have those properties, particular things that are contingent upon the world that God chose to create.[97] Notice that for de Volder, there is no way for humans to know if God could have created this existing world in different ways, that is, if God had the option of choosing among different sets of first principles, all of which would have resulted in this exact physical world. We cannot know this through reason or experience—it is simply beyond human purview.[98] However, for the purpose of natural philosophy, which is concerned only with this world, it is of no matter.

Now we can see several key differences between de Volder and Descartes. De Volder does not distinguish between the certainty achieved through mathematics and experiments in terms of (1) the possibility of doubt or (2) the domain in which that knowledge is to be used. Rather, mathematical certainty pertains to first principles, the causes of all things, and provides an understanding of how all bodies relate. Experiment provides us with a certainty in the existence of particulars, which through natural philosophy we can come to understand to instantiate the laws of nature. Both *scientia* and experiment are essential components of natural philosophy (which de Volder takes to include medicine).

For de Volder, experiments can show that a theory has misconstrued the phenomena it attempts to explain. De Volder, as a modern philosopher, recognizes that the world is not always what it seems. Experiment can help reveal realities that seem paradoxical, counterintuitive, or otherwise might go unnoticed, particularly against the background of longstanding scholastic assumptions. This is where experiment becomes an invaluable pedagogical tool. In *Questiones Academicae De Aëris Gravitate* we see an example of how experiments can help students see problems with an incorrect theory, in this case the incorrect scholastic view that air has levity. De Volder wrote this 1681 work as a syllabus for students and it should be

---

[96] De Volder 1698, 17: "Nam, praepter hoc commune omnium de corporibus disciplinam vitium, quod absque sensuum & experientiae auxilio determinare nequeant; num quae ratione vel exactissima conclusa sunt, ulli in rerum natura corpori competunt; multa alia sunt, quae in singulis haram ad certam desunt scientiam."

[97] Letter from Leibniz to de Volder dated May 13, 1699 in Leibniz 2013, 84–85.

[98] De Volder 1681a, 20. He is echoing the view captured in Descartes' famous clock metaphor, AT IXb 327, CSM I 289.

understood in this pedagogical context. As Klever points out, by this time the gravity of the air was accepted by the scientific community, however:

> because this conviction is opposed to the whole (Aristotelian) tradition of philosophy and to everyday belief, according to which the air is light, the subject lends itself extremely well to didactic purposes and to instruct 'tyronum animos' in order to make them sensitive for the paradoxical laws of natural science.[99]

Here experiment is used to secure among his students belief in the particular phenomena that not only cannot be explained by the debunked Scholastic theory, but could not happen were it correct. Through the experiments of Galileo, Torricelli, Roberval, Pascal, Guericke, Boyle, and Huygens, de Volder demonstrated to his students the certainty of the phenomena of the gravity of air.[100]

The distinction between de Volder's two types of certainty helps us make sense of the request for the Physics Theatre. Recall that it claims that experiment will demonstrate the truth and certainty of what is learned through theoretical physics. Experiment provides certain belief that physical theories apply to this world and are not mere logical possibilities available to the creator.

In conclusion, de Volder presents an interesting case of pre-Newtonian experimental physics. His use of experiment is best understood within the context of the teaching traditions at Leiden. Given that context, combining a Cartesian commitment to a priori reasoning with a commitment to the demonstration of physical experiments was a completely natural thing to do.

Acknowledgements This work was supported with an American Council of Learned Societies Fellowship, an American Philosophical Society Franklin Research Grant, an Obermann Center Fellowship at the University of Iowa and research support from Grinnell College. I would like to thank Mihnea Dobre, Paul Lodge, Gary Hatfield, and Sujeev Wickramasekara for their helpful comments on earlier drafts of this paper.

References

Ahnert, Thomas. 2002. *The culture of experimentalism in the Holy Roman Empire: Johann Christoph Sturm (1635–1703) and the Collegium Experimentale.* Sammelpunkt Elektronisch archivierte Theorie. http://sammelpunkt.philo.at:8080/308/. Accessed 14 Apr 2010.

Ariew, Roger. 2011. The new matter theory and its epistemology: Descartes (and late scholastics) on hypothesis and moral certainty. In *Vanishing matter and the laws of motion: Descartes and beyond*, ed. Dana Jalobeanu and Peter R. Anstey, 31–46. London: Routledge.

Boas Hall, Marie. 1966. *Robert Boyle on natural philosophy. An essay with selections from his writings*. Bloomington: Indiana University Press.

Bos, E.P., and H.A. Krop (eds.). 1993. *Franco Burgersdijk (1590–1635): Neo-Aristotelianism in Leiden*. Amsterdam: Rodopi.

Cardoso, Adelino. 1996. Mathesis Leibniziana. *Philosophica: Revista do Departamento de Filosofia da Faculdade de Letras de Lisboa* 8: 51–77.

---

[99] Klever 1988, 205.

[100] De Volder 1681b, 12.

Cook, Harold J. 1992. The new philosophy in the low countries. In *The scientific revolution in national context*, ed. Roy Porter and Mikulas Teich, 115–149. Cambridge: Cambridge University Press.
Cook, Harold J. 2007. *Matters of exchange: Commerce, medicine, and science in the Dutch Golden Age*. New Haven/London: Yale University Press.
Crommelin, Claude August. 1926. *Physics and the art of instrument making at Leyden in the 17th and 18th centuries*. Leiden: A.W. Sijtoff.
Crommelin, Claude August. 1935. Die Holländische Physic Im 18 Jahrhundert Mit Besonderer Berücksichtigung Der Entwicklung Der Feinmechanik. *Für Geschichte der Medizin und der naturwissenschaften* 28(3): 129–142.
Crommelin, Claude August. 1951. *Descriptive catalogue of the physical instruments of the 18th century (including the collection 'S Gravesande-Musschenbroek) in the Rijksmuseum Voor De Geschiedenis Der Natuurwetenschappen*. Leiden: National Museum of the History of Science at Leiden.
Daumas, Maurice. 1972. *Scientific instruments of the seventeenth and eighteenth centuries*. Trans. and ed. Mary Holbrook. New York: Praeger Publishers, Inc.
De Clercq, Peter. 1988. The 'S Gravesande Collection in the Museum Boerhaave, Leiden. *Nuncius/ Istituto Museo Di Storia Della Scienza Firenze: Nuncius: Annali di Storia della Sceinza* Anno III: 127–137.
De Clercq, Peter. 1989. *The Leiden Cabinet of Physics*. Leiden: Museum Boerhaave.
De Clercq, Peter. 1997a. *At the sign of the oriental lamp: The Musschenbroek workshop in Leiden, 1660–1750*. Rotterdam: Erasmus Publishing.
De Clercq, Peter. 1997b. *The Leiden Cabinet of Physics: A descriptive catalogue*. Leiden: Museum Boerhaave.
de Volder, Buchard. 1664. *Disputatio Medica, Inauguralis de Natura*. Leiden: Severinum Mattiae.
de Volder, Burchard. 1679. *Oratio Funebris in Obitum Consultissimi Viri Siberti Coeman, Philosophiae & U. Juris Doctoris, hujusque in Acad. Lugduno Batava Professoris*. Leiden.
de Volder, Burchard. 1681a. *Disputationes Philosophicae Sive Cogitations Rationales de Rerum Naturalium Principiis*. Middelburg: Remigii Schreverii.
de Volder, Burchard. 1681b. *Quastiones Academicae De Aëris Gravitate*. Middelburg: Remigii Schreverii.
de Volder, Burchard. 1682. *Oratio de Conjugendis Philosophicis et Mathematicis Disciplinis*. Leiden: Apud Jacobum Voorn.
de Volder, Burchard. 1698. *Oratio de Rationis Viribus, Et Usu in Scientiis*. Leiden: Fridericum Haringium.
de Volder, Burchard. 1705. *Oratio Qua Consentientibus Illistr. Acad. Curatoribus, Urbisque Leidensis Coss. Sese Laboribus Academicis Abdicavit. Habita*. Leiden: Cornelium Boutestein.
[de Volder, Burchard, Christoph Wittichius] and Adriaan Heidanus. 1676. *Consideratien, over Eenige saecken onlanghs voorgevallen in de Universiteyt binnen Leyden. Den Derden Druck is achter by-gevoegt een Extract uyt de Resolutie van de Curateuren over de un iversiteyt en Burgermeesteren de Stadt Leyden*. Leiden: Aernoudt Doude.
Descartes, René. 1970. *Oeuvres de Descartes*, ed. Charles Adam and Paul Tannery. Paris: J. Vrin.
Descartes, René. 1985. Principles of philosophy. In *The philosophical writings of Descartes*, vol. I. Trans. John Cottingham, Robert Stoothoff, and Dugald Murdoch, 177–292. Cambridge: Cambridge University Press.
Dobre, Mihnea. 2013. Knowledge and certainty in the foundation of Cartesian natural philosophy. *Revue Roumaine de Philosophie* 57(1): 95–110.
Gori, Giambattista. 1972. *La Fondazione Dell 'Esperienza in 'S Gravesande*. Firenze: La Nuova Italia Editrice.
Gronovio, Jacobo. 1709. *Laudatio Burcheri de Volder Philosophiae & Matheseos Professoris LAUDATIO ab Jacobo Gronovio Peracta Ex decreto Rectoris Magnifici & Senatus Academemici*. Leiden: Cornelium Boutestein.
Heilbron, J.L. 1982. *Elements of early modern physics*. Los Angeles: University of California Press.

Jalobeanu, Dana (ed.). 2013. The creative role of experimentation in early modern science. Special Issue of the *Journal of Early Modern Studies* 2(1).

Jesseph, Douglas Michael. 1998. Leibniz on the foundations of the calculus: The question of the reality of infinitesimal magnitudes. *Perspectives on Science* 6(1): 6–40.

Kegal-Brinkgreve, E., and A.M. Luyendijk-Elshout. 1983. Introduction. In *Boerhaave's orations*. Trans. E. Kegal-Brinkgreve and A.M. Luyendijk-Elshout. Leiden: Brill and Leiden University Press.

Klein, Ursula. 2003. Experimental history and Herman Boerhaave's chemistry of plants. *Studies in History and Philosophy of Science Part C: Studies in History and Philosophy of Biological and Biomedical Sciences* 34(4): 533–567.

Klever, Wim. 1988. Burchard De Volder (1643–1709)–a Cryptospinozist on a Leiden Cathredra. *LIAS* 15(2): 191–241.

Klever, Wim. 1997. *Mannen rond Spinoza*. Hilversum: Uitgeverij Verloren.

Knoeff, Rina. 2002. Herman Boerhaave: Spinozist? In *Herman Boerhaave (1668–1738): Calvinist chemist and physician*, ed. R. Knoeff, 21–51. Amsterdam: KNAW (Koninklijke Nederlandse Akademie van Wetenschappen).

Krop, Henri. 2003. Medicine and philosophy in Leiden around 1700: Continuity or rupture? In *The early enlightenment in the Dutch republic*, ed. Wiep van Bunge, 173–196. Leiden: Brill.

Le Clerc, Jean. 1709. Eloge de feu Mr. De Volder Professeur en Philosophie & aux Mathématiques, dans l'acadmémie de Leide. *Bibliothèque Choisie* 18: 346–401.

Leibniz, Gottfried Wilhelm. 2013. *The Leibniz-De Volder correspondence: With selections from the correspondence between Leibniz and Johann Bernoulli*, ed. and trans. Paul Lodge. New Haven: Yale University Press.

Leomker, Leroy E. 1946. Leibniz's judgments of fact. *Journal of the History of Ideas* 7(4): 397–410.

Lindeboom, Gerrit Arie. 1972. Burchard De Volder. In *Mensen Om Boerhaave*. The Hague: Geneeskundige gids.

Lindeboom, Gerrit Arie. 1974. *Boerhaave and Great Britain*. Leiden: Brill.

Lodge, Paul. 1998. The failure of Leibniz's correspondence with De Volder. *Leibniz Society Review* 8: 47–67.

Lodge, Paul. 2001. The debate over extended substance in Leibniz's correspondence with De Volder. *International Studies in the Philosophy of Science* 15(2): 155–165.

Lodge, Paul. 2004. Leibniz's close encounter with Cartesianism in the correspondence with De Volder. In *Leibniz and his correspondents*, ed. Paul Lodge, 162–192. Cambridge: Cambridge University Press.

Lodge, Paul. 2005. Burchard De Volder: Crypto-Spinozist or Disenchanted Cartesian? In *Receptions of Descartes: Cartesianism and anti-Cartesianism in early modern Europe*, ed. Tad M. Schmaltz, 128–145. London: Routledge.

McGahagan, T.A. 1976. *Cartesianism in The Netherlands, 1639–1676. The new science and the Calvinist reformation*. Unpublished PhD, University of Pennsylvania, Pennsylvania.

Molhuysen, P.C. 1918. *Bronnen Tot De Geschiednis Van De Leidsche Universteit. 'Sgravenhage, 1913–1924*, vol. III. 's Gravenhage: Marinus Nijhoff.

Newton, Isaac. 1982. Further Newton correspondence. *Notes and Records of the Royal Society of London* 37(1): 7–34.

Niceron, Jean Pierre. 1733. *Mémoires pour servir á l'histoire des hommes illustres dans la République des Lettres. Avec un catalogue raisonné de leurs ouvrages*, vol. XXII. Paris: Briasson.

Nyden, Tammy. In press. Living force at Leiden: De Volder, 's Gravesande, and the reception of Newtonianism. In *Newton and Empiricism*, ed. Zvi Biener and Eric Schliesser. Oxford: Oxford University Press.

Nyden-Bullock, Tammy. 1999. Radical Cartesian politics: Van Velthuysen, De La Court, and Spinoza. *Studia Spinozana* 15: 35–65.

Nyden-Bullock, Tammy. 2007. *Spinoza's radical Cartesian mind.* London: Continuum.
Orr, Robert R. 1967. *Reason and authority: The thought of William Chillingworth.* Oxford: Clarendon Press.
Otterspeer, Willem. 2001. The University of Leiden: An eclectic institution. *Early Science and Medicine* 6(4): 324–333.
Otterspeer, Willem. 2008. *The bastion of liberty: Leiden University today and yesterday.* Leiden: Leiden University Press.
Remer, Gary. 1992. Humanism, liberalism, & the skeptical case for religious toleration. *Polity* 25(1): 21–43.
Rey, Anne-Lise. 2009a. L'ambivalence de la notion d'action dans la dynamique de Leibniz: La correspondance entre Leibniz et de Volder. Première partie. *Studia Leibnitiana: Zeitschrift Fuer Geschichte Der Philosophie Und Der Wissenschaften* 41(1): 47–66.
Rey, Anne-Lise. 2009b. L'ambivalence de la notion d'action dans la Dynamique de Leibniz: La correspondance entre Leibniz et De Volder. IIe Partie. *Studia Leibnitiana: Zeitschrift fuer Geschichte der Philosophie und der Wissenschaften* 41(2): 157–182.
Ross, George. 1986. Leibniz and de Volder on the infinitely small in metaphysics. *Studia Liebnitiana. Supplemata: im Austrage der Gottfried-Wilhelm-Leibniz-Gesellschaft E.V* 14: 169–177.
Ruestow, Edward G. 1973. *Physics at seventeenth and eighteenth-century Leiden: Philosophy and the new science in the University.* The Hague: Martinus Nijhoff.
Sassen, F.L.R. 1970. The intellectual climate in Leiden in Boerhaave's time. In *Boerhaave and his time (Analecta Boerhaaviana VI)*, ed. G.A. Lindeboom, 1–16. Leiden: Brill.
Schuurman, Paul. 2004. *Ideas, mental faculties and method: The logic of ideas of Descartes and Locke and its reception in the Dutch Republic, 1630–1750.* Leiden/Boston: Brill.
Siegenbeek, Matthijs. 1832. *Geschiedenis der Leidsche Hoogeschool, van Hare Oprigting, in den Jare 1575, tot het Jaar 1825*, vol. II. Leiden: J. Luchtmans.
Smith, Pamela. 2004. *The body of the Artisan: Art and experience in the scientific revolution.* Chicago: University of Chicago Press.
Spaans, Jo. 2007. The Alphen pig war. In *Iconoclasm and Iconoclash: Struggle for religious identity*, ed. Willem van Asselt, Paul van Geest, Daniela Müller, and Theo Salemink, 331–352. Leiden: Brill.
Suringer, G.C.B. 1865. De Leidsche Hoogleraren in De Natuurkundige Wetenschappenn… *Ned. Tijdschrift voor Geneeskunde* 2nd Ser. I:275–306.
Thijssen-Schoute, C. Louise. 1954. *Nederlands Cartesianism.* Amsterdam: Verh. Kon. Ned. Akad. Wetensch.
Turner, G.L.E. 1998. *Scientific instruments: 1500–1900, an introduction.* Berkeley: The University of California Press.
van Berkel, Klaas. 1985. *In Het Voetspoor Van Stevin: Geschiedenis Van De Natuurwetenschap in Nederland 1580–1940.* Amsterdam: Boom Meppel.
van Berkel, Klaas, Albert van Helden, and Lodewijk Palm (eds.). 1999. *A history of science in the Netherlands: Surveys, themes and reference.* Leiden: Brill.
Van Bunge, Wiep. 2001. *From Stevin to Spinoza: An essay on philosophy in the seventeenth-century Dutch Republic.* Leiden: Brill.
van Leeuwen, Henry G. 1963. *The problem of certainty in English thought 1630–1690.* The Hague: Martinus Nijhoff.
van Poelgeest, L. 1984. *De Wis–En Naturrkundesfaculteit*, De Leidse Hoggleraren En Lectoren 1575–1815, vol. IV. Leiden: Werkgroep Elites Rijksuniversiteit.
Van Miert, Dirk. 2003. Bie, Alexander de (1623–1690). In *The dictionary of seventeenth and eighteenth-century Dutch philosophers*, ed. Wiep van Bunge, Henri Krop, Bart Leeuwenburgh, Han van Ruler, Paul Schuurman, and Michael Wielema, I:102–103. Bristol: Thoemmes Press.
Van Miert, Dirk. 2009. *Humanism in an age of science: The Amsterdam Athenauem in the Golden Age, 1632–1704*, Trans. Michiel Wielema and Anthony Ossa-Richardson. Leiden: Brill.

Vermij, Rienk. 2003. The formation of the Newtonian philosophy: The case of the Amsterdam mathematical Amateurs. *British Journal for the History of Science* 36(2): 183–200.

Vinson, Carolus. 1676–1677. *Experimenta Anatomica et Philosophica*. Manuscript. Sloane Collection 1676–1677, The British Library.

Wiesenfeldt, Gerhard. 2000. The virtues of new philosophies or: How the Leiden philosophical faculty survived the crises of 1676. In *19th International Congress of Historical Sciences* Oslo, Norway. http://www.oslo2000.uio.no/AIO/AIO16/group%203/Wiesenfeldt.pdf. Accessed 8 Oct 2012.

Wiesenfeldt, Gerhard. 2002. *Leerer Raum in Minervas Haus: Experimentelle Naturlehre an der Universität Leiden, 1675–1715*. Amsterdam: Royal Netherlands Academy of Arts and Science.

Wiesenfeldt, Gerhard. 2003. Senguerd, Wolferd (1646–1724). In *The dictionary of seventeenth and eighteenth-century Dutch philosophers*, vol. II, ed. Wiep van Bunge, Henri Krop, Bart Leeuwenburgh, Han van Ruler, Paul Schuurman, and Michael Wielema, 911–914. Bristol: Thoemmes Press.

Chapter 11
The Cartesian Psychology of Antoine Le Grand

Gary Hatfield

Abstract In the Aristotelian curriculum, *De anima* or the study of the soul fell under the rubric of physics. This area of study covered the vital ("vegetative"), sensitive, and rational powers of the soul. Descartes' substance dualism restricted reason or intellect, and conscious sensation, to human minds. Having denied mind to nonhuman animals, Descartes was required to explain all animal behavior using material mechanisms possessing only the properties of size, shape, position, and motion. Within the framework of certainty provided by the metaphysical foundations of his physics, he posited such mechanisms in accordance with appropriately lessened standards of certainty. As Cartesianism (or the Cartesian revolution) spread, adherents offered survey textbooks or treatises of physics to replace the corresponding Aristotelian curriculum. These books typically discussed the role of experience in physics and the appropriate standard of certainty. A comprehensive Cartesian natural philosophy needed to mechanize the offices of the Aristotelian sensitive soul, including sense perception, memory, and cognitive and appetitive responses to the environment. Descartes' *Treatise on Man* (1664) offered an initial explanatory program. This chapter examines the role of experience in the natural philosophy and mechanistic psychology of Descartes' English follower Antoine Le Grand. He offered detailed accounts of the sensory and motor mechanisms shared by human and nonhuman animals (for which he claimed "physical certainty"). These frame his Cartesian psychology.

G. Hatfield (✉)
Department of Philosophy, University of Pennsylvania,
Philadelphia, PA, 19104-6304, USA
e-mail: hatfield@phil.upenn.edu

M. Dobre and T. Nyden (eds.), *Cartesian Empiricisms*, Studies in History and Philosophy of Science 31, DOI 10.1007/978-94-007-7690-6_11,
© Springer Science+Business Media Dordrecht 2013

11.1 Introduction

Antoine Le Grand (1628/1629–1699) can lay claim to being the most productive and influential follower of Descartes in England during the second half of the seventeenth century. He extended the original contributions of Descartes into a full philosophical program that might replace the corresponding subjects in the college curriculum, running the gamut from metaphysics and natural theology to physics broadly construed so as to include all natural phenomena from fundamental extended matter to organic bodies (living things). Since he intended to be faithful to the thought of Descartes, his work offers an opinion on interpretive questions about Descartes' philosophy. And since he elaborated topics that Descartes had discussed and also expanded the range of topics covered, his work offers insight into how Descartes' philosophy might be extended.

In this chapter, I focus on Le Grand's extensions of Descartes' physiology and psychology of animal and human behavior. Le Grand accepted and defended Descartes' program to mechanize the sensitive and motor functions of the Aristotelian animal and human souls (or their Galenic counterparts). The Aristotelian sensitive soul (or sensitive power) was responsible for several functions: sense perception, memory, and cognitive and appetitive responses to beneficial and dangerous aspects of the environment. The Cartesian program required that these functions be mechanized. That is, they must be explained through bodily mechanisms that mediate between sensory stimulation and motor response, in a way that produces situationally appropriate behavior—that is, behavior that (on the whole) avoids harms and obtains benefits. In this context, "mechanization" means constructing explanations that are couched wholly in terms of the sizes, shapes, positions, and motions of Cartesian bare extended matter. In a way to be explained, this program amounts to a Cartesian mechanistic psychology.

The chapter begins with a discussion of Descartes' philosophical program in metaphysics and natural philosophy (or physics), including the role of observation and experience and the lowered standard of certainty he accepted in positing particular mechanisms to explain natural philosophical phenomena. Having noted the rapid spread of Descartes' natural philosophy in the decades after his death, which amounted to a "Cartesian revolution,"[1] the chapter turns to Le Grand in relation to Descartes, including Le Grand's discussion of the status of knowledge in natural philosophy. I claim that Le Grand was closer to Descartes himself on the matter of certainty in natural philosophy than were some other Cartesians (Rohault and Régis)[2] and that he helpfully distinguished between "physical" and "metaphysical" certainty. The chapter further considers Descartes' program to mechanize the powers of the Aristotelian sensitive soul and Le Grand's extension of it. Le Grand accepted and defended Descartes' conceptions that all animal functions must be explained mechanistically and that a great majority of human behaviors can be

[1] Hatfield 1985, 160; 1996, 504.

[2] On Jacques Rohault (1618–1672) see Chap. 9 by Dobre. On Pierre-Sylvain Régis (1632–1707) see Chap. 6 by Joly.

explained in the same manner. This program of explanation amounts to a Cartesian mechanistic psychology. Consciousness, intellectual understanding, and will remain outside the ambit of such explanations.

11.2 Descartes and Appeals to Experience

For much of the past century, Descartes was best known for his metaphysical and epistemological works and arguments. Of his works, the *Meditations on First Philosophy* and the *Discourse on the Method* were and remain the most widely read. In these works, certain topics and arguments have been privileged over others. Understandably, these are the metaphysical and epistemological arguments that seem "philosophical" by recent standards. They include the method of doubt, the cogito reasoning, the quest for certainty, proofs for the existence of God, arguments for the existence and general nature of an external world of extended matter, and mind–body dualism along with mind–body union and interaction. These discussions have emphasized Descartes' quest for certainty and his appeal to "reason" or the "intellect" as a source of knowledge independent of the senses, that is, as a basis for a priori knowledge.[3] It is because he recognized the pure intellect as a self-sufficient means for some kinds of knowledge that Descartes has legitimately been labeled a "rationalist."

In recent decades, historians of philosophy have given renewed attention to the fact that Descartes was a "philosopher" in the seventeenth-century sense of that word: a person who sought organized knowledge in the various "philosophical" disciplines. At the most generic level, these disciplines included logic, metaphysics, physics or natural philosophy,[4] and ethics. These constituted the core disciplinary structure of seventeenth-century philosophical learning, and were expected to be covered in textbook surveys of philosophy.

Descartes in some sense offered materials in each of these areas. Although he rejected Aristotelian logic, in the early *Rules for the Direction of the Mind* and in the *Discourse* he discussed the rules for proper reasoning and his discussions influenced the later Port Royal logic, as well as the various logics by Cartesian textbook writers. The *Passions of the Soul* was his most sustained foray into topics that touched upon ethics,[5] and it too was used by Cartesian textbook writers who constructed a full-blown Cartesian ethics.

[3] Prior to Descartes' day, the standard meaning of the term "a priori" was "known through its cause." Descartes may have had a hand in facilitating a later meaning, "known independently of the senses," since much of the metaphysical knowledge he claimed to know through its cause was known independently of the senses, as in his appeal to God's immutability in his discussion of the laws of motion (God being a cause known independently of the senses). On the philosophical interest of Descartes' fluid use of terms such as "a priori," see Buchdahl 1969, Chap. 3.

[4] The terms "physics" and "natural philosophy" were used synonymously by Descartes and his followers. In this context, "physics" is the science of nature in general, including organic bodies. Some Cartesians even included mind–body interaction under the rubric of physics (see Hatfield 2009).

[5] An edition of Descartes' *Passions* was issued in 1707 under the title *La Philosophie morale de Monsieur Descartes*.

Of the traditional philosophical disciplines, Descartes devoted by far the greatest effort to metaphysics and natural philosophy, as is apparent from the *Discourse* and its essays on *Dioptrics* and *Meteorology*, the *Meditations*, *Principles of Philosophy*, the post-humously published *World, or Treatise on Light* and *Treatise on Man*, and the *Passions* (which he approached "as a physicist").[6] In these works, he sketches his metaphysics in the *Discourse*, presents its main tenets in the *Meditations*, and reprises and extends those tenets in Part One of the *Principles*. The *Discourse* also contains an overview of his program in natural philosophy, including his intent to cover the workings of animal and human bodies; the *Meteorology*, portions of the *Dioptrics*, *Principles*, and *World* cover the inorganic universe; *Man*, portions of the *Dioptrics*, and the *Passions* examine the workings of animal and human bodies, focusing especially on sensory and motor mechanisms and the production of situationally suitable behavior. From this listing of his major works, we see that Descartes wrote considerably more on natural philosophy than he did in metaphysics.[7]

In agreement with his Aristotelian predecessors, Descartes recognized metaphysics and physics as distinct disciplines. Although adherents of Aristotelian philosophy disagreed on whether metaphysics supports physics, Descartes was clear that his metaphysics provided the first principles of his physics. This includes the conclusion that the essence of matter is extension, that the relevant properties of extended matter, including its smallest parts, are size, shape, position, and motion, and presumably also the laws of motion.[8]

There has been scholarly disagreement over how much of his physics Descartes believed he could derive from his metaphysics.[9] For his physics included not only general claims, such as that matter is extended substance, the denial of a vacuum (or the assertion of a plenum), and the laws of motion, but also the description of specific microstructures, such as branchy particles as constituents of oily liquid, eel-shaped particles for water, and corkscrew shaped particles in motion as the basis for magnetic phenomena.

In some passages, such as a notorious one in the Sixth Part of the *Discourse*, Descartes seems to make ambitious claims about the extent of a priori deduction. But I believe that a careful reading of even these passages, as well as the

---

[6]That is, "en Physicien" (AT XI 326). There is some possibility that Descartes meant "as a physician" (medical doctor) as opposed to "as a physicist," since "physique" was used in the sense of "medicine" into the seventeenth century. Rey 1992, II, 1509.

[7]This fact accords with Descartes' remarks to Princess Elizabeth that one ought spend only an hour *a year* on metaphysics (especially after Descartes has shown the way to the true one) and several hours *a day* on things that engage the senses and imagination, which includes natural philosophy (AT III 692, 695, CSMK 227, 228). All the same, Descartes valued his metaphysics highly, as is evident from the effort he expended on the *Meditations* and from his making metaphysics the roots of the tree of knowledge and the foundation of physics (in his Author's Letter to the French edition of the *Principles*, AT IXb 1–20, CSM I 179–190).

[8]For discussion of the role of metaphysics in relation to physics in Descartes and others, see Hatfield 1990.

[9]Consider the range of positions found in Clarke 1982; Hatfield 1985; Garber 1992; and Machamer and McGuire 2009.

Meditations and *Principles*, reveals that Descartes held the following position from the time of the *Discourse* or before: metaphysics provides the general principles of physics or natural philosophy, as sketched above; these general principles then provide constraints on any acceptable mechanistic explanation, including micromechanisms; experience and observation reveal many of the particular phenomena that must be explained; more than one micromechanism may be imagined that is consistent with the general principles, so that experience and observation are called in to decide among them; and sensory experience provides less certainty than does a clear and distinct deliverance of the pure intellect. Some of these attributions are more widely accepted than others. I focus initially on Descartes' statements in the *Discourse* (and contemporary letters) about the role of and need for observation and experience in natural philosophy (and the lowered standard of certainty).[10]

In the *Discourse* Descartes says, speaking of his program in natural philosophy: "I also noted, regarding observations, that the further we advance in our knowledge, the more necessary they become."[11] At the same time, he speaks of "the principles or first causes of everything that exists or can exist in the world," which must include "fundamental principles of physics" that he subsequently mentions, which are "almost all so evident that they need only to be understood to be believed."[12] What is the relation between these fundamental principles and the need for observation? Here the text becomes more difficult to interpret. Descartes first considers "the first and most ordinary effects deducible from these causes," and in this way says he "discovered the heavens, the stars, and an earth; and, on the earth, water, air, fire, minerals, and other such things which, being the more common of all and the simplest, are consequently the easiest to know."[13] Does he here claim to "deduce" these things a priori from his evident first principles? Or has he taken his own advice, from a few lines back, that one should, at the beginning of an investigation into nature, "resort only to those [observations] which, presenting themselves spontaneously to our senses, cannot be unknown to us if we reflect even a little"?[14] I read him as following his own advice, and so as saying that, in relating his a priori explanatory categories to the things he "discovered" from his first principles, he is fitting what his first principles allow to the most general phenomena (such as the heavens, stars, and earth) as already known from experience.

---

[10] Some scholars hold that Descartes first accepted a reduced standard of certainty for claims about particular micromechanisms only with the *Principles* in 1644 and see it as a kind of defeat (e.g., Garber 2001; Ariew 2011), whereas I find this reduced standard already present in the *Discourse* and related correspondence and find Descartes suggesting it would be inappropriate to seek absolute certainty (as found in metaphysics and mathematics) about such mechanisms (see Hatfield 1985).

[11] AT VI 63, CSM I 143.

[12] AT VI 63–64, 68, CSM I 143, 145–146.

[13] AT VI 64, CSM I 144.

[14] AT VI 63, CSM I 143.

Be that as it may, the next stretch of the *Discourse* makes clear that Descartes, at this early date, considered observations and experiential test as absolutely required in natural philosophy. Having mentioned the "most common" and "simplest" things (as just listed, the heavens, etc.), he reports:

> Then, when I sought to descend to more particular things, I encountered such a variety that I did not think the human mind could possibly distinguish the forms or species of bodies that are on the earth from an infinity of others that might be there if it had been God's will to put them there. Consequently I thought the only way to make these bodies useful to us was to progress to the causes by way of the effects and to make use of many special observations. And now, reviewing in my mind all the objects that have ever been present to my senses, I venture to say that I have never noticed anything in them which I could not explain quite easily by the principles I had discovered. But I must also admit that the power of nature is so ample and so vast, and these principles so simple and so general, that I notice hardly any particular effect of which I do not know at once that it can be deduced from the principles in many different ways; and my greatest difficulty is usually to discover in which of these ways it depends on them. I know no other means to discover this than by seeking further observations whose outcomes vary according to which of these ways provides the correct explanation.[15]

As has been noted before, the picture that Descartes presents here is that of hypothesis testing, where the range of hypotheses is constrained by his first principles (for instance, he must appeal only to matter in motion, and not to other sorts of explanatory entities, such as substantial forms).[16] Leaving aside the details of how to reconstruct his account, this passage makes clear that even in 1637 Descartes was prepared to rely heavily on observations in deciding which of the many possible posited micro-mechanisms he should affirm. There is also evidence that he accorded less certainty to such posits than to, say, geometrical demonstration.[17]

---

[15] AT VI 64–65, CSM I 144.

[16] Clarke 1992, 263–265. On hypothesis in Descartes, see also Laudan 1966 and Buchdahl 1969, Chap. 3. I do not go so far as Clarke 1982 in viewing Descartes as an outright empiricist, if an "empiricist" is someone who holds that all knowledge comes through the senses. (Such an empiricist need not, of course, deny that reason operates, but they hold either that it operates on sensory materials, or, if there are cases it does not, that such operations are equivalent to definitional relations that do not establish substantive truths about existing things.) The notion of "rationalism" is broad enough to include valuing the senses and the empirical, if one recalls that Descartes is a rationalist because he recognized intellectual knowledge that comes independent of the senses; that does not require that he hold sensory knowledge and empirical observation in contempt (as in a Platonic rationalism). On Descartes' scientific rationalism, see Plutynski 2011.

[17] A year after the publication of the *Discourse*, Descartes wrote to Mersenne: "You ask if I regard what I have written about refraction as a demonstration. I think that it is, in so far as one can be given in this field without a previous demonstration of the principles of physics by metaphysics, and so far as it has ever been possible to demonstrate the solution to any problem of mechanics, or optics, or astronomy, or anything else which is not pure geometry or arithmetic. But to require me to give geometrical demonstration on a topic that depends on physics is to ask the impossible. And if you will not call anything demonstrations except geometers' proofs, then you must say that Archimedes never demonstrated anything in mechanics, or Vitellio in optics, or Ptolemy in astronomy. But of course nobody says this." AT II 141–142, CSMK 103. The fact that Descartes refers to "demonstrating the principles of physics by metaphysics" does not obviate the point that particular arguments in physics cannot reach the certainty of geometrical demonstration.

It is less controversial that, by the time of the *Principles* in 1644, Descartes acknowledged the need for experience and the fact that arguments resting on experience do not possess the absolute certainty that he demanded in metaphysics.[18] He introduces the distinction between "moral" and "absolute" certainty, the first being good enough for "ordinary life," and the second pertaining to metaphysics and mathematical demonstration. He then claims "more than moral certainty" for his natural philosophy. The ensuing passage is nuanced, but taking the surrounding context into account, he seems to be suggesting that the general features of his physics concerning "the universe and earth" approach absolute certainty but do not quite attain it.[19] One might add that in the *Meditations*, as well, Descartes acknowledges the need to appeal to experience in some of his arguments. In the Sixth Meditation, he appeals to the facts that we receive sensations, and that we naturally believe them to come from external bodies, as grounds for inferring (by appeal to God's benevolence) that bodies exist. He also indicates that it is only by means of the senses that we can know the particulars of these bodies, such as that the sun is of a certain size.[20] And he acknowledges that, even though things known by the senses can be known very well if we have time to be careful in observing, we are nonetheless subject to error in such cases in a way that we are not with the cogito (or any other simple thing known by the clear light of the intellect).[21]

The historical Descartes devoted greater effort to natural philosophy than to metaphysics—still granting of course that his work in metaphysics was important to him and remains important in the history of philosophy. As it happens, in the second half of the seventeenth century, Descartes was more known for his natural philosophy than his metaphysics—again, acknowledging the strong response to his metaphysics in prominent figures such as Spinoza, Malebranche, and Leibniz.

It does suggest that the arguments will more closely approach the standard of geometrical demonstration if the first principles of physics have been certified by metaphysics.

[18] See Chap. 2 by Ariew. Ariew 2011 argues that Descartes held to an absolute standard of certainty in natural philosophy even at the time of the *Discourse*. He reads an important letter to Jean-Baptiste Morin of July 13, 1638 as containing an explanation by Descartes that he intended to "demonstrate" all of his hypotheses and "deduce" them from first principles. I read the same letter as indicating that Descartes will elsewhere demonstrate his "single supposition that all bodies are composed of parts" (AT II 200, CSMK 107). I take this to be the content of the suppositions he speaks of at the beginning of the *Meteors* (at AT VI 233), which he claims (in a letter to Antoine Vatier in 1638: AT I 563, CSMK 87) to be able to "demonstrate" a priori (but not without exposing his entire physics and, presumably, its metaphysical foundations). This leaves open that the particular explanations (of salt, snow, etc.) in the *Meteors* cannot be demonstrated a priori but are posited as part of a mechanistic program that receives empirical support from the fact that it proposes unified and plausible causes of the effects. On the care needed in interpreting Descartes' use of the term "deduce," see Clarke 1982 and Hatfield 1985.

[19] *Principles* IV 203–206. AT VIIIa 325–329, CSM I 288–291.

[20] AT VII 80, CSM II 55.

[21] On the roles of purely intellectual knowledge vs. sensory knowledge in the *Meditations* and related writings, see Hatfield 1986 (which contrasts with Clarke 1982).

11.3 Descartes and the Cartesians: Le Grand

During the latter years of his life and in the decades following his death, Descartes' philosophy faced a hostile reception in many quarters. It was subject to many condemnations and prohibitions. Numerous foes took up the pen against it.[22] All the same, his philosophy, and especially his natural philosophy, transformed the philosophical landscape with such rapidity and scope that it is appropriate to speak of a "Cartesian revolution."[23] There were legions of Cartesian followers pursuing general programs of Cartesian natural philosophy, as well as Cartesian physiology and medicine and Cartesian mechanistic psychology. Prominent followers were found in the Netherlands, France, Germany, Italy, and England.[24] The two dozen or so Cartesians discussed in the present volume make up but a fraction of the diverse body of Cartesian philosophers and physicians.

These Cartesians showed varying degrees of acceptance of Descartes' principles, methods, and conclusions. Some had doubts about his rendering of animals as mere insentient machines.[25] There were various attitudes toward whether Descartes was an occasionalist about body–body and mind–body interaction and also anent whether occasionalism is the right doctrine to hold.[26] There was disagreement over the need for Cartesian doubt as a methodological tool.[27]

At the same time, I have found broad agreement among Cartesians on the claim that some truths can be established by the pure intellect (or reason, or "the light of nature") operating independently of the senses. To name only two, this is found in Régis and Rohault.[28] Nonetheless, I find these same authors accepting the need for empirical observation in natural philosophy to tell among competing proposals for micromechanisms.[29]

---

[22] See Chap. 2 by Ariew.

[23] Bouillier 1842, I, 439; Ryan 1935, 227; Cohen 1985, 44, 49; Hatfield 1985, 160; 1996, 504.

[24] For some accounts of various Cartesians, see Trevisani 1992; Schmaltz 2002; Lennon 2003; and other chapters in this volume.

[25] Rosenfield 1968, Appendix B. Régis 1691, II, 506 expressed reservations, but ultimately accepted the hypothesis of animal automatism on theological grounds (that is, to avoid attributing animals an immaterial and hence immortal soul, a reason also given by Descartes). See also Hatfield 2000, 2005.

[26] Clatterbaugh 1999.

[27] See Chap. 10 by Nyden.

[28] Régis 1691, I, 70–71, 127, 160; Rohault does "conclude from that Knowledge which we have by Reason, that the *Essence* of Matter consists in *Extension*," something he derives from "the Light of Nature" (Rohault 1987, I, 24–25). He also lays down certain "Axioms of Natural Philosophy" as "self-evident," including that "Every Effect presupposes some Cause" (Rohault 1987, I, 18–19). In my experience, Anglophone interpreters of Descartes in the second half of the twentieth century rarely acknowledged that Descartes embraced a kind of knowledge that was *completely* independent of the senses, although from the 1980s this aspect of his philosophy has become more widely acknowledged.

[29] Régis 1691, I, 275–278 (on 276 he in effect paraphrases Descartes' letter to Morin, which was published in 1657 in the first volume of Descartes' *Lettres*, see AT II 196); and Rohault 1987, I, 13–14.

Among the large body of seventeenth-century Cartesians, I am focusing on the Franciscan divine and professor of philosophy Antoine Le Grand. Not a great deal is known about Le Grand's life.[30] He was born in Douai (then part of the Spanish Netherlands) in 1628 or 1629 and trained there as a Franciscan, affiliated with the second English Province, a group devoted to reconverting England to Catholicism. He was ordained in 1655 and taught philosophy briefly before being sent on a mission to England in 1656, where he lived until his death in 1699. He spent most of those years in London, but moved to Oxford as a private tutor in 1695. In England, he served as professor of philosophy and held various church offices. As a Catholic priest in England during this time, he was in some danger and indeed was briefly arrested in 1692.[31] Hence, his teaching occurred in private homes or in rooms rented to or otherwise made available to Franciscan missionaries.

Le Grand was the author of several works of philosophy and a smaller number in Catholic theology.[32] He wrote all of his significant works after moving to England. His first three books, which appeared in the 1660s, were ethical in nature and did not show the influence of Descartes. Then in 1671 he published a "digest" of Descartes' philosophy, followed by an expanded "institution" of philosophy in 1672, entitled *Institutio philosophi, secundum principia domini Renati Descartes*. These works were clearly intended as textbooks and as such were reasonably priced.[33] The more complete *Institutio* was well received and underwent frequent revision, going through several London editions as well as being printed in Geneva and Nuremberg, into the eighteenth century (e.g., 1711). It was soon supplemented by two related and more specialized books, the *Historia naturae* (1673), which was dedicated to Robert Boyle, and *Dissertatio de carentia sensus*

The interpretations offered by these authors are consonant with Descartes having expressly accepted that experience is needed in natural philosophy from the time of the *Discourse*, as these authors cite the need to know causes by effects (e.g., Régis 1691, I, 277; Rohault 1987, I, 13–14), which paraphrases *Discourse* VI, AT VI 76, CSM I 150. They may also mention (e.g., Rohault 1987, I, 14) the well-known "code" passage from *Principles* IV 205, AT VIIIa 327–328, CSM I 289–290, but this appears along with wording reminiscent of the *Discourse*.

[30] The most complete source remains Ryan 1935. See also Easton 2012. Thaddeus 1898 provides much information on the activities of the Franciscans in England, with biographical notice of Le Grand.

[31] Thaddeus 1898, 231.

[32] Ryan 1935 lists Le Grand's publications.

[33] Ryan 1935, 239. Le Grand 1671 was immediately and favorably reviewed in the *Philosophical Transactions* (Oldenburg 1671), although the notice gave greater attention to Jacques Rohault's 1671 *Traite de physique* (see Rohault 1987), which went further into natural philosophy than Le Grand's initial brief digest. Le Grand 1671 drew positive comment from John Beale (F.R.S. and a promoter of the "Moderne Philosophy") in a letter to Henry Oldenburg (June 12, 1671, Hall and Hall 1971, 92, 95). Slightly later, Beale (to Oldenburg, June 24, 1671, Hall and Hall 1971, 122) approvingly reported a plan to translate Le Grand 1671, but a translation appeared only in 1694 (of a subsequent edition, described below). In 1671, Descartes' works were well known in England and had their admirers, often critical, including Kenelm Digby, Thomas Hobbes, and Henry More (see Rogers 1985). Robert Boyle was also an admirer, only sometimes critical (see Boyle 1674). Descartes' *Discourse* appeared in English translation in 1649, his *Passions* in 1650, his *Compendium* in 1653, and the *Meditations* in 1680; Latin editions were, of course, also available.

& cognitionis in brutis (1675), which was dedicated to Henry Jenkes (F.R.S.). There were also theological works and works defending Descartes against the charge of atheism and responding to other attacks.[34]

The three Latin works just mentioned are an early exposition of a complete Cartesian philosophy, covering logic, metaphysics, natural philosophy, and ethics (the *Institutio*), together with defenses of portions of Cartesian natural philosophy. The three works were later translated through the London publisher Richard Blome—who specialized in fancy volumes supported by subscription—in a sumptuously illustrated edition of 1694, entitled *An Entire Body of Philosophy, According to the Principles of the Famous Renate Des Cartes*;[35] the translation was reportedly made by a member of the Royal Society.[36] My interest here is not to chart Le Grand's development,[37] but to examine his most extensive statement of Cartesian philosophy for what it can show us about how Descartes was read, interpreted, and defended by a devoted expositor who intended to be faithful to Descartes' opinions.[38]

The first book of the *Entire Body*, the Institution, purports to survey all of philosophy, in ten parts. Le Grand incorporated the usual four-part scheme, adjusting it to the new Cartesian ontology. He treats logic as a propaedeutic "instrument" of philosophy; his first part is a brief introduction to logic, featuring clear and distinct

[34] The attacks came from Samuel Parker and John Sergeant, on which see Easton 2012.

[35] This work, advertised on the title page to contain "large Additions of the Author, never yet Published," is in the passages relevant herein often but not always a translation of Le Grand 1680 (*Institutio*), 1673, and 1675 (I regularly cite the Latin original, including 1672 where the earlier passage is sufficiently similar). Samuel Pepys is listed among some 100 subscribers (or "benefactors"), as is Godfrey Kneller, the German-born English portrait artist. Blome himself is best known for his cartographic publications. The broadsheet proposal survives (Blome 1692).

[36] Ryan 1935, 241. Le Grand dedicated many of his Latin works to Catholics or Catholic-friendly patrons. Blome dedicated the translation to Henry of Sydney, one of the Immortal Seven and indeed author of the letter inviting the Protestant William of Orange to take the throne from the Catholic James II in the Glorious Revolution of 1688. Hence, Le Grand's Catholicism (and Descartes') might be balanced by Henry's Protestant credentials. Blome's dedicatory letter presents Henry as "Master of the whole Body of PHILOSOPHY" and suggests that Henry might regard philosophy as needing no protection, on the grounds that "Works of the Learned, like the Actions of the Noble, ought only to be judged by their Peers; nor can I think your Lordship, or any Learned Person will say, that a virtuous desire of Knowledge (and such I humbly conceive is Philosophy) needs or wants a Protection from any, since 'tis an experienced Truth, That Virtue carries her own Safeguard, as well as her own Reward" (Blome in Le Grand 1694, front matter). Apparently, philosophy is to be judged as philosophy, no matter the creed of its author. Le Grand's Catholicism does manifest itself, as when he assigns to princes and priests respectively the obligations to "suffer no Innovation in matters of Worship" and to "avoid all novel Doctrins." Le Grand 1694, I, 400, 402.

[37] The question might arise of whether Le Grand's accommodation of experience was influenced by his British context, including the publication of Locke's *Essay* in 1690. No strong influence is evident. I coordinate my citations of Le Grand's methodological remarks in 1694 with the Latin editions from the 1670s and 1680.

[38] Le Grand 1694, I, preface iv; 1672, preface xv. Le Grand did not follow Descartes on everything. In natural theology (Le Grand 1694, I, Part 2), he offers a proof for the existence of God from the order of created things (not found in Descartes), and he discusses God's providence in ways not found in Descartes (befitting Le Grand's role as a divine). These are not the only departures, which I note only as needed.

mental perceptions, a mitigated aspiration for certainty,[39] syllogisms, and method. He reports that philosophy itself is "commonly divided into "Metaphysics, or Natural Theology," "Physiology, or Natural Philosophy," and "Moral Philosophy, or Ethicks."[40] In accordance with Descartes' dualistic ontology of thinking and extended substance, he expands metaphysics to cover all immaterial substances, grouped under the rubric "Pneumatica."[41] He treats these topics in the second, third, and ninth parts, respectively on God, angels, and the human mind (the latter part labeled "psychology"). In fact, the ninth part, on the human mind, falls within the portion of the book devoted to natural philosophy and includes discussion of bodily mechanisms that cause the passions, as well as mind–body union and interaction. The tenth part pertains to ethics, leaving the bulk of the work, parts 4–9, to cover natural philosophy. He divides these parts in a manner inspired by Aristotelian treatments (with some adjustments): general natural philosophy, covering extended body and its general properties; special natural philosophy, concerning the world and heaven; the four great bodies, earth, water, air, and fire,[42] with minerals and meteorological phenomena; living creatures in general and plants and animals in particular; and man (human beings) in relation to his body, which includes anatomy, the physiology of nutrition, growth, and blood; the senses, imagination, and memory; and the mode of operation of medicaments. The ninth part, on the human mind, includes natural philosophical aspects of mind–body union as noted, but also touches on the metaphysics of immortality. In any event, we have the traditional four parts of philosophy, with emphasis on natural philosophy and then metaphysics.

The work contains two additional books, each briefer than the previous. The second book is a "History of Nature." It is billed as an application of Cartesian natural philosophy to well-known or problematic phenomena that other "sects" of philosophers might claim to handle better.[43] The topics overlap the parts of the first book on natural philosophy. This second book comprises: body; qualities; the world and heaven; the four bodies; fossils; meteors; plants; animals; and man, including his body and passions. Finally, the third book is a brief "Dissertation of the Want of Sense and Knowledge in Brutes," in which Le Grand defends what he considers to be Descartes' thesis that beasts are "meer Engins or Machins."[44]

---

[39] In "science" nothing dubitable is to be accepted, but for everyday life a weaker standard is needed; the senses are dubitable and are not in themselves instruments of truth, but nonetheless may be used in the investigation of nature; the senses do not themselves err, but in sensory perception the mind may err in judging. See Le Grand 1694, I, 5–7, 30; 1680, 9–11, 65.

[40] Le Grand 1694, preface iv; 1680, iv. The English translates *physica* as "physiology," which means the science of nature (the *logos* of *physis*) and does not have the medical connotation of covering the natural functioning of living organisms, which was also extant. See Hatfield 1992.

[41] Le Grand 1694, I, 77; 1680, 159.

[42] Le Grand accommodates the traditional four elements in his textbook (Le Grand 1694, I, 179; 1680, 359), which he calls types of "body" and explicates using the three elements or types of particle found in Descartes, which Le Grand introduces earlier in his general physics (Le Grande 1694, I, 99–100; 1680, 203).

[43] Le Grand 1694, I, preface ix; compare 1673, preface ii–iii.

[44] Le Grand 1694, I, preface xii.

11.4 Le Grand on Certainty and Experience in Natural Philosophy

Le Grand is an avowedly faithful follower of Descartes, and he knew Descartes' works very well. In comparison with Rohault and Régis, and perhaps with Descartes himself, he claims a high standard of certainty for his Cartesian natural philosophy, calling it "a Species of Science" that is "conversant about things that are True and Necessary."[45] He gives as examples the doctrine that body is extended and hence divisible. All the same, he by no means held that the claims of "special natural philosophy," concerning the natural world and its particular structures, can be known by reason alone, or with the same certainty as accrues to metaphysics.

In first considering the range of things he says about certainty and the need for sense perception in natural philosophy, a reader may suspect that his position is unformed or confused. Thus, under general natural philosophy, he responds to the objection that "the Certainty which is required of Science, cannot be had in Natural things, seeing that Bodies and all their Attributes are only perceived by the Senses" by rejecting this claim and asserting that "Bodily things are not known by the Senses, but by the Understanding alone."[46] But in the History of Nature, he writes:

> Tho' our Senses are often mistaken, and being deluded by false Representations, do circumvent our Mind; yet have we no safer way by which we may arrive to the Knowledge of Corporeal Things, than with their assistance we measure their Dimensions, and by Signs conveyed through our Eyes, discover their Existence: For it is by their Advertisements we come to know, that Matter consists of 3 Dimensions; and that the Parts of it are capable of various motions, have different forms, and do diversely affect the Organs of our Senses.[47]

He seems to take back in the latter passage what he had asserted in the former. It may nonetheless be possible to reconcile the two passages, by considering their respective functions and placement in his work as a whole, and in this way to gain a fuller understanding of his attitudes toward sense experience and toward absolute certainty.

The first passage denies that the attributes of bodies are "only perceived by the senses" and affirms that bodily things are known not by the senses "but by the Understanding alone" (echoing Descartes' passage on the wax in the Second Meditation). The denial that bodies are perceived only by the senses is consistent with some things about bodies being found out by the senses. And the claim that body is known "by the Understanding alone" is consistent with the position that the essence of each body can be grasped only by the understanding or intellect, even if the understanding can discover the particular size and shape of a body only with the aid of the senses. The first passage occurs near the beginning of the part on "general physicks,"

[45] Le Grand 1694, I, 91; 1672, 116.

[46] Le Grand 1694, I, 92; 1672, 116–117.

[47] Le Grand 1694, II, 1; 1675, 1. Here and throughout, I don't replicate the use of italics from the Blome edition, but use italics to mark technical words and to indicate propositions that are mentioned.

in which Le Grand ultimately establishes that body is extended substance.[48] But he also makes clear that sense perception can be a source of knowledge if joined together with reason or understanding. He recounts that when touch and vision give conflicting appearances on whether a stick in water is bent, it is "by our Touch we know the Stick to be straight"; but it is by reason that we decide to "give credit to the report of our touch."[49] Hence, the senses can give knowledge, but only in connection with the understanding, a position closer to the second passage.

In another passage prefatory to the History of Nature, Le Grand declares that this book will test the principles achieved in the first book:

> For seeing that the Truth of the Principles of any Science is made manifest by the Evidence of its Deductions, and that their Certainty is looked upon as Indubitable, if those Things that are Inferr'd from them, do wholly depend upon the knowledge of them; I was desirous to try, whether several Appearances of Nature, or all those Things which our Senses perceive to be in Bodies, did comport with the Principles laid in my Institution of Philosophy, and whether there be such a Connection between them, as that tho' the latter may be apprehended without the former, yet the former can never be Understood without the latter.[50]

Several points might be made. First, in the History, Le Grand is starting from the facts of sense (as befits a history).[51] Second, he intends these facts of sense to confirm and to (help) render indubitable the principles achieved in the Institution. Thus, this passage need not be read as asserting that, in the order of knowledge, we know that bodies have three dimensions because we perceive this through the senses. But we may first learn this fact by the senses, and then come to know it "philosophically" or by way of "science" when we arrive at it through the understanding, in general physics.[52] Third, although we can cognitively grasp or "apprehend" the appearances without aid of the principles (as when we perceive a square tower in the distance as round), we cannot understand the appearances without the principles. In his example of the tower,[53] this understanding amounts first to the realization that there must be a thing of some sort that we are seeing (hallucination appears to be ruled out), even if we should be mistaken about some of its properties. Fourth, the notion that we start from the appearances is

---

[48] Le Grand 1694, I, Part 4, Chap. 3; compare 1672, 123.

[49] Le Grand 1694, I, 92, echoing Descartes' discussion in the Sixth Replies (*Meditations*. AT VII 439, CSM II 296).

[50] Le Grand 1694, I, preface vii; 1673, preface i. One may note an echo of Aristotle's distinction between fact and reasoned fact, and also of Descartes' discussion in *Discourse* VI, AT VI 76, CSM I 150. That the things "inferred" from the principles do "wholly depend on the knowledge of them" seems not to mean that their existence (e.g., of the phenomena of the heavens) depends on "knowledge" (scientific derivation) of them, but rather that an understanding of the things that appear depends on those principles. As Descartes says in the cited passage, his causes don't prove the effects (which are known by the senses) but rather serve to explain them.

[51] Le Grand's strategy is to use a detailed natural history to show that Cartesian natural philosophy can indeed account for the great diversity of effects that other systems of philosophy claim to cover. Of course, the Institution was not itself without appeal to facts and observations.

[52] Le Grand 1694, I, 94; 1672, 123.

[53] Le Grand 1694, II, 1; 1673, 2.

consistent with Le Grand's methodological claim that the "object" of natural philosophy is to answer questions in which "the Cause is searched out by the Effects."[54] He gives the example of reasoning from the effect that water dissolves salt and sugar to the motion of the "insensible" parts of water as the cause.[55]

The use of the effects to confirm (render "indubitable") the principles may provide a hint for interpreting the apparently contradictory passages from above, on the certainty of natural philosophy as a science or demonstrative body of knowledge. At the beginning of the parts on natural philosophy, Le Grand writes:

> that Physiology is a Species of Science, and is conversant about things that are True and Necessary, appears from the Demonstrations that are made of Natural things; the Certainty whereof depends on the stability of the Things that are defined, and supposeth their determinate Essence.[56]

His example is the demonstration that body is divisible from the claim that body is extended, which depends for its necessity on bodies being by nature extended. So it would appear that Le Grand holds that natural science—or at least its general part—reaches necessary truths through knowledge of unchangeable essences.

Other passages qualify the certainty of natural philosophy. In the section on method in the Logic, Le Grand admonishes: "Yet are not we to conceive that all things that are true, are of the same certainty; for some things are only Contingently true, that is, such as are taken to be true by us, tho' indeed they may be false."[57] As an example, he describes the possibility of inferring someone's piety from church attendance and good behavior, an inference that is "morally" certain but may nonetheless be false. He contrasts such contingent truths with those that are "altogether Certain and True, as are all Propositions of Eternal Truth," such as "Twice four make Eight." But within these, he again finds different degrees of certainty:

> Some *Attributes* are said to be joynd with their Subjects, by a *Physical* certainty, when according to the order of Nature, it is impossible but they must be joyned to them; as when we say, that *a Man hath two Feet*; because, tho' a Man may be conceived without Feet, yet Naturally Man is never without them. Other things are called true, by *Metaphysical* certainty, when an *Attribute* is so indissolubly attributed to its Subject, that it cannot be conceived to be otherways: As when we say, *Three is a number*.[58]

---

[54] Le Grand 1694, I, 45; 1672, 61.

[55] The passage clearly affirms that the effect is known by the senses (supplemented by the understanding, of course): "it is apparent that Water, and all Liquid things, have their Parts in Motion: For we see that Water dissolves Salt and Sugar, which would not be, if the insensible parts of the Water by their continual motion, did not run against the Salt and Sugar" (Le Grand 1694, I, 46; 1680, 97).

[56] Le Grand 1694, I, 91; 1672, 116.

[57] Le Grand 1694, I, 44; 1672, 57.

[58] Le Grand 1694, I, 44; 1672, 58. Boyle 1674, 139–143, distinguishes metaphysical, physical, and moral certainty in a manner similar to Le Grand. Ariew (2011, 40) quotes an earlier distinction among moral, physical, and metaphysical certainty in the Jesuit philosopher Rodrigo Arriaga. The distinction is similar but may not be exactly the same as in Le Grand, since Arriaga's example of physical certainty is seeing that someone is running (perception of a particular act) whereas Le Grand uses a general claim about the nature of the human body (it has two feet). Shapiro 1983

The denial of a metaphysical truth is inconceivable. But the certainty of science rests on the stability of essences in nature, or what "naturally" occurs. We might ask how the certainty of essences of various kinds of natural things contrasts with the eternal truths, whether Le Grand is invoking a *hypothetical necessity* (given that the essence of human beings is to have two feet, then all humans must naturally possess them), a discovered universal generalization about the *normal* state of human beings (and hence admitting exceptions), or some other status.

Le Grand himself refers us to his earlier discussion of judgment and propositions for clarification. But this discussion offers only moderate help. He there distinguishes metaphysical universality from moral universality, the first being exceptionless, the second admitting exceptions. "Every body is extended" has metaphysical universality, whereas in daily life we act on maxims that are true only for the most part ("moral" universals).[59] The claim about extension falls into the metaphysical certainty of the first principles of natural philosophy. But what about the claim that human beings have two feet? Le Grand wants such claims to reach the standard of science and so to be in some sense necessary. But he also acknowledges that a human being might conceivably have no legs, as with a double amputee.[60]

Interestingly, in this discussion on universals, we do not find a counterpart notion to physical necessity, which would be termed a "physical universal." But some such notion seems to be required to fit Le Grand's concession that physical truths about particular essences (such as the essence of a human being) do not accrue the same certainty as metaphysical truths (including those that serve as the foundations of natural philosophy). It may be too much to ask for a systematic account of the status of knowledge of particular essences, as this knowledge surely must depend on observation to some extent, and to ask for a systematic account would be to ask Le Grand to address the inductive problem of supporting universal claims through observations.

Le Grand sought to reconcile certainty with lack of metaphysical necessity. His efforts are not completely effective. But he did make the interesting claim that the certainty of physics lies between the metaphysical certainty of eternal truths and the moral certainty of daily exigency. In this he reflected Descartes' general ordering of certainties. Indeed, his "physical certainty" might address Descartes' assertion that aspects of his natural philosophy, though not quite reaching the certainty of metaphysics, have "more than moral certainty."[61] Le Grand

---

reviews the concepts in British culture more generally; however, her treatment does not recognize "moral certainty" in Descartes and does not include Le Grand, who lived in England for more than 40 years.

[59] Le Grand 1694, I, 28; 1672, 38.

[60] Le Grand 1694, I, 266 subsequently affirms that the "bare essence" of human beings or the "perfection of Man as such" consists in mind–body union, so that an amputee is no less a human being.

[61] Descartes says that his general results on "the universe and the earth" will "perhaps" be admitted to the class of absolute certainties, offering as grounds that they can "hardly" be explained otherwise than he has suggested (*Principles* IV 206. AT VIIIa 328–329, CSM I 290–291). Le Grand's "physical certainty" might just fit the bill. It purports to describe how nature *in fact* acts, not how it *must* act. Only the metaphysically secured aspects of Descartes' physics fit the latter proscription.

may well be closer to Descartes concerning the level of certainty ascribed to natural philosophy than are Régis and Rohault, who offer a significantly lowered standard.[62]

11.5 Mechanizing the Sensitive Soul: Descartes and Le Grand

In the *Discourse*, Descartes recommended that in natural philosophy one begin with observations that are "common" and usual. In Parts 3 and 4 of the *Principles*, he frequently refers to common phenomena, such as the observed paths of the planets against the stars, or the behavior of the magnet (commonly known among natural philosophers of his day).

We can look deeper for the common phenomena that Descartes accepted. For indeed in many branches of natural philosophy and what were known as mixed mathematical sciences, such as optics, there was a wealth of description of phenomena either offered as such, or embedded in the very vocabulary of description. An example of the former is observations about the rainbow. An example of the latter is the vocabulary in which the offices of the Aristotelian soul were described: vegetative, sensitive, and rational, with various sub categories, such as generative, nutritive, and motor powers.

Put another way, Aristotelian and Galenic accounts of the soul and its powers offered both a description of the phenomena of living things and an explanatory framework.[63] Explanation of the vegetative power might invoke subpowers, such as a nutritive or generative power. Descartes rejected these explanations but retained the basic list of phenomena to be explained. Thus, he accepted that animals reproduce, are nourished, grow, have sense organs, which guide them as they move about, and that this system of sensory guided motion allows them to approach beneficial objects and avoid harmful ones (on the whole).

Scholars have generally read Descartes as seeking to replace the Aristotelian account of sentient animals with the unfeeling, mindless and soulless animal machine. In recent decades this orthodoxy has been challenged. Although I do not

[62] Régis 1691, I, 275: speculative physics (knowledge of causes) is "problematic" and "uncertain." Rohault 1987, I, 14: "we must content ourselves for the most part, to find out how Things may be; without pretending to come to a certain Knowledge and Determination of what they really are." On experience and certainty in Rohault, see Chap. 9 by Dobre.

[63] In the sixteenth and seventeenth centuries, authoritative accounts of life and living things (such as Fernel 2003) included a synthesis of Aristotelian and Galenic concepts. Le Grand discussed a number of previous philosophers in his work, but Aristotelian philosophy provided the chief point of comparison. He did not include Galen in his list of historically important philosophers (Le Grand 1694, I, preface), but he did occasionally refer to Galen in discussing living things. Galen was part of the general context, even if Le Grand (by comparison with Descartes) did not invoke him prominently.

find the challenges convincing,[64] my focus here concerns Le Grand's interpretation of Descartes on animal sentience.

Le Grand ascribes to Descartes the view that animals are unfeeling machines.[65] In that way, he offers an interpretation of Descartes' intended position (the common interpretation of Descartes in his time). But beyond offering a take on Descartes' views, Le Grand elaborates the program of mechanization. Descartes offered a strong beginning. In the *Treatise on Man*, he described a mindless human body that could also serve as a model of an animal body, and he explained how it could carry out various sensory and motor functions, including many that occur in human beings without the guidance of thought.[66] The *Dioptrics* offers some account of the anatomy and physiology of sense perception (including mind–body interaction), and the *Passions* describes the machine of the human body. To this picture Le Grand adds a wealth of detail at every level, from an extensive discussion of what makes something a living being, to treatment of various kind of plants and, among animals, discussions of fourfooted beasts, birds, fishes, and insects, a discussion of what constitutes health in a mere machine, and a detailed account of the anatomy and physiology of the human body.

Although Le Grand's treatment is remarkable for its conceptual and descriptive intricacy, here we must settle for examining only a few instances. To begin with, it is noteworthy that in this treatment of living things, as elsewhere in the work, he followed the strategy (found in other Cartesians) of using language similar to the Aristotelian philosophy and so familiar from other textbooks. A general example is the use of the terms "matter" and "form" to mean, not prime matter that is "informed" by a substantial form or real accident, but bare extended matter that takes on various forms (spatial configurations). In opposition to Aristotelian orthodoxy, extension is not a mode or universal accident, but the essence of matter, and the "forms" of such extended matter are the various modes of extension: magnitude, figure, situation, motion, and rest.[67]

Le Grand used a similar strategy of terminological repurposing in his definition of living things. In dividing things into living and nonliving, he offers the following definition:

> Now we call those things *Living Bodies*, that have received from the Author of Nature such a Disposition of an Organical Body, as that by innumerable passages and conveyances it hath, the Alimentary juice, being by motion thrust into them, is by the Soul every way dispersed and distributed, for their Nourishment, Growth and Conservation. They are called Living Bodies from the Life they possess, and Animate Bodies from their Soul.[68]

---

[64] A challenge to the previous orthodoxy: Cottingham 1998; a response: Hatfield 2007.

[65] Le Grand 1694, II, 250–252; 1675, 141–145.

[66] Allusion to mechanistically produced human responses, which occur without the direction of thought or the intervention of the soul, occur in the Fourth Replies (AT VII 229–230, CSM II 161) and the *Passions* (AT XI 338–339, 358, CSM I 333–334, 342–343). Le Grand acknowledged such processes (e.g., Le Grand 1694, II, 243; 1675, 104–105).

[67] Le Grand 1694, I, 94–96, 106; compare 1672, 123–125, 148.

[68] Le Grand 1694, I, 229; compare 1680, 446. The seventh part of the English translation, on living creatures, is considerably revised from the 1680 *Institutio*.

On the surface, this seems like an appeal to an Aristotelian sensitive soul (or its Galenic counterpart), in which vital processes are governed by the vegetative power of the animal and human soul. But the appearance is deceiving. The "soul" turns out to be nothing more than Descartes' "fire without light" that gives motion to the blood in animals. As described by Le Grand, "This Soul, a fit disposition of parts being first supposed, chiefly consists in the Innate or Inborn-heat, which is a Heating, but not a shining Fire."[69] From this heat the alimentary juice in plants and the blood in animals is distributed to the parts of the organic body. Life consists in the fact that an organic body has heat and moisture to move the needed fluids that provide nourishment and growth; death occurs when cold and dryness predominate and fluids don't flow.[70]

A second thing to note about this passage is that Le Grand has taken over the functional description found in previous works on living things. He speaks of "nourishment, growth and conservation," which are standard descriptions. Indeed, he generally accepts previous functional descriptions. Thus, not only nourishment, growth, and reproduction figure in, but the senses, imagination, and memory, motor capacities, and instincts for the preservation of life.

Although Descartes does not specifically speak to what counts as the "health" of a mere machine, and hence the health of plants or nonhuman animals, he does discuss the conditions for death. His account of death, even for the ensouled human being, is that it occurs when the body is broken.[71] The complementary notion of health should then be the well-functioning of bodily mechanisms. And this is exactly the notion that Le Grand supplies in his elaborated Cartesian theory of living things, in discussing health and disease. He writes:

> *Sanity* or *Health* is a certain disposition of the Body, by which it is rendered capable to perform its Offices. So that when-ever the temperaments of the parts chance to be altered, or its Organs to be intercepted, the Order or Course of Nature being hereby inverted, the Creature must needs be deprived of Health, as falling from its primitive State of Body.[72]

Health is a disposition of the body so that it can carry out its offices; that is, for an animal body, its various vegetative and sensitive functions. The notion of "performing an office" implies a standard of functionality. If, as Le Grand suggests, living things are the product of a divine cause, then this finality can be seen as God's handiwork. For a Cartesian who accepted Descartes' conjecture that life might develop out of the chaos of particles by natural causes, organic finality might be harder to account for. But Le Grand invokes divine origin for plants and animal bodies.[73]

---

[69] Le Grand 1694, I 229; compare 1680, 443, 446, which highlights heat and moisture but does not lead off with the notion of a "Soul" consisting of a non-shining fire. Descartes' "fire without light": AT XI 123.

[70] Le Grand 1694, I, 231; compare 1680, 452.

[71] Descartes, *Passions*, Art. 6. AT XI 330–331, CSM I 329–330.

[72] Le Grand 1694, II, 198; compare 1673, 373.

[73] Le Grand 1694, I, 60, 240; II, 235. Le Grand affirms the Cartesian doctrine that "The Ends of God are not be enquired after in Natural Philosophy" (1694, I, 132) and interprets this as an injunction

Throughout his discussion of the particular capacities of animals and of the human body, Le Grand appeals to common descriptions (from previous writers) and to common observations as a starting point in considering the "offices" of the animal machine. Thus he cites Edmé Mariotte on the phenomena of milky plants, Virgil on setting twigs to root, and Marcello Malpighi and Nicolas Steno on the organs of touch.[74] He describes, in his discussion of the human body, both the organs of sense and the things that they allow us to perceive, including, for vision, the familiar list of light and color, situation, distance, magnitude, figure, and motion or rest.[75] In relation to these phenomena, he describes various causal mechanisms to account for them, sometimes adhering closely to things Descartes has said, sometimes elaborating on topics not covered by Descartes, and sometimes offering an alternative doctrine. In describing how we see distance, Le Grand suggests that to accommodate the lens system for various distances, the whole shape of eye is altered, whereas Descartes had conjectured that the shape of the lens changes.[76]

In defending the view that animals are mere machines, Le Grand was required to explain mechanistically not only the internal processes of nutrition and the like but also the phenomena of the sensitive powers of the Aristotelian soul. These phenomena form a subpart of Aristotelian psychology (*Peri psyches*), or *De anima* studies (in accordance with the odd practice of citing Aristotle's works with their Latin titles). In this Aristotelian psychology, the powers of the sensitive soul included sense perception, motor action, and cognitive operations that mediated between the two. In a common example, when the sheep sees the wolf, it perceives not only its shape and color, but also the fact that it is an enemy. To perceive shape and color is the office of the sense of sight; to perceive the enmity of the wolf is the office of what was called, in various contexts, the estimative or cogitative power.[77]

In discussion prior to Le Grand's writing, various opinions prevailed on whether animals perform such operations only by instinct or by a combination of instinct and learning. Le Grand (as Descartes before him) had to allow for both, for he was aware that animals can be trained.[78] As had Descartes, Le Grand appeals to changes in corporeal memory, that is, in the brain, to explain how beasts might be trained.

---

into asking why God created the things he did, and what purpose he might have had in making them. This may well allow seeing creatures as having an internal functionality that is God's handiwork. On the relation between external and internal finality, see Laporte 1928, Simmons 2001, and Hatfield 2007 (the last of these also considers Descartes' conception that living things might have arisen from the primordial chaos, as did stars and planets).

[74] Le Grand 1694, I, 243, 245, 287; in Le Grand 1680 I find only the reference to Virgil (472–473).

[75] Le Grand 1694, II, 298, 300; 1680, 560, 565.

[76] Le Grand 1694, I, 301; 1680, 568. Descartes, *Dioptrics*, third and sixth discourses (AT VI 108, 137), in which Descartes' language suggests that he might have allowed a change in the shape of the whole eye as well. In the *Treatise on Man* (AT XI 156) he clearly posits that only the lens changes shape.

[77] On the notion of Aristotelian and mechanistic psychologies, see Hatfield 2009.

[78] For details on Descartes' mechanistic accounts of animal learning and animal instinct in context, see Hatfield 2012.

He refers to the explanation, also found in Descartes' *Man*,[79] of inner pores in the brain that, when they have once been opened together by the flow of animal spirits as directed by the senses, gain a tendency to open together again, even if only a part of the sensory pattern is repeated. This forms a kind of material associative memory. In this way, staying at a general level of description, Le Grand explains how hawks and hunting dogs may be trained, among other things.[80]

Le Grand explains the more elaborate patterns of animal behavior as instincts. The notion of instinct was widely discussed in the seventeenth century. For Le Grand, an instinct in a nonhuman animal (and one based in the human body) is a mechanical disposition to produce a certain result given some sensory situation. Le Grand lists some (supposed) behavioral generalizations concerning animals: that camels prefer to drink muddy water, that rabbits ("conies") dig burrows, that hares shot with arrow seek the herb dittany, that cats skip and dance when given the herb valerian, and that hens cluck at the sight of a kite overhead. He then explains that these are all the product of instinct:

> The cause of all these Actions can be referred to nothing else but Natural Inclinations and Instinct, by which Animals are instigated to such and such motions, and resolve upon what things are convenient for them: So that this, or that whole Brutal Species is carried with one propension, and there is the same force, and the same impulse found in all of that Kind. So every Hare is a like fearful, and by the method of subtilty declines all Dangers and Inconveniences.[81]

The language here may be deceiving. He speaks of the "sight of a kite" in the case of the hen; animals "resolving" upon a path of action that is "convenient"; and the hare avoiding "dangers and inconveniences."

These seem to be instances in which perceptual and cognitive language is applied to animals, and they are. But this perceptual and cognitive language is used to describe the behavior of the hen and hare, not to characterize the processes that produce the behaviors. As descriptions of behavior patterns, they are couched in a psychological vocabulary; but they do not entail that sentient and cognitive processes underlie these behavior patterns. Indeed, Le Grand is repeatedly explicit that animals do not feel, sense, "cogitate," or reason. They are mere machines:

> The Faculty therefore of Sense and Motion (which by some is called the Sensitive Soul) in Animals consists in a due disposition of the Parts, viz. their Nerves, Muscles, Spirits, Fibres, Joints and of their other Organs; by the help whereof Animals become differently affected from outward and inward objects, and are carried from one place to another. For all animals (Man excepted) are a kind of Watches or Clocks, which by a fit adaptation of their parts, having a Bodily Principle of Motion in themselves, as long as they are well disposed, and have whatsoever is required to perform and exert the several actions to which they are design'd. For all the Effects we perceive in Animals (Man excepted) have no other cause or Principles but the Body.[82]

---

[79] AT XI 178–179.

[80] Le Grand, 1694, II, 248, 258–260; 1675, 132–133, 181–189.

[81] Le Grand 1694, II, 250; 1675, 137–138.

[82] Le Grand 1694 I, 253; 1680, 487. From the Dissertation Concerning Brutes: "From this similitude of the Vectis [lever] with the Machin of an Animal, and from the manner by which we have

The role played by the sensitive soul (and its organs) in animals is, in Le Grand's reckoning, actually played by suitably arranged bare Cartesian matter. As he also says: "Nor am I more inclinable to the Opinion of the Peripateticks, who attribute Sense to Brute Animals, and are perswaded that they See, Hear, Smell, &c. in the same manner as we do."[83]

Le Grand fully recognized that the project of mechanizing the offices of the sensitive soul was extensive. It included the many behaviors of animals that have the effect of benefitting them or preserving them from harm. These offices are in the first instance carried out by "Natural Instinct," due to "corporeal dispositions" and the "local motion" of matter. Le Grand ranged the fundamental behaviors of animals under three headings: (1) avoiding hurtful and troublesome things; (2) propagating their kind; and (3) self-preservation through acquiring food and also laying it up for the winter.[84] These phenomena of animal behavior are well known. Le Grand claims that they can be accounted for by mere internal mechanisms. He is short on details of exactly how these mechanism work, beyond the economy of the "flow" of animals spirits and their diversion to the various muscles. For our purposes, the important point is that he recognizes broad categories of animal behavior, which vary in realization from species to species, each of which must be explained by hypothesized subvisible mechanisms.

11.6 Le Grand: Theory and Practice

In theory, Le Grand held that Cartesian natural philosophy must hold to a standard of physical certainty that is quite high. Officially, he repeated often that the senses are not to be trusted, or at least are not to be trusted without scrutiny of sensory appearances by the intellect.

In practice, in all of the areas of particular natural philosophy, including the study of living things, he referred frequently to phenomena that were known by observation, including the habits and behaviors of animals. In the book on the History of Nature, Le Grand recognized a role for sensory observation (monitored by the intellect) in testing the principles of natural philosophy, and in the Dissertation Concerning Brutes he recounts many reported and common observations of animal behavior.

above made out, that the functions of the Body are performed; it is clear enough evinc'd, that there is no necessity of a Soul in Beasts, for the producing of a Pulse of the Heart and Arteries, the Concoction of Meats, Nutrition, Respiration and Procreation of Spirits; but that they are produced without any Cogitation, and consequently that the parts of the Body in Animals are moved from place to place, in the nature of Pneumatick or Hydraulic Automata" (Le Grand 1694, II, 242; 1675, 100–101).

[83] Le Grand 1694, II, 244; 1675, 110.

[84] Le Grand 1694, II, 251; 1675, 144–145.

In this way, Le Grand's stance on certainty and natural philosophy reflects a tension that has existed in the interpretation of Descartes' works. Many readers of Descartes have attributed to him an across-the-board desire for absolute certainty, even though he distinguished the certainty he claimed for first principles from that he claimed for his account of the structure of the universe and his postulated micromechanisms. Le Grand also had to reconcile his expressedly high standard of certainty with the actual practice of natural philosophy. These efforts led him to invoke a notion of physical certainty, lying between merely moral certainty and the absolute certainty of metaphysics. Perhaps this physical certainly captures the more-than-moral certainty that Descartes himself sought for his account of the actual structure of the universe, beyond the metaphysically insured foundations of his natural philosophy.

Finally, there is Le Grand's larger significance, beyond his contribution to the development of Cartesian methodology, his comprehensive elaboration of Descartes' natural philosophy (conceptually and empirically), and his vociferous representation of Cartesianism in England. His larger significance may well lie in a fact about him that is unexceptional. Le Grand granted a large role to experience in the conduct of natural philosophy, in filling out explanatory schemes framed by metaphysics. In this way, he was, as a Cartesian, unexceptional. For as we have seen, Descartes himself saw no contradiction between his own claim to found natural philosophy on an intellectualist (rationalist) metaphysics and his pursuit of a natural philosophy that responds to and is guided by experience in forming and testing conceptions of the particular mechanisms of nature. Le Grand's body of work sustains the proposition that pairing a Cartesian intellectualist metaphysics with a Cartesian reliance on sensory experience in natural philosophy is not contradictory. His own portrayal of this path comes closer to capturing the spirit of Descartes than did other prominent Cartesians, including Rohault and Régis. Le Grand was in this sense a particularly acute interpreter and defender of Descartes' philosophy, especially his natural philosophy and, in particular, his mechanistic psychology.

References

Ariew, Roger. 2011. The new matter theory and its epistemology: Descartes (and late scholastics) on hypothesis and moral certainty. In *Vanishing matter and the laws of motion: Descartes and beyond*, ed. Dana Jalobeanu and Peter R. Anstey, 31–46. London: Routledge.

Blome, Richard. 1692. *Proposals for the printing an entire course or body of philosophy, according to the principles of the famous Renate Des Cartes*. London: R. Blome.

Bouillier, Francisque. 1842. *Histoire et critique de la révolution cartésienne*. Lyon: Boitel.

Boyle, Robert. 1674. *The excellency of theology compar'd with natural philosophy*. London: Henry Herringman.

Buchdahl, Gerd. 1969. *Metaphysics and the philosophy of science, the classical origins: Descartes to Kant*. Cambridge, MA: MIT Press.

Clarke, Desmond M. 1982. *Descartes' philosophy of science*. Manchester: Manchester University Press.

Clarke, Desmond M. 1992. Descartes' philosophy of science and the scientific revolution. In *Cambridge companion to Descartes*, ed. John Cottingham, 258–285. Cambridge: Cambridge University Press.

Clatterbaugh, Kenneth. 1999. *The causation debate in modern philosophy, 1637–1739*. New York: Routledge.

Cohen, I. Bernard. 1985. *The Newtonian revolution*. Cambridge: Cambridge University Press.

Cottingham, John. 1998. Descartes's treatment of animals. In *Descartes*, ed. John Cottingham, 225–233. Oxford: Oxford University Press.

Descartes, René. 1649. *A discourse of a method for the well guiding of reason, and the discovery of truth in the sciences*. London: Thomas Newcombe.

Descartes, René. 1650. *The passions of the soule*. London: J. Martin, and J. Ridley.

Descartes, René. 1653. *Excellent compendium of Musick*. London: Humphrey Moseley.

Descartes, René. 1707. *La Philosophie morale de Monsieur Descartes*. Brussels: Foppens.

Descartes, René. 1680. *Six metaphysical meditations wherein it is proved that there is a god and that mans mind is really distinct from his body*. Trans. William Molyneux. London: Benjamin Tooke.

Descartes, René. 1984–1991. *The philosophical writings*, 3 vols. Trans. J. Cottingham, R. Stoothoff, D. Murdoch, and A. Kenny (for the third volume). Cambridge: Cambridge University Press.

Descartes, René. 1996. *Oeuvres de Descartes*. 2nd ed, ed. Ch. Adam and P. Tannery. Paris: Vrin.

Easton, Patricia. 2012. Antoine Le Grand. In *The Stanford encyclopedia of philosophy (Fall 2012 edition)*, ed. Edward N. Zalta. http://plato.stanford.edu/archives/fall2012/entries/legrand/. Accessed 28 Nov 2012.

Fernel, Jean. 2003. *Physiologia*. Trans. J.M. Forrester. Philadelphia: American Philosophical Society. (Originally published in Latin in 1567.)

Garber, Daniel. 1992. *Descartes' metaphysical physics*. Chicago: University of Chicago Press.

Garber, Daniel. 2001. Descartes on knowledge and certainty: From the *Discours* to the *Principia*. In *Descartes embodied: Reading Cartesian philosophy through Cartesian science*, 111–129. Cambridge: Cambridge University Press.

Hall, A. Rupert, and Marie Boas Hall (eds.). 1971. *The correspondence of Henry Oldenburg: Volume VIII, 1671–1672*. Madison: University of Wisconsin Press.

Hatfield, Gary. 1985. First philosophy and natural philosophy in Descartes. In *Philosophy, its history and historiography*, ed. A.J. Holland, 149–164. Dordrecht: Reidel.

Hatfield, Gary. 1986. The senses and the fleshless eye: The *Meditations* as cognitive exercises. In *Essays on Descartes' Meditations*, ed. Amélie Rorty, 45–79. Berkeley: University of California Press.

Hatfield, Gary. 1990. Metaphysics and the new science. In *Reappraisals of the scientific revolution*, ed. David Lindberg and Robert Westman, 93–166. Cambridge: Cambridge University Press.

Hatfield, Gary. 1992. Descartes's physiology and its relation to his psychology. In *Cambridge companion to Descartes*, ed. John Cottingham, 335–370. Cambridge: Cambridge University Press.

Hatfield, Gary. 1996. Was the scientific revolution really a revolution in science? In *Tradition, transmission, transformation*, ed. Jamil Ragep and Sally Ragep, 489–525. Leiden: Brill.

Hatfield, Gary. 2000. Descartes' naturalism about the mental. In *Descartes' natural philosophy*, ed. Stephen Gaukroger, John Schuster, and John Sutton, 630–658. London: Routledge.

Hatfield, Gary. 2005. Rationalist theories of sense perception and mind–body relation. In *Blackwell companion to rationalism*, ed. Alan Nelson, 31–60. Oxford: Blackwell.

Hatfield, Gary. 2007. Animals. In *Companion to Descartes*, ed. John Carriero and Janet Broughton, 404–425. Oxford: Blackwell.

Hatfield, Gary. 2009. Psychology in philosophy: Historical perspectives. In *Psychology and philosophy: Inquiries into the soul from late scholasticism to contemporary thought*, ed. Sara Heinamaa and Martina Reuter, 1–25. Dordrecht: Springer.

Hatfield, Gary. 2012. Mechanizing the sensitive soul. In *Matter and form in early modern science and philosophy*, ed. Gideon Manning, 151–186. Leiden: Brill.

Laporte, J. 1928. La finalité chez Descartes. *Revue d'Histoire de la Philosophie* 2: 366–396.

Laudan, Larry. 1966. The clock metaphor and probabilism: The impact of Descartes on English methodological thought, 1650–65. *Annals of Science* 22: 74–104.

Le Grand, Antoine. 1671. *Philosophia veterum, e mente Renati Descartes more scholastico breviter degesta*. London: J. Martyn.

Le Grand, Antoine. 1672. *Institutio philosophiae, secundum principia domini Renati Descartes: novo methodo adornata & explicata: in usum juventutis academicae*. London: J. Martyn.

Le Grand, Antoine. 1673. *Historia naturae, variis experimentis & ratiociniis elucidata: secundum principia stabilita in Institutione philosophiae edita ab eodem authore*. London: J. Martyn.

Le Grand, Antoine. 1675. *Dissertatio de carentia sensus & cognitionis in brutis*. London: J. Martyn.

Le Grand, Antoine. 1680. *Institutio philosophiae, secundum principia D. Renati Descartes: nova methodo adornata, & explicata: in usum juventutis academicae*, 4th ed. London: J. Martyn.

Le Grand, Antoine. 1694. *An entire body of philosophy according to the principles of the famous Renate Des Cartes*, 2 vols, ed. R. Blome. London: Samuel Roycroft. (Reprint, New York: Johnson Reprint, 1972.)

Le Grand, Antoine. 1711. *Institutio philosophiae, secundum principia domini Renati Descartes: nova methodo adornata, & explicate: in usum juventutis academicae*. Nuremberg: J. Ziegeri.

Lennon, Thomas M. (ed.). 2003. *Cartesian views: Papers presented to Richard A. Watson*. Leiden: Brill.

Locke, John. 1690. *An essay concerning human understanding*. London: Bassett.

Machamer, Peter, and J.E. McGuire. 2009. *Descartes's changing mind*. Princeton: Princeton University Press.

Oldenburg, Henry. 1671. An accompt of some books. *Philosophical Transactions* 6: 2136–2142.

Plutynski, Anya. 2011. In defense of rationalist science. In *Science at the frontiers: Perspectives in history and philosophy of science*, ed. William Krieger. Lanham: Lexington Books.

Régis, Pierre-Sylvain. 1691. *Cours entier de philosophie, ou systeme general selon les principes de M. Descartes*, Latest edition, 3 vols. Amsterdam: Huguetan. (Reprint, New York: Johnson, 1970).

Rey, Alain (ed.). 1992. *Dictionnaire historique de la langue française: contenant les mots français en usage et quelques autres délaissés*, 2 vols. Paris: Dictionnaires Le Robert.

Rogers, G.A.J. 1985. Descartes and the English. In *The light of nature: Essays in the history and philosophy of science presented to A. C. Crombie*, ed. J.D. North, J.J. Roche, and A.C. Crombie, 281–302. Dordrecht: Nijhoff.

Rohault, Jacques. 1987. *System of natural philosophy, illustrated with Dr. Samuel Clarke's notes, taken mostly out of Sir Isaac Newton's philosophy*. New York/London: Garland Publishing. (Originally published in French in 1671.)

Rosenfield, Leonora Cohen. 1968. *From beast-machine to man-machine: Animal soul in French letters from Descartes to La Mettrie*, 2nd ed. New York: Octagon Books.

Ryan, John. 1935. Anthony Legrand (1629–1699): Franciscan and Cartesian. *New Scholasticism* 9: 226–250.

Schmaltz, Tad M. 2002. *Radical Cartesianism: The French reception of Descartes*. Cambridge: Cambridge University Press.

Shapiro, Barbara J. 1983. *Probability and certainty in seventeenth-century England: A study of the relationships between natural science, religion, history, law, and literature*. Princeton: Princeton University Press.

Simmons, Alison. 2001. Sensible ends: Latent teleology in Descartes's account of sensation. *Journal of the History of Philosophy* 39: 49–75.

Thaddeus, Father. 1898. *The Franciscans in England, 1600–1850: Being an authentic account of the second English Province of Friars Minor*. London: Art and Book Co.

Trevisani, Francesco. 1992. *Descartes in Germania: la ricezione del cartesianesimo nella Facoltà filosofica e medica di Duisburg (1652–1703)*. Milan: Angeli.

Chapter 12
Mechanical Philosophy in an Enchanted World: Cartesian Empiricism in Balthasar Bekker's Radical Reformation

Koen Vermeir

Abstract Balthasar Bekker is seen as one of the seminal thinkers sparking off the early (radical) Enlightenment, the battle against superstition and the 'disenchantment' of the world. The secondary literature has interpreted him a Cartesian rationalist, focussing on his a priori treatment of theology and metaphysics. In this article, I stress the importance of Bekker's Cartesian empiricism instead, which will allow me to reassess the traditional historiography. I show that Bekker was not a forerunner of the enlightenment, but instead aimed at radicalizing the reformation. He did not battle superstition in the enlightenment sense of the term, but inveighed against what he considered corrupted forms of religion. Furthermore, he did not disenchant the world in the sense of freeing it from occult and magical powers, powers which Bekker accepted and explained in natural terms. For Bekker, instead, disenchantment meant denying all demonic activity in the world. He argued that belief in the action of the devil was a pagan remnant in Christianity, which had to be weeded out in order to purify Protestantism. In this article, I argue that not only Bekker's Cartesian metaphysics or hermeneutics, but especially his Cartesian empiricism buttressed his project of disenchanting the world. His theological and philosophical empiricism was necessary in order to shield his system from otherwise fatal criticisms. In particular, I show here how the mechanical philosophy provided him with the tools to develop his empirical approach to natural philosophy. Even if he did not initiate the Enlightenment, Bekker's work did play a crucial role in early modern discussions of Cartesianism, reformed theology and the radical reformation, and some of his ideas would be taken up by later Enlightenment thinkers.

K. Vermeir (✉)
CNRS, Paris, France
e-mail: koen.vermeir@paris7.jussieu.fr

M. Dobre and T. Nyden (eds.), *Cartesian Empiricisms*, Studies in History and Philosophy of Science 31, DOI 10.1007/978-94-007-7690-6_12,
© Springer Science+Business Media Dordrecht 2013

12.1 Introduction: Questioning Demonic Action

In 1695, a printed pamphlet from Saxony reported a mysterious disembodied voice. This pious ghost manifested itself by reciting catechism questions and singing church songs at people's kitchen doors. When visiting foreigners suggested that this might just be a hidden clergyman instead of a ghost, the disembodied voice reacted strongly, by giving a vehement diatribe against pietists as well as unbelievers. This story fitted perfectly the expectations of critical thinkers in the tradition of Baruch de Spinoza (1632–1677), Anthonie Van Dale (1638–1708), Bernard le Bovier de Fontenelle (1657–1757), and Balthasar Bekker (1634–1698), who claimed that priests had abused popular belief in ghosts and demons for centuries in order to enforce their power over the credulous populace. When the Dutch journal editor Pieter Rabus (1660–1702) reported on this case in his journal, *De Boekzaal van Europe*, he wrote: "[Such ghosts] do not have to appear to me, or they may know, that I will deal with them humanly, without charging this buffoonery to the devil."[1]

Balthasar Bekker was the most prominent voice in this debate (see Fig. 12.1), and his place is assured in today's narrative of the beginnings of the early Enlightenment, the battle against superstition and the "disenchantment" of the world.[2] As Jonathan Israel has indicated in his grand story of the radical Enlightenment, Bekker's work caused the largest controversy in Early Modern Europe, with thousands of copies sold in the first 2 months of its publication and with more than 300 texts published for or against.[3] On the other hand, the so called "Enlightenment" is a complex phenomenon that cannot be reduced to a linear story or a few hero's, such as Spinoza, Diderot and Voltaire. In this article, I will focus on Bekker, critically assessing his role in the tolerant Dutch republic and the rise of Enlightenment ideas. In particular, a study of Bekker allows us to reassess a classical question from the history of ideas: how much of the Enlightenment is due to Descartes and Cartesianism?[4] Balthasar Bekker was well known as a Cartesian theologian, and he was part of the long struggle of the reception of Descartes in the Dutch Republic during the seventeenth century. In this paper, I will argue that not only his Cartesian metaphysics but especially his Cartesian empiricism was crucial to his project of disenchanting the world.[5]

[1] Rabus, September–October 1695, 359–364: "Zy behoeven by my niet te verschijnen, of ze mogen weten, dat ik menselijkerwijze met hen zoude omgaan, zonder de potsen ten lasten van den duivel te leggen."

[2] See e.g., Hazard 1961; Israel 1996, 2001, 2012. See also van Ruler 2000 on disenchantment.

[3] Israel 2001, 382. For more on the reception of Bekker and the controversy he provoked, see van Bunge 1998; Fix 2013; Monfils 2004a; Nooijen 2009; Sluis 1994.

[4] Attfield 1985; Fix 1993, 1996; Israel 2001; van Ruler 2000.

[5] Both "Cartesian" and "empiricism" should not be understood here in a too restrictive sense (see also Chap. 1 by Dobre and Nyden), and the concept is chiefly useful as a corrective to rationalist approaches to Bekker's work. It is important to note, however, that Bekker contrasted reason to Scripture rather than to experience. In fact, for him, taking into account experience was part of a rational approach. Nevertheless, he sometimes opposed experience and reasoning (see e.g., the references in note 45), and his major work has more "empiricist" and more "rationalist" parts, as I show below.

Fig. 12.1 Portrait of Bekker from Bekker (1739), Book 1, on folded leaflet located between the advertisement to the reader and the first book. Courtesy of Maurits Sabbe Library, Faculty of Theology and Religious Studies, KU Leuven: GBIB; 46 N; BEKK 1739

Studying Bekker's work in detail will also allow me to reassess his role in the early Enlightenment. Instead of promoting the Enlightenment, Bekker used Cartesianism to reform the Reformation, by weeding out superstitious beliefs.

12.2 The Friesian Hercules

Balthasar Bekker was born on March 20, 1634 in a typical Friesian village, located on an artificial mound to protect it against the high tides from the Wadden Sea. He studied philosophy at the University of Groningen, where he witnessed the ongoing disputes about Cartesianism, and theology at the old University of Franeker, receiving his doctorate there in 1665. He later worked as a reformed pastor in Friesland and Gelderland, before settling in Amsterdam as a city preacher in 1679. When the Calvinist conservatives pursued action against the Cartesians at Franeker University, Bekker openly took sides and published a defense of Cartesianism. In his *De Philosophia Cartesiana admonitio candida et sincera* (1668), he argued that the

Cartesian philosophy posed no risk to the Reformed Church, because philosophy and theology had their own domains. He also defended specific Cartesian ideas that had been attacked by orthodox critics. Bekker's effort was successful and the synod of Friesland rejected the anti-Cartesian objections.[6]

In the same period, Bekker published two children's catechisms, but it was his adult catechism *Vaste Spyse*, published in 1670 that provoked a controversy.[7] The catechism made clear that Bekker opposed a strict and literal interpretation of the bible and adhered to Johannes Coccejus' (1603–1669) more liberal exegesis. For Bekker, God's words could only be properly understood when interpretation took into account that God aimed His message at particular people in a specific context. Bekker's adult catechism encountered stiff opposition from his colleagues and from the Church hierarchy. The theological faculty accused him of socinianism, and the States of Friesland forbade the printing of the catechism. The controversy escalated and Bekker would eventually flee Franeker for more tolerant cities such as Amsterdam. Nevertheless, Bekker took a strong stand in the many controversies that were to follow, an attitude that earned him the epithet 'the Friesian Hercules'.[8]

Bekker's early publications demonstrate two strands of ideas that would define the rest of his career. He adhered to Cartesianism combined with an anti-dogmatic but radicalized reformed theology.[9] At the time, these two approaches were to some extent independent, of course, but they could often not be disentangled. Both Cartesian and anti-confessional thinkers were seen as unorthodox and were criticized together.[10] On the one hand, Cartesian theologians often had heterodox views as the new philosophy became involved in many theological disputes at the end of the seventeenth century. Also for Bekker, despite his protestations that philosophy and theology were separate realms, Cartesianism had a marked impact on his exegetical assumptions and theological ideas.[11] On the other hand, many Dutch academic Cartesians professed religious orthodoxy, and argued for separating philosophy and theology, in order to shield Cartesianism from further attacks by theologians. Bekker's work should be understood in this context of Dutch Cartesianism. In particular, Dutch academic Cartesians focussed on physics rather than metaphysics, in order to avoid conflict with orthodox Reformed critics.

As Andrew Fix has shown, Bekker played an important role in what he called the struggle between confessionalism and anti-confessionalism, taking place at the time

---

[6] Bekker 1668; De Vleeschouwer 1939. For Bekker's life, see Fix 1999; van Sluis 1994. For a discussion of the various attacks and condemnations of Cartesianism, with a focus on the French context, see Chap. 2 by Ariew.

[7] Bekker 1670.

[8] This expression was used by Johannes Duikerius in his politico-religious novel *Het Leven van Philopater* of 1691 (Duikerius 1691); see Fix 1999, 79.

[9] See Fix 1999 for a general analysis.

[10] Cf. Samuel Desmarets, who attacked Cartesians and moderate Calvinists, grouping them together as a "Cartees-Loevesteyns-Remonstrantsche" faction (Desmarets 1672).

[11] van Bunge 1993.

in the Dutch Republic.[12] All over Europe, the centrifugal forces triggered by the Reformation had left people reeling, and the various protestant churches that had sprung up tried to stake out their territory and define their identity. One way to do this was by 'confessionalization': strict adherence to doctrine, church law and institutional hierarchies became of paramount importance.[13] Furthermore, religious confessionalism went hand in hand with the creation of the modern state. Indeed, the new states could draw from the social control and intellectual discipline that characterized confessionalism in order to form and strengthen their identity, coherence and expanding power. Threats to these newly won identities, such as the constant danger of war, increased the need for discipline and conformity even more. For the new Dutch Republic, the constant threat of invasion by Spanish, French, English and German troops played in the advantage of the Orangists and confessionalists, who tried to diminish civil, religious and intellectual liberties and heterodoxies. It was in this context that Bekker wrote his theological works, promoting civil and religious liberties, and provoking major opposition and discord.[14]

In 1680 and 1681, just after Bekker's arrival in Amsterdam, large comets appeared in the sky and Bekker gave a sermon to calm his parishioners. He elaborated on his natural philosophical and exegetical arguments to make clear that comets were no portents. He published these arguments as the *Ondersoek van de Betekeninge der Kometen*, a book in which his Cartesian inclinations came again to the fore. Bekker argued that we do not know enough about the nature of comets, so we cannot conclude anything about their meaning, let alone make predictions from them. In particular, the empirical basis for the study of comets is problematic. On the one hand, Bekker's empiricism makes him stress the importance of personal observation and experience for natural knowledge. On the other hand, he shows that we can only have a very limited experience of comets, because they occur so rarely and meteorological conditions usually hamper good observations. This means that knowledge of comets is very difficult to obtain. Nevertheless, he proffers the best theories of comets available to make clear that these are just natural phenomena. Descartes' theory of vortices gets pride of place, but he also discusses what he considers the derivative 'Cartesian' theories of Hevelius and Bernoulli. Furthermore, by means of an extensive bible exegesis, he explains that comets are not mentioned in the bible, and there is thus no scriptural legitimation for treating them as portents.[15]

Bekker's work on comets is usually interpreted as a work against popular superstition, a project that would eventually culminate in his magnum opus, the *Betoverde Weereld* (translated as the *Enchanted World*, or the *World Bewitched*). In this book, he

---

[12] Fix 1999, 127 ff.

[13] The term 'confessionalism' was first used in the nineteenth century. Although it is somewhat anachronistic to apply it to the early modern period, its use here does not obscure historical understanding.

[14] For more on the Dutch context of Cartesianism, see e.g., Chap. 4 by Van Bunge and Chap. 10 by Nyden.

[15] Bekker 1683; Fix 2000.

argues that the devil cannot act in this world. This does not mean that the devil does not exist, but he is a powerless and pitiful creature, thrown in Hell by God. Many scholars have argued that Cartesian rationalism, with Bekker as a powerful exponent, was responsible for the rejection of superstition and the flourishing of an early Enlightenment.[16] More recently, the central importance of Cartesianism for Bekker's Enlightenment project has been questioned, and the focus has shifted towards the importance of a new kind of biblical exegesis for Bekker's project.[17] In this chapter, I will revise or nuance some of these claims. In fact, both Cartesianism and liberal biblical exegesis were crucial for Bekker's thought. The relation between reason and scripture was a difficult problem he needed to confront in all his work. Furthermore, the important role of Bekker's empiricism, for his Cartesianism as well as for his bible interpretation, has been neglected in the literature. It is also not enough appreciated that Bekker's arguments were not put into service of a straightforward 'Enlightenment project'. Bekker did not aim to be a forerunner of the Enlightenment, and certainly not a 'radical enlightenment' of any sort. With his works, Bekker wanted to reform religion, to continue and fulfill the Reformation. A theologian and practicing pastor, his work was not directed against religion, nor was it opposed to vulgar beliefs or occult sciences *in se*. Bekker wanted to purify religion from superstitious elements.[18] He remained heir to the reformation rhetoric against rituals and ceremonies, but he radicalized it to include belief in demonic action in this world. Devil and demon beliefs were alien to true Christianity, he argued. These were pagan elements that had crept in, and led people to superstition, which meant that it led people to false religious beliefs and worship.[19] His understanding of superstition was thus very different from later Enlightenment thinkers. His motivation also differed significantly from those who would later launch a philosophical critique of superstition, which would eventually also include religion.

Bekker's works should be read in the context of a more general late seventeenth century movement of religious reformation that took place in Catholic as well as in Protestant regions. In the years that Bekker was working on his *Betoverde Weereld*, for instance, the French Cardinal Étienne Le Camus (1632–1707) was traveling through his diocese to weed out malpractice and superstition. Le Camus argued that superstition, i.e., excessive religion, was not less dangerous than libertinage or irreligion, and he ordered his priests to abolish it as much as possible. They should teach the people the legitimate use of sacred images, not to adore relics without permission, not to use these objects in superstitious practices, and not be involved in divination, sorcery or magic.[20] As a result of such campaigns, the meaning of superstition itself was transformed, and in the course of the eighteenth century, it changed from denoting an excessive religious cult to signifying vulgar credulity.[21] For Le Camus and his pupils, an excessive cult meant in many cases idolatry, worshiping the devil or being involved in demonic practices. For these Catholic

[16] Attfield 1985; Fix 1989; Knuttel 1906.

[17] Buisman 1998; Fix 1999; Jorink 2006; Nooijen 2009.

[18] Cf. van Ruler 2000.

[19] To clarify my point: today's popular Christmas trees would be a major superstition for Bekker.

[20] Le Brun 1702; Le Camus 1690; Thiers 1697.

[21] Dompnier 1998.

authors, belief in the devil was *essential* to a correct understanding of superstition. For Protestants too, superstition had the same meaning of false religious worship. They only extended its range by including Catholic practices, which involved what they considered to be excessive cults, idolatry, and, as some would even claim, a pact with the devil. In this period, believing in hell and the devil was not only thought to be necessary for upholding Christian morality, which needed the threat of eternal punishment;[22] it was also necessary for Christian belief in general, because doubting evil spirits and the devil could lead one to doubt God. Furthermore, devil beliefs were necessary for defining the limits of legitimate religious practice. When he denied demonic activity, Bekker thus went against a general consensus that was shared between Catholics and Protestants.

Recently, scholars have stressed that neither the Scientific Revolution nor the Enlightenment inaugurated the end of the witch trials. Indeed, in most of Western Europe, these trials had generally ended some time before the mid seventeenth century.[23] Bekker himself commented on this, and it formed part of his argument. He explained that the waning of witch trials and demonic magic was a result of the Reformation, which had purified religion, even if its full promise had not yet been fulfilled. In other countries—especially the Catholic ones, Bekker argued—belief in witchcraft was still rife.[24] Indeed, even if witch prosecutions in Holland had effectively ended by the time Bekker wrote the *Betoverde Weereld*, the Scandinavian and eastern European witch hunt only reached its peak. Reports of demonic possession, witchcraft and the actions of spiritual beings were also still at the forefront of the intellectual life in Holland and elsewhere, and the account of a Saxon ghost mentioned at the start of this paper was not an exception. In 1693, reports were published about a case of demonic possessions in Køge, Denmark.[25] In 1694, the journal editor Pieter Rabus wrote with compassion about an older case in Flanders, when an old man was executed for attending witches' Sabbaths, for enchanting animals and people, and for assuming the appearance of a werewolf.[26] In 1695, he also noted with indignation that Jean Bodin's (1580–1596) *Démonomanie* had recently been used in Paris to justify the death sentence of peasants accused of demonic magic.[27] Bekker and likeminded authors were appalled by the horrors inflicted on what they thought were innocent victims. He belonged to a strong Dutch tradition of authors, such as Johannes Wier (1515–1588), Johannes Grevius (1584–1622), Daniël Jonctijs (1600–1654), Gerhard Tuining (1566–1610) and Abraham Palingh (1588/1589–1682), who had criticised the witch trials throughout the sixteenth and seventeenth centuries.

The most famous case of demonic activity, perhaps, was the 1634 case of demonic possession at Loudun, which still evoked strong reactions at the end of the seventeenth century. In this case, Father Urbain Grandier was convicted of the crime

[22] E.g., Walker 1964.

[23] Ankarloo et al. 2002; Bever 2009; Golden 2006; Levack 2006. For a dissenting voice, see Wasser 2008.

[24] Bekker 1693, 268.

[25] Bekker 1693, Book 4; Brunsmands 1693.

[26] Pieter Rabus, March–April 1694, 333–337.

[27] Pieter Rabus, November–December 1696, 555–558.

of sorcery and evil spells. He was deemed responsible for the possessions visited upon the Ursuline nuns, and the sentence was based on the testimony of the possessed demoniacs. For progressive thinkers, the conviction of Grandier became the paradigm case of dogmatism, superstition and intolerance.[28] In 1693, Nicolas Aubin, a Huguenot preacher exiled in the Netherlands and a native of Loudun, still struggled to come to terms with the event. He published a book in which he argued that the possessions were a fraud.[29] The story of Loudun was also one of the key examples in Bekker's *Betoverde Weereld*. In chapter 11 of book 4, he gave a full analysis of the case, with a full description of the facts and citations from the court case records, arguing that the phenomenon had nothing to do with de devil. Bekker closely collaborated with Aubin, and he had advised him when publishing his critical analysis.[30] For Bekker, this was one of the many crucial analyses that made up de groundwork of his empirical approach. This empiricism, based on concrete analyses of particular examples supported the main thesis of his work: that these devil stories were superstitious and should be rejected.

The beginning and end of the witch craze are extremely complex phenomena, and cannot be attributed to a few simple causes. Bekker's role in the demise of the witch trials and in the disappearance of beliefs in witchcraft and other demonic activities is still contested. It is now accepted that Cartesian rationalism was not a major cause in the decline of witchcraft ideas, if only because the trials were already on their decline, but also because prominent Cartesians defended the existence of witches and evil spirits. Henry More (1614–1687) comes to mind, but also lesser known Dutch Cartesians defended witchcraft.[31] Nevertheless, Cartesianism had a crucial role to play in Bekker's disenchantment of the world, that is, in his project of freeing the world of the belief in evil spirits. Although scholars have generally focussed on his Cartesian metaphysics, I will show that Bekker's Cartesian empiricism was a major resource for his program of disenchantment. I will argue that it was the striking explanatory power of Cartesianism which allowed him to explain any imaginable phenomenon as natural, and hence counter the plethora of empirical proofs proffered for demonic phenomena.

12.3 The Enchanted World

In the late 1680s, Bekker wrote a book on ecclesiastical history, a commentary on the Book of Daniel and a work on child baptism. His book on comets had set him thinking, however, and had prompted reflections about other superstitions. Bekker was convinced that belief in sorcery and witchcraft, just like interpreting comets as

[28] There is a lot of literature on the Loudun case; most famously, see De Certeau 1970. See also Pieter Rabus, July–August 1693, 56–61 and Bekker 1693, Book 4.

[29] Aubin 1693.

[30] Bekker 1693, Book 4, 85.

[31] On Henry More, see e.g., Vermeir 2012.

portents, implied wrong ideas about God and religion, which could be detrimental to one's spiritual welfare. Originally planned as an appendix to a new edition of his comet book, his research into witchcraft and devil beliefs grew quickly into a four-volume work.[32] The work had a very complex publication history, however, and Bekker encountered many problems in getting it printed. Communication with his printer in Leeuwarden was difficult, weather conditions caused delays, and critics who had been informed about his work already started to denounce him. A controversy was well on its way even before the first volumes had appeared. He quarrelled with his Leeuwarden printer and moved the printing process to Amsterdam. Incomplete copies circulated early and different versions—more or less authorised—appeared in Leeuwarden and Amsterdam in the spring of 1691. Rumours had gathered pace and the first 8,000 copies were immediately sold. In the meantime, Bekker continued to make corrections and changes, even reacting to criticisms, which resulted in a myriad of different versions of the work printed in the course of 1691.[33] Only Books 1 and 2 were published in 1691, with books 3 and 4 to follow in 1693, but his critics did not wait until all four volumes were available.

In the first book of the *Betoverde Weereld*, Bekker presents an empirical historical study into the origins of the common opinions of beliefs of devils and demons by looking into ancient texts from all over the world. He concludes that our ideas about the devil originated in paganism, not in Holy Scripture. In his controversial Book 2, Bekker takes a philosophical and theological a priori approach to the study of the action of the devil. From June 1691 onwards, several commissions and synods drafted objections to Bekker's book. Some attempts at compromise were made, and Bekker was asked to retract some of his controversial views, but church councils from other towns intervened with further criticisms. Appalled theologians opposed especially the arguments in book 2, where Bekker argued on philosophical and theological grounds that the devil had no power here on earth. He was accused of atheism, socinianism and Spinozism (even though he explicitly opposed Spinoza's monism).[34] Many scholars wrote books and pamphlets against Bekker's magnum opus in the 2 years after the first two parts appeared in print (see Fig. 12.4). Johannes van der Waeyen (1639–1701) and Florentius Costerus (1635–1703), for instance, provided long and substantial critiques of book 2 of the *Betoverde Weereld*. Some, like Everardus van der Hooght (1642–1716) and Henricus Groenewegen (1640–1692), focused only on Bekker's exegesis, while many others, such as Henricus Brink (1645–1723), Melchior Leydekker (1642–1721), Johannes Verrijn (1672–1698), Jacobus Koelman (1632–1695), Johannes Aalstius (1660–1712) and Paulus Steenwinkel, (1662–1740) took issue with Bekker's Cartesianism.[35] Next to

[32] See Bekker 1689; see also Simoni 1979.

[33] See Monfils 2004b; van Sluis 1994.

[34] On Bekker's affinity to Spinozism, however, Van Bunge writes: "The irony of Bekker's predicament seems to have been that his position would have been far more consistent, had he had the stomach to subscribe to Spinozism in its entirely. The best way to have defended himself against the accusation of being a Spinozist would probably have been to become one" (van Bunge 1993, 79).

[35] See Israel 1996, 2001; Knuttel 1906; Nooijen 2009.

Spinoza's *Tractatus*, the *Betoverde Weereld* became the most notorious theological work of the century. As a consequence, at the height of the controversy, the Amsterdam church council stripped Bekker of his post, forbade him to preach and barred him from taking part in communion.[36]

Bekker felt misunderstood. His readers had missed the force of his arguments in book 1, he thought, and had misinterpreted book 2. Book 1 was not just a collection of historical studies, but it provided empirical proof aimed at showing that the belief in the power of the devil is of pagan origin and has no biblical warrant at all. He established that beliefs about demons are pagan impurities that had infiltrated Christian religion. In book 2, despite all the allegations, Bekker did not argue that spirits cannot act on bodies or on other spirits. Nor did he claim that the devil did not exist. He only argued that the devil did not have power over our world. He felt much of the attacks were misdirected, but most importantly, he felt that the aim of his project had been insufficiently understood. Whilst dealing with this controversy, Bekker prepared the publication of the next volumes, which he thought essential for the understanding of his argument. He did not want to attack religion. To the contrary, he wanted to bolster piety and counter idolatry, religious error and sin, and fulfil the purification process that the reformation had begun. In the introduction to book 3 and 4, which he managed to publish only in 1693, he expressed his hope that reading the entire work would make his aims clearer. Book 3 was an extension of book 2, and focussed specifically on the pacts and communication between man and devil. He argued that demonic magic, witchcraft and demonic pacts did not exist. In book 4, he put together empirical evidence for his conclusion that there is no proof for the operation of the devil in our world. It shows concretely that each allegedly demonic phenomenon was in fact a hoax or a natural phenomenon. One should reject these profane fables, he advises, and exercise godliness.

Bekker is still misunderstood. Because of the current interest in Cartesianism and in methods of exegesis for explaining the origins of modernity, scholars have focussed almost exclusively on book 2, just like Bekker's contemporary critics. There is no

---

[36] Bekker was not decried by everyone, and many pamphlets appeared to support him. Many authors, however, took care not speak too explicitly in his favour, or published anonymously. In 1691, for instance, the *Nagt-Gezigte*, an anonymous satire, was published in Bekker's support. In this satire, Molinaeus, who had preached against Bekker, was poked fun at. At the end of the pamphlet, a portrait of Bekker was printed, and a poem by Pieter Rabus was put underneath it. Soon, another satire was published as a response. In it, Pieter Rabus was identified as the author of the *Nagt-Gezigte*. Rabus was accused of being a notorious *Bekkerian* and was dragged through the mire, and Molinaeus staged an *ad hominem* attack on Rabus. Driven into a corner, Rabus had to respond. On the one hand, he denied all accusations, especially the authorship of the *Nagt-Gezigte*. Given his position as a teacher and his intention to begin a learned journal (under censure by the church council), Rabus did not want to become too involved in the Bekker controversy. On the other hand, he also made clear that he agreed with Van Dale and Bekker that the devil's power was overestimated, that he cannot do supernatural acts, and that he did not prophecy through the pagan oracles. For the connection between Rabus and Bekker, see De Vet 1980.

Fig. 12.2 Frontispiece of the pamphlet, Hilarides (1691), in which a bust of Bekker is shown, underneath of which we find the devil confined in Hell. On top of the image, Bekker is described as a monotheist, because he did not want to attribute supernatural power to anyone but God. The rest of the pamphlet contains an interpretation of the image, or rather, a panegyric to Bekker. This image would also be used as the frontispiece of early editions of Bekker's *Betoverde Weereld*. Image Library, Creative Commons Licence, Universiteitsbibliotheek Ghent, BIB.MEUL.006770

significant discussion of book 1, and there is hardly mention of book 4.[37] As I will argue, however, these books are central to Bekker's argument. In what follows, I will look especially at book 4, in which Bekker's Cartesian empiricism comes to the fore most clearly. In order to show its importance in Bekker's general argument, however, I will briefly review his better known book 2. Here, Bekker argues from reason and scripture that the devil is confined in hell and has no power over the world. In the first part, the argument concerns nature and proceeds by reason.[38] He argues from first principles, inspired by Cartesian philosophy, which show that there are no demons and that the existence of angels is uncertain. Reason teaches us that God is perfect, unique and infinitely transcendent. God's perfection implies that there are no intermediaries, such as demons, angels, spirits, emanations, Platonic ideas, etc., which mediate between God and the world. God's uniqueness also implies that there are no demigods like demons. Bekker was a strict monotheist, and he was unwilling to hand over any of God's power to the devil (see Fig. 12.2).

[37] Scholars analyzing Bekker's *De Betoverde Weereld* do not mention book 4, or they treat it in a few lines and dismiss it as a collection of curiosities. See e.g., Fix 1999, 74.

[38] Note that 'nature' meant something different here than we might think, because this section also included discussion of God, angels and spirits. Bekker tried to separate the realms in which reason and scripture could be applied. Nevertheless, he admitted that reason and scripture could support each other mutually, and the distinction between their domains was not always so clear-cut.

In his metaphysics, Bekker follows Descartes' dualism when he defines spirit and body. Spirit is an immaterial substance, body is a material substance, and both are entirely distinct. It is clearly not the case that spirits do not exist, Bekker argues. Indeed, we have clear personal knowledge of one spirit: our own soul. This is also the only source of knowledge we have about spirit. From experience, we know that our spirit acts on other souls or bodies exclusively through the intermediary of bodies. Angels and demons are said to be spirits, that is, purely immaterial beings without bodies. This implies, Bekker argues, that it is impossible for them to act on other spirits or on the material world. The question whether angels or demons exist cannot even be answered with certainty by means of a philosophical analysis. At least, we cannot have any direct experience of them. So far, we do not know if they really exist, we have no idea what their essence is, and so it makes no sense to speculate about their actions. Man, however, is special in that God connected his mind with his body. This body needs to be complete and well formed in order for the connection to take hold. This confirms our experience, but this special connection is also impossible for us to understand.

The second part of book 2 proceeds not from the book of nature but from scripture, which Bekker interprets by means of his own particular hermeneutics. From scripture, we do learn about angels and the devil, Bekker admits, and so we have to accept their existence. We can read in the bible that angels are the ministers of God, but we do not learn anything more about their essence. Bekker's interpretation makes clear that angels do not operate by their own virtue, but they are the mouthpieces of God. Their 'action' is not really theirs; it is rather God's action through them. It is similar to how holy men performed miracles in patristic times. Miracle working is not part of their own essence and power but the result of God working through them. The case of devils is more complicated, however, because the relevant biblical texts are more difficult to interpret. Bekker argues that there are specific interpretative problems with the languages in which the bible was written, as well as with the accepted translations. Supposed corruptions of the bible text together with the use of figurative speech aggravate this situation even more. Bekker argues that the term 'devil' was often a name given to evil men. Bekker establishes his conclusions by interpreting key biblical passages metaphorically (often on linguistic grounds). The principle of accommodation, to which Bekker ascribed, held that God's Word in the revealed scriptures was accommodated to human capacity, so that common people at the time would be able to understand it. For Bekker, this allowed interpreting the bible in a liberal way. Bekker interpreted passages referring to the appearance of devils as referring to human messengers, for instance. Possessions were not caused by spirits but by diseases, Bekker maintained, and Christ therefore did not expel demons but he cured disorders. Christ himself was not really tempted by the devil, as scripture seems to suggest, but only in a vision, while he was struggling with himself.[39] Sometimes the bible had to be read literally instead of metaphorically. The serpent in Genesis was a real serpent, not a devil in disguise. Some biblical episodes remained problematic, however, such as the fight between

[39] On Bekker's hermeneutics, see e.g., van Bunge 1993; Fix 1999; Nooijen 2009.

Archangel Michael and the devil, but Bekker thinks that these passages are too obscure to draw any conclusions from them. Bekker thus concludes that the empire of the devil is a chimera, because there was no biblical warrant to substantiate it.

It should be clear by now that Bekker's work should not be read as an Enlightenment work, nor is it about superstition in the current sense of the word. In fact, Bekker believed in magic, that is, in its natural form. He only denied the efficacy of demonic magic. In the late seventeenth century, superstition meant bad religion, and even if this often referred to popular practices (such as using the host in fertility rituals for cattle), other superstitious practices were not necessarily confined to the vulgar. Indeed, contemporary tomes on demonology show that devil beliefs were as much part of elite discourse as they were part of popular practices.[40] Bekker inveighs against vulgar and elite beliefs alike, when they contradicted what he thought was the purer form of Christian religion. Bekker was in the first place a theologian, and philosophy was useful for him in so far as it helped to buttress his theological aims. His central aim was to be a good pastor and to save the souls of his parishioners. In order to do so, he wanted to fulfill the reformation by purging religion of elements that could obstruct the road to salvation.[41] Therefore, he attacked not only Catholic superstitions but also those Protestant beliefs and practices that he considered to be dangerous pagan remnants within Christianity. Bekker used Cartesianism not for an Enlightenment programme *avant la lettre* but in the service of the further purification and reformation of protestant religion. As he himself expressed it: "It can be seen that there is rather much work to do—because so much is still turned upside down—to further clear out Protestant Christendom (Fig. 12.3)."[42]

12.4 Bekker's Empirical Theology

Bekker was aware that the philosophical and theological arguments in books 2 and 3 were not enough. He knew that it was possible to make different interpretations of scripture, and that his method of exegesis was controversial. He himself had accused his opponents of selecting the interpretation that suited them best, and he knew he was vulnerable to the same allegation. Of course, his own bible interpretations were not made in a void. They were buttressed by reason and the abstract principles of Cartesian philosophy. But again, this was not enough, as these Cartesian principles were at least as controversial as his exegesis. He knew that it was possible to start with other principles, and indeed, the philosophy of the schools was still favoured by

[40] For an overview of demonological discourse, see Clark 1997.

[41] Bekker argued that belief in witchcraft and the devil was dangerous for piety and for one's salvation, taking both a rationalist (esp. in Bekker 1693, Book 2, Chaps. 35–36 and Book 3, Chaps. 21–23) and an empiricist approach (esp. in Bekker 1693, Book 4, Chaps. 34–35).

[42] Bekker 1693, 268: "Het is dan wel te sien dat vry veel werk te doen is/daar so veel noch over hoop leid/om de Protestantische Kristenheid voorts op te schonen…."

Fig. 12.3 Frontispiece of Bekker (1739). Book 1. Courtesy of Maurits Sabbe Library, Faculty of Theology and Religious Studies, KU Leuven: GBIB; 46 N; BEKK 1739

most theologians. Orthodox theologians especially inveighed against the application of Cartesian principles to bible interpretation.[43] Crucial for his position, Bekker thought, was empirical evidence that could prove that his interpretation was the right one. This is why books 1 and 4 are so essential to his overall argument.

Book 1 provided empirical evidence from history. It shows that beliefs in demons and the action of the devil originated in pagan sources, which had infiltrated and corrupted Christendom. Bekker explains that his method is a priori (‘*van voren*’) and not a posteriori (‘*van achteren*’), following Scholastic terminology. This means that Bekker’s method progresses from causes to effects, and not from effects to causes.[44] Indeed, he starts with reading the most ancient texts (causes) and studies

[43] It should be understood that these debates took place in the wake of vigorous confessionalist attacks directed at Baruch Spinoza (1632–1677), Lodewijk Meyer (1629–1681), Lambert Velthuysen (1622–1685) and others who argued that philosophical analysis could interpret obscure and difficult passages in the bible. Bekker believed that the bible is a higher source of knowledge than reason, but he also thought that reason needed to *precede* Scripture. He took the controversial stance that the bible needs to reveal and make clear to ‘healthy reason’ that it is of a divine origin, before we can take it seriously. Bekker 1693, Section “Naakte Uitbeeldinge,” 8.

[44] Bekker 1693, Section “Naakte Uitbeeldinge,” 2. Note the important difference with the common modern usage of a priori (as deductive) and a posteriori (as inductive).

how these opinions influenced later beliefs about witchcraft (effects). This is important, because he needed to forestall criticism that he read the sources in order to find confirmation of his own prejudices. If one wants to confirm a pre-established principle (such as: current devil beliefs come from pagan sources), it is often possible to select historical texts that provide evidence for this principle. In contrast, Bekker claims to take an a priori, i.e., an open-ended empirical approach. He compares all texts he can find, from all over the world, written by Pagans, Jews, Islamites, Catholics and Protestants, before drawing conclusions. This method of intellectual history is objective, Bekker believes, because he did not impose his judgement on the texts, nor did he set out to confirm a pre-established idea. In this way, he shows that beliefs in the 'action of the devil' were alien to Christianity itself. Bekker believed that interpretation of Scripture should be consistent with this historical fact. The empirical data he found lent very strong support for his exegesis in book 2, and provided evidence that showed this was the right interpretation. Without book 1, books 2 and 3 stood on shaky grounds indeed; but given the historical evidence in book 1, Bekker claimed, his reading of the bible was much more plausible than that of his critics.

Bekker's empiricism was not only textual; it was also based in natural philosophy. He accepted the authority of experience ('*ondervindinge*' or '*ervarentheid*'), and he knew his whole analysis in books 1–3 would crumble if his opponents were able to garner convincing empirical evidence for spirit activity in this world. Bekker even states that "experience is the mistress of all things," and claims that all the rational explanations he presents will fall on deaf ears if people are confronted with a profusion of counterexamples.[45] He therefore had to take very serious experiential reports of demonic phenomena, apparitions and witchcraft, which he analyses in book 4. How could he deal with people's personal strong experiences of demonic phenomena, and with the reports of authoritative and credible witnesses? As a pastor, he worked with his parishioners on an everyday basis, and was aware of their strong personal experiences and convictions. He also knew that philosophical and theological speculation were not an adequate response to these direct and extraordinary experiences. Furthermore, a central part of his opponent's tactics was, and had always been, to flood the public with credible examples of demonic activity. One of the most impressive works in this genre was Joseph Glanvill's (1636–1680) *Saducismus Triumphatus* (1681), a work Bekker discussed in detail. According to Glanvill, to deny such a large collection of witchcraft stories would amount to destroying "the credit of all humane testimony, and to make all men liars."[46] Plenty of other pamphlets circulated as well, describing cases of possession, witchcraft and demonic magic. Glanvill admits that many of these stories can be dismissed as fraud, but he stresses that "one relation, wherein no fallacy or fraud could be suspected," is enough to make his case.[47] Therefore, Bekker wanted to forestall objections of dogmatism (to be '*hardnekkig*') and to honour the respectability of those

[45] Bekker 1693, Book 4, 1 and Section "Naakte Uitbeeldinge," 27.

[46] Glanvill 1681, 28.

[47] Glanvill 1681, 32–33.

who had had such experiences.[48] He did not want to dismiss their accounts, but needed to take them seriously, and this is what book 4 of the *Betoverde Weereld* is about. The title of the book appropriately reads: "Fourth book, in which the proof that is taken from experience is explored in depth."

In the beginning of book 4, Bekker sets up guidelines for assessing personal experience and how it should guide one's beliefs about sprits and devils. In order to acquire knowledge, preference and priority should be given to personal experience, even if it is particularly difficult to have personal experiences of extraordinary phenomena (cf. Bekker's analysis of comets, discussed above). The empirical basis of preternatural phenomena will to a large extent be based on testimony, but one should first look at one's own experiences before accepting stories by others as true. One's personal experience and judgment does not deliver a direct road to the truth, however, because it is riddled with prejudice. Bekker develops a subtle theory of prejudice, based on Descartes' philosophy.[49] He argues that our opinions are influenced by literature, stories, one's own circumstances and expectations, and these biases affect knowledge and even perception. Prejudices also affect morality and charity, as for instance in witchcraft accusations, which are biased towards old women. Bekker stresses that one has to be free of fear and other emotions in order to make a careful judgment of a phenomenon. Such emotions hinder the thorough examination of a case and help prejudices to take hold. They take away sound judgment, fill the imagination with false images, and distort perception.[50] In order to judge testimony, one should examine in detail its credibility, the circumstances in which it is made, and the reliability of those who made it. One should always do a thorough research of the phenomenon itself before judging it. And even then, one will often lack the means, the possibility, and the courage to learn everything sufficiently for assessing the phenomenon properly.

In general terms, Bekker is confronted with the following problem. Imagine that you experience or hear of a wondrous, extraordinary phenomenon. Either you deny it, because you are sceptical about your perception (possible reasons are prejudices, illnesses, or the illusions of imagination) and the testimonies heard (maybe the witnesses are deceitful, or they are not rational enough to make a proper judgment); or you try to explain this phenomenon. In early modern Europe, there were usually four causal explanatory categories available. Phenomena could be caused by divine intervention, by demonic action, by natural causes or by artificial means. Divine intervention was not a plausible option for many Protestants, among whom we should count Bekker, because they believed that God had not intervened directly in worldly affairs

---

[48] Bekker 1693, Book 4, 1.

[49] Bekker 1693, Book 4, 3: "Descartes has taken this as a theme of his philosophical thought, and he himself has shown us the way" ("Dit heeft Des-cartes voor een hoofdstuk sijner filosofische gedachten waargenomen: en self daar in den wegh gewesen").

[50] See especially Bekker 1693, Book 4, Chap. 3 for an extensive analysis of the distortions of sense perception and imagination. In fact, nature also hides the essences of things, so that it is very difficult to know the real causes of phenomena. Nature sometimes even plays with man, like a teacher who can make a fool of a student, to sharpen our judgment and to test us. Many extraordinary phenomena are therefore very difficult to penetrate.

since patristic times.[51] The second option, demonic causality, was not a possible explanation for Bekker either, because this was exactly what his work tried to deny. He had already shown that this was an impossibility for philosophical and theological reasons, but he still needed to counter empirical accounts of demons and devils. This means that he was left with two kinds of explanations: either the phenomenon was natural, or it was artificial.[52] If he were able to explain all extraordinary events by means of these two causal structures—natural or human action—he could exclude demonic causality. Reference to demons would become unnecessary and even implausible. In order to do this, he needed to study the limits and powers of nature, and natural philosophy became of central importance to him.

Artificial phenomena usually referred to phenomena created by human ingenuity, which include machines, special effects or human trickery and fraud. Bekker explained many allegedly demonic phenomena by showing how they might be the result of trickery by a conman. Nevertheless, Bekker considered it impossible to explain all extraordinary phenomena as artificial effects created by tricksters. It was difficult to dismiss credible and prominent witnesses, for instance, and sometimes one might have a convincing personal experience of an extraordinary occurrence. The category of the artificial was thus not reliable and powerful enough to explain away all extraordinary phenomena. The brunt for explaining the most difficult cases, which could not so easily be dismissed as human trickery, was therefore on natural philosophy. It was crucial to make the explanatory power of "the natural" as comprehensive as possible. Bekker had paid close attention to other authors who

[51] Bekker 1693, Book 4, 15: "Neither Moses, educated in the wisdom of the Egyptians, (Acts. 7:22) nor such a heathen king had in this case the least suspicion of sorcery; concluding immediately (as expressed clearly about the latter) that it must be something divine. Even the magicians, when they saw clearly that the louses generated by Moses and Aaron were real louses, were forced to say frankly that this was [the work of] God's finger (Exod. 8: 19–20). If in our own times, when miracles have ceased, such a thing is seen or heard; and we for certain reasons should believe that this is not a special wonder of God, something that has been prophesied to manifest his holiness by such a proof of his power; and if the phenomenon comes across as impossible; then we should mistrust our own ears and eyes" ("Noch Moses, in de wijsheid der Egyptenaren onderwesen, (Hand. 7:22) noch zulk een heidensch koning hadden hier het minste achterdenken op de Toverkunst; terstond besluitende (gelijk het van den laatsten klaar staat uitgedrukt) dat dit wat goddelijk moest zijn. Self ook de Toveraars, wanneer se klaarlik sagen, dat de Luisen door den dienst van Moses en Aaron voortgebraght in waarheid luisen waren, so wierden se geperst rond uit te seggen dat dit Gods vinger was (Exod. 8: 19–20). Doch nu ter tijd dat so geen meer Mirakelen geschieden so iet gesien wordt of gehoord dat wy om redenen geloven moeten geen besonder wonderwerk van God te zijn op zulken wijse als 't voorseide was, om sijne heiligheid door sulk bewijs van sijne mogentheid te openbaren; en 't komt ons al so seer onmogelijk te voren; so moeten wy onse eigene ooren en gesight mistrouwen").

[52] Bekker 1693, Book 4, 8: "What is nothing else but the effect of nature or of human brains: and therefore can be found out by those who know the secrets of nature or the tricks of impostors; or by those who have been tricked many times and are therefore less apt to be cheated again" ("het gene anders niet en is dan een uitwerksel der Natuur of menschelike herssenen: en daarom ook wel uit te vinden van sodanige die de verborgentheden der Natuur verstaan of op bedriegeryen afgereght; of menigmaal bedrogen en daarom niet so licht meer te bedriegen zijn"). In Chaps. 2 and 3 of Book 4, Bekker described how extraordinary phenomena could be explained naturally; in Chap. 4, he treats of imposture and trickery.

had already naturalised seemingly demonic phenomena, and he had studied the writings of Giambattista della Porta (1535–1615), Gaspar Schott (1608–1666) and Johann Christian Frommann (1662–1719) on natural and artificial magic.[53] He finds the best cultural and intellectual resources in the mechanical philosophy, however, and he refers to Sir Kenelm Digby (1603–1665) and René Descartes (1596–1650) as his main sources of inspiration. These authors show that phenomena that appear to be caused by spirits or demons can in fact be explained by the flux of very subtle and indivisible substances.

For Bekker, the systems of Descartes and Digby were similar and complementary.[54] It is especially Descartes' and Digby's corpuscular theory that had attracted Bekker's attention, because they were able to explain so many wondrous phenomena with it, from the phenomenon of light to the sympathetic cure.[55] After explaining the central propositions of Digby's system, Bekker claims that it is not his task to defend or refute them. The only thing he needs to do, he argues, is to show that a philosophy of very fine particles can naturalize everything that usually was attributed to magic, witchcraft or the devil.[56] Whether the particular theory is Digbian, Cartesian or of another origin does not matter much. Bekker is ready to defend three central and abstracted principles, however, on which he thinks everyone can agree: (1) these very subtle particles are present in each body; (2) these particles, entering, leaving and permeating all bodies, are the cause of all changes (3) these particles are connected with each other in long chains, so that one body can act on another even if it is far away.[57] The existence of these little particles could not be doubted anymore, Bekker claims, because of the many wonders discovered by the microscope, recent studies of generation, and other new developments in seventeenth–century natural philosophy. The chains of connected particles were confirmed in every day practice, for instance by hounds able to find traces of game over a distance of many miles. Such dogs are even able to recognise and find a stone touched by a particular man, which proves that we emit particular little particles continuously.

Bekker keeps his distance from doctrinal struggles in natural philosophy, and does not commit himself to Digbian principles. His own three principles are the abstracted core of what a number of different natural philosophies have in common. These principles give a sufficient fit with Digby's and Descartes' theories, but they

[53] Della Porta 1589; Frommann 1675; Schott 1659. For an analysis of Schott's work, see esp. Bekker 1693, Book 1, Chaps. 19 and 20.

[54] In fact, Digby and Descartes were on friendly terms, exchanged ideas and appreciated each other's work. Digby was one of the first to introduce Descartes to an English public, notably in a letter to Hobbes from 1637 (See Nicolson 1929).

[55] Digby 1658.

[56] Bekker 1693, Book 4, 10: "I only need to prove that this knowledge of the most subtle particles and their very quick and wide-ranging movements opens our eyes to see that everything that is often attributed to witchcraft or demonic action can be natural" ("Ik heb alleenlik te betonen dat die kennisse van d'allerfijnste stofjes en derselver oversnelle en seer verre henen strekkende veelvuldige beweginge ons d'ogen opent om te zien hoe't alles kan natuurlijk zijn het gene veelmaals aan de Tovery of Duivels werkinge word toegeschreven").

[57] Bekker 1693, Book 4, 10.

also fit older theories of action at a distance. Action at a distance had always been a problem for Aristotelian philosophy, which only accepted natural action by contact. Medieval and early modern Aristotelians had already found a solution, however, by explaining interactions by means of subtle vapours flowing between bodies.[58] Bekker's own natural philosophical explanations in book 3 takes the theory of subtle corpuscles to its extreme by explaining sympathy, divination and other occult phenomena by means of the flow of little particles. He even goes so far as to admit that human intention can influence and direct vapours emitted by the imagination. This is the controversial theory of the power of imagination, in which small particles are emitted from the imagination of one person and can affect animate or inanimate objects outside the body.[59] He refers to this external power of imagination, as well as to a recent phenomenon of divination with a divining rod, as the two most difficult phenomena to explain in natural terms. Nevertheless, these phenomena can be explained by means of his theory of the exhalation of little particles.[60]

To some extent, Bekker's natural philosophy is in agreement with the mechanical philosophy, most importantly because he prefers to explain all phenomena in terms of corpuscles. Bekker is not in principle against explanations that refer to sympathy and antipathy, but he thinks all phenomena of sympathy and antipathy can be explained in mechanical terms.[61] On the other hand, the theoretical details and exact explanations of atomism, the rules of movement and interaction of these little particles do not matter for him. Only the fact that there is a plausible natural explanation of these phenomena is important. This hybrid mechanical philosophy was basically limitless in its explanatory power, and the most extraordinary phenomena could easily be explained by its means. It is this limitless explanatory power that Bekker needed in order to naturalize extraordinary phenomena and dismiss all demonic agency.[62] In any case, for Bekker, it was evident that one needs first to look for a natural explanation, before one can start to invent different kinds of spirits which would act in some incomprehensible

---

[58] On how the mechanical philosophy reinterpreted occult phenomena in mechanical terms, see Henry 1986; Hutchison 1982; but see Vermeir 2004 for similar explanatory models in Aristotelianism.

[59] Bekker 1693, Book 3, 177–179. For a discussion of such theories of a strong imagination, see Vermeir 2004.

[60] Bekker 1693, Book 4, 13–14.

[61] In older traditions, the theory of vapors was only one explanatory model among others (such as natural place, occult qualities, etc.), but the mechanical philosophy made explanations by means of subtle particles universal.

[62] Bekker 1693, Book 4, 14: "Now we shall see whether there ever happened something in the world that one calls ghostly, sorcery or such like, which would not be explicable by these atoms; these atoms can move, divide and assemble in so many ways that they can act on vision, hearing and on the movements in and around humans [and create phenomena] as are [usually] ascribed to ghosts and to the devil in particular" ("Nu sullen wy eens sien of iets ter weereld ooit of ooit gebeurd zy datmen Spook of Tovern of diergelijke noemt, het gene niet aan dese Atomi zy toe te schryven; dewelke veelsins sich beweegende sich scheidende of tsamenpakkende sodanigen gesight, gehoor, beweeginge in en ontrent de menschen werken konnen als aan de Geesten en besonderlik den Duivel toegeschreven worden").

Fig. 12.4 Frontispiece of a book of one of Bekker's critics, van Rusting (1694). Courtesy Leiden University Library, Special Coll. Reading Room GM. Call Number 1199 G 19

way on bodies.[63] But if everything is explicable by natural means, his opponents had no legitimate motive to resort to explanations involving demonic causality. An abstracted version of the mechanical philosophy, with eclectic and hybrid roots, thus allowed for a comprehensive disenchantment of the world (Fig. 12.4).

[63] Bekker 1693, Book 4, 14: "In the examples that follow, we shall first refer to the rule of bodily divisions and movements before we will take recourse to ghosts. Because it is without doubt appropriate to search [the explanations] of our bodily sensations in corporeal causes and in the nature of bodies; rather than to invent ghosts, which act by means of corporeal or even inconceivable way on bodies; and to use [these ghosts] as an alternative explanation for occult qualities. Because this is like the heathens who were all too ready to ascribe phenomena to the devil when they could not understand the cause or nature of something" ("D'exempelen hier na te melden en te ondersoeken sullen wy eerst aan dien regel der lichamelike deelinge en beweeginge beproeven eer dat wy aan de Geesten komen. Want buiten twijfel is 't behoorlijk 't gene wy lichamelik verneemen eerst lichameliker wijse en in de nature der lichamen te gaan soeken: eer dan sulke Geesten te versinnen die lichamelik of immers onbedenkeliker wijse op een lichaam werken; en also dien toevlught in de plaats te stellen van d'occultae qualitates de verborgene hoedanigheden. Want dat is gelijk den heidenen wanneer sy d'oorsaak of de wijse van een ding niet meghtig waren te begrijpen het seggen seer gereed was dat een Daemon 't dede").

12.5 Demons, Impostors or Corpuscles?

How does Bekker's empiricism work in practice? In many of the specific examples of witchcraft and demonic involvement he discusses, he takes a sceptical stance. He uses historiography and source criticism to deny a demonic interpretation of the phenomenon or to deny the phenomenon itself and expose it as the result of deceit. In some difficult cases, however, when he is confronted with indisputable accounts by credible witnesses, or when his personal experience is involved, Bekker resorts to a natural explanation. In this section, I will explore one of the most difficult cases, which he discusses in book four.[64] It regards a remarkable feat that happened exactly at the time when he was finishing the book 3 and 4 of the *Betoverde Weereld*, in 1692–1693. It is an excellent example that shows how Bekker's Cartesian empiricism comes to the fore when he needs to resolve challenging examples of demonic activity in the world.

The story, recorded in many contemporary accounts, is as follows (see Fig. 12.5).[65] One day in 1688, Jaques Aymar, a Dauphiné peasant, went out to search for water with his divining–rod. When he felt his rod turn strongly, he was sure that he was standing over an underground spring. While they dug at the appointed spot, instead of water, they found the remains of a woman. Indeed, a woman from the village had been missing for 4 months, and Aymar went to the house where she had once lived. He directed his rod, in turn, upon each person there, and it moved when he directed it to the widower. The man immediately fled, so proving his guilt, and as a consequence people bestowed Aymar with the capacity to trace murderers. It was not until 1692, however, that Aymar was called upon to solve the difficult case with which he was to acquire international fame. Thieves had broken into a wine shop in Lyons, stolen the money and killed the owners. When called for by the police, Aymar was able to follow the trail of the murderers for hundreds of miles. Some of the murderers were able to escape over the border, but by means of his divining rod, Aymar could track down one of them before he could flee. The criminal confessed his crime and was executed shortly afterwards.

The first accounts of the case are two letters by the Royal Prosecutor in Lyons,[66] which were later published in the *Mercure Galant*, a popular journal.[67] The case

---

[64] Bekker calls it a "very special story" (Bekker 1693, Book 1, "Openinge van het IV boek," 29), or a "most singular and ingenious adventure" in the French edition (Bekker 1694, Book 1, "abrégé du livre 4"); yet even this phenomenon can be attributed to natural causes (Bekker 1693, Book 4, 14). Note that the first cited passage in this footnote is lacking in some copies, because the Aymar story was happening during the writing of the book, and the passage was inserted only in later print runs of the book. Also Rabus notices Bekker's special attention in his review of the *Betoverde Weereld*: "In't stuk van wichelary komt den Heere Bekker onder't schrijven nog wel't bedenkelijkste voor, dat hy ooit gelezen heeft" ("In the course of writing, Mr. Bekker finds the piece on dowsing as yet the most dubious that he has ever read") (Rabus 1693–1696, 353).

[65] This story has been recounted a number of times; see especially Figuier 1860 for more details on the story. For a close study of the Dutch reception of this controversy, see Vermeir 2011.

[66] [Vanini] 1692.

[67] *Mercure Galant*, August 1692 issue, 114–128 and the September 1692 issue, 226–237.

Fig. 12.5 Image of a dowser, in Bekker (1694). Book 4, 476. 7A6372; BIBC; BRES; Digilab, Universiteitsbliotheek KULeuven

sparked a big controversy, first in Lyons and then in Paris, which led to various tests, experiments and attempts at explanation. Provincial physicians such as Garnier, Panthot and Chauvin proffered explanations of the phenomenon, based on Cartesian medicine and the mechanical philosophy. Others reacted to this audacity with indignation. The Oratorian priest Pierre Le Brun (1661–1729), who had already done research on the matter, declared that de divining rod could only work with the help of the devil.[68] Le Brun's judgment was supported by the famous Cartesian philosopher Nicolas Malebranche (1638–1715), who confirmed that the devil had to be the cause of the movements of the divining rod.[69] This was a great case for Bekker to examine: a controversy that contrasted natural and demonic causality, occurring exactly at the time when he was finishing his book. It would allow him to test his ideas and to convince his readership of the power of his philosophical and theological views.

[68] Le Brun 1693.

[69] Malebranche 1693.

As was often the case, four different explanatory models were proposed. Some of the dowsers claimed that they had a special divine gift, but they were quickly rebuffed by the theologians examining the case.[70] For their part, elite commentators were convinced that it was all a hoax. After successful tests on the famous dowser in Lyons, Henri-Jules, Prince de Condé (1643–1709), a famous Cartesian and one of the highest ranking noblemen in France, brought Aymar to Paris. He performed experiments with him, partly as a philosophical project, but partly also as a salon entertainment for the aristocracy. The prince wanted to find out whether Aymar could really find springs, hidden treasures and criminals. Many experiments were performed in de Condé's mansion and gardens in Paris and at his castle in Chantilly. The first tests seemed promising, but later Aymar failed at almost all of the experiments, and he was debunked as an impostor. The Prince gave much publicity to his conclusions, and his followers published journal articles and books in which they unmasked the divining rod.[71]

Many commentators, however, remained convinced that the divining rod phenomena were real. Bekker was one of them. It would have been easy for him to dismiss Aymar as a fraud, because credible persons such as the Prince de Condé gave ample authority to do so. In making his judgment, however, Bekker carefully weighed all the testimonials he had access to. He found credible persons who believed in Aymar as well as honourable persons who wrote that Aymar was a fraud. Other savants such as Pierre Le Lorrain, abbé de Vallemont (1649–1721), for instance, had also done experiments with Aymar, and he came to conclusions opposite to those of the Prince de Condé. If we cannot even confirm the facts of this episode, Bekker remarks warily, it does not make much sense to look for the causes. But Aymar was not the only dowser around. Other credible men also turned out to possess the gift of dowsing. Bekker mentions Monseigneur Galet, the bishop of St Jean de Morienne and a great astronomer; Tonnelier, an apprentice apothecary in Paris; Grimaut, an officer of the douane in Lyons; and Besson, a young prosecutor in Lyons.[72] In the case of Tonnelier, experiments had been performed in the garden of the *Académie des Sciences* in Paris and in the houses of officials and noblemen. At first, these experiments sometimes failed, but in the end, some accounts pronounced them convincing and successful, even when his examiners tried to trick Tonnelier into making mistakes.[73] Furthermore, Bekker knew that many people went searching for metals in the mountains with a divining rod. Would these all be impostors? And why would they do such a thing? Finally, and crucially for making up his mind, Bekker had personal friends who had the capacity of dowsing. One of these friends can be identified as the publisher and bibliographer Cornelis Van Beughem (c. 1637–c. 1710). Bekker calls him a man of great

[70] Vermeir 2013.

[71] Buissière 1694; "On publie qu'Aimar est un fourbe. Lettres à ce sujet," 1693.

[72] Bekker could have read about the first two in the April 1693 issue of the *Mercure Historique et Politique* (p. 434 ff.) and of the latter two in a letter (reprinted in Vallemont's book) written to the abbot Bignon, a famous intellectual in Paris.

[73] On Tonnelier, see Comiers' article in the *Mercure Galant*, June 1693, 91; and Comiers 1693, 59.

experience, judgment and reason, and he recounts how the man had discovered his own gift of dowsing when he had accidentally found a treasure in his youth.[74] Bekker therefore had firsthand experience with the practice of dowsing, and it is plausible that this was decisive in his judgment that dowsing with a divining rod was a real and natural phenomenon.

For those, like Bekker, who believed the divining rod worked but rejected divine intervention, only two causal explanatory principles remained available: demons and nature. Some of the theologians argued that this was clearly a demonic phenomenon.[75] The phenomenon was erratic, because only a few people had the capacity for dowsing, and others had not. The tests were sometimes successful, at other times they failed. It seemed impossible to find the necessary regularities behind the phenomenon in order to attribute it to natural causes. Furthermore, finding criminals with a divining rod seemed to mix up natural and moral causes in impossible ways. From a physical point of view, a criminal was not different from a morally upright person. The difference between them was only moral and could not be measured by a physical instrument. The divining rod could therefore not be natural, they argued, because it was necessarily an instrument guided by a deceitful higher intelligence, i.e., the devil. These theologians had a strong case. This was exactly why Bekker was, of course, interested in the phenomenon at all, because it allowed him to show that even such difficult cases could be explained in a natural way.

News of Aymar's spectacular feats reached the northern Netherlands through a variety of channels, amongst which, most prominently, published books, journal issues and personal contacts between members of the republic of letters. Bekker referred to the accounts written by the provincial abbot De Lagarde and the Lyons physician Garnier, but his major source was the *Physique Occulte*, a widely read book written by the theologian and courtier Pierre Lorraine de Vallemont, printed in many editions in Paris, Amsterdam, and The Hague.[76] The book appeared just before Bekker finished books 3 and 4 of his *Betoverde Weereld*, and he based his judgment of divining rod in particular on Vallemont's descriptions and explanations. It is clear why Bekker was attracted to Vallemont's Cartesian explanation of dowsing. Nevertheless, Bekker did not naïvely accept Vallemont's account. Indeed, Bekker was clearly aware of the publications that exposed Aymar as a fraud. Bekker thinks that there could have been specific, as yet undiscovered causes that could explain why the experiments with Aymar did not work in Paris. A failed experiment did not imply that people were frauds. Indeed, one should not be so naive as to think that every experiment should work invariably. Experiments with magnets did not always work either. In his judgment, the circumstances of the Aymar episode, as well as the experiments with other diviners, had been too well researched into, to

[74] Despite his reticence and magnanimity, Van Beughem would later become embroiled in his own controversy on dowsing. Note that in the French edition of Bekker's book (Bekker 1694, 489, 497), some of this material is replaced by a discussion of Claude François Milliet Deschales' (Deschales 1674, 190–191) experiments with divining rods.

[75] Vermeir 2013.

[76] de Vallemont 1693.

maintain a rational suspicion of fraud. The solution to the dilemma should be sought in a natural explanation.

The natural explanations that were proffered were medically inspired and focused on the bodily states of the dowser. When Aymar was on the trail of the murderers, not only did the rod react, but his body as well, and he experienced severe affections. His temperature and heart pulse increased and he became feverish, he felt faint and sometimes had to vomit blood.[77] When he returned with the criminal he had tracked down, he found that he could not walk close to him without suffering from severe heart spasms. Aymar stated that he felt those violent agitations only when tracing criminals, but similar cases were known, in which sensitive people experienced violent emotions when walking above certain metals.[78] Other symptoms were headaches, fatigues and muscle spasms. According to the Cartesian physicians and natural philosophers, every object transpires specific vapors and Aymar claimed that he could discern the exhalations of springs, of metals and even of different murderers, by means of the distinctive emotions he experienced. Because of these violent emotions, he did not even need to use the divining rod; it only served to enhance the effect when the trace was feeble.[79] Some people are bestowed with a special and almost inexplicable capacity for feeling these exhalations, these physicians claimed, and they have very susceptible bodies which interact with the surrounding environment.

The physician Chauvin made the connection between the passions experienced in dowsing and the union of body and soul. Few, he says, know the condition of the union between mind and body, of which God alone is the cause. He writes:

> all the ideas of the soul, which regard the conservation of the body, as are those that are accompanied by sentiments and passions, will always be followed by a movement of the animal spirits, which will be the most suitable for the execution of the desires of the soul, and for the conservation of the unity of spirit and body, which constitutes man.[80]

---

[77] de Vallemont 1693, 434: "One knows that his pulse increases as in a serious fever" ("On sait que son pouls s'éléve alors comme dans une grosse fiévre...").

[78] de Vallemont 1693, 446: "I know a man who finds gold that is hidden in the ground without a divining-rod. The metallic exhalations permeate him so strongly that he feels his pulse increasing, & his hart weakening until it makes him vomit with terrible violence" ("Je connois un homme qui trouve sans Baguette l'argent qu'on a caché dans terre. Les seuls écoulements métalliques l'imprégnent si fort, qu'il sent son pouls s'élever, & son coeur s'affoiblir jusqu'à le faire vomir avec des violences terribles").

[79] de Vallemont 1693, 436: "It is this sudden change that occurs so violently within himself, which informs him that he is in the atmosphere of vapors, exhalations & fumes. When this internal disturbance is great & clearly perceptible, it directs Jaques Aymar sufficiently, & thus he does not need the divining-rod, which only serves him when he is internally affected in a weak and equivocal manner" ("C'est ce subit changement qui se fait si violemment en dedans de luy-même, qui l'avertit qu'il est dans l'atmosphére des vapeurs, des exhalaisons & des fumées. Quand ce dérangement intérieur est grand & bien sensible, il dirige suffisamment Jaques Aymar, & alors il n'a pas besoin de la Baguette, qui ne luy sert que quand il n'est émû intérieurement que d'une maniére foible & équivoque").

[80] Chauvin 1692, 9: "Toutes les idées de l'âme, qui regardent la conservation du corps, telles que sont celles qui sont accompagnées des sentiments & des passions, seront toûjours suivies du mouvement des *esprits animaux*, qui sera le plus propre pour l'éxécution des désirs de l'âme, & pour la conservation de l'union de l'esprit avec le corps, ce qui constitue l'homme."

The murderer and the victim must both have experienced strong internal agitations together with passions of self-preservation, fear, hate and revenge. This causes irregular motions of the animal spirits which flow together with the blood. These violent emotions, according to Chauvin, cause the body to perspire little corpuscles of a determinate figure.[81] Another physician, Garnier, gives a similar interpretation of Aymar's performance. Doctrinal differences were not so important in this debate, and theories of vapors could find a place in both Aristotelian and Cartesian philosophies. Even Vallemont, an avowed Cartesian, was not too worried about ideological distinctions between Gassendists and Cartesians. He even cited the Jesuit Gaspar Schott (1608–1666), who had stated that exhalations of subtle particles were the best way to explain phenomena that had previously been attributed to occult qualities.[82] Nevertheless, those adhering to Cartesian mechanical philosophy were especially amenable to explain the curious action at a distance of the divining rod by means of fluxes of special corpuscles.

Vallemont had extensively experimented with Aymar when the latter stayed in Paris. He was especially concerned with explaining how the divining rod could act at a distance. If one presupposed a circulation of corpuscles between the murderer and the divining rod, Vallemont still had to explain how it was possible for him to find the trace of a murderer days, weeks, sometimes even years after the facts. Vallemont had to demonstrate that the subtle particles were not mingled or blown away by the wind. He argued that those particles were so subtle that they did not interact with the coarser air particles, and that the whole world was filled with subtle particles which did not modify each other. This explanation would have been dubious if Vallemont had not been able to demonstrate that there existed particles with exactly such a behavior. These subtle corpuscles were not dispersed by the wind, just like light rays were not scattered. Another, perhaps more convincing analogy, was that between the divining rod and a hound, which could smell scents even if the trace was already cold for some days. Vallemont described many experiments that—by analogy—were able to show that the wind did not blow away the subtle

[81] Chauvin 1692, 10: "This movement cannot occur without some little and specifically shaped corpuscles, which pass through the miliary glands and are pushed and pulled outside by transpiration" ("ce mouvement ne peut se faire sans qu'il ne se sépare au travers des glandes miliaires, quelques petits corpuscules d'une certaine figure déterminé, qui sont poussés & entraînes au dehors par la transpiration").

[82] de Vallemont 1693, 142. "Finally, I explain the sympathy of the divining-rod with metals & other things toward which it bends, by the pouring out, & the flux of the subtle matter, which emanates from all bodies, & which disperses in the air; & the Jesuit Father Schott declares that this is the right manner to explain those effects that one has until now attributed to occult qualities" ("Enfin j'explique la sympathie de la Baguette de coudrier avec les métaux, & les autres choses surqoy elle s'incline, par l'écoulement, & le flux de la matiére subtile, qui se transpire de tous les corps, & qui se répand dans l'air; & le Pere Schott Jésuite déclare que c'est la bonne maniere de développer les effets, qu'on a jusques icy attribuëz à des qualitez occultes"). See also the title of Ch. 3: "Nature has only one mechanism in all its operations: & the philosophy of corpuscles is the only theory that can make sense of the wonders of sympathy, & of the movement of the divining-rod" ("La nature n'a qu'un seul mécanisme dans toutes ses opérations: & la Philosophy des corpuscules est la seule, qui puisse rendre raison des merveilles de la sympathie, & du mouvement de la Baguette Divinatoire").

transpirations exhaled by the murderer. Vallemont's particular Cartesian empiricism presented analogical demonstrations that were meant to confirm his central hypothesis.[83] Bekker did not oppose this kind of reasoning, as he used analogical arguments himself when he claimed that "experience proves entirely what I say."[84] One of his analogical arguments goes as follows. We know the imagination works by means of subtle particles. The nature of such subtle particles has been well established by the study of magnetism. By means of these subtle particles, magnetism can act at a distance. Therefore, experience proves that the imagination can work outside the body.

Bekker probably did not experiment directly on diviners, nor did he present actual analogical demonstrations to prove his point.[85] His explorations were sheer textual. Nevertheless, this time, he wanted to present an unambiguously Cartesian explanation of the divining rod, and following Vallemont, he claimed to follow the rules of movement set down by Descartes. For him, this was the best way to show that divining is a natural phenomenon and should not be attributed to the devil. According to Bekker, the murderer leaves subtle particles behind, wherever he goes. These particles are different from those exhaled before the murder, because murder is usually committed in a passionate state, full of fear and terror. These passions change the blood, which in their turn affect the vapours exhaled by the body. These small corpuscles, exhaled by the criminal, can enter the skin of the diviner if there is a special fit between the exhaled particles and the pores in the skin of the diviner. This is the reason why only some people have the capacity to find specific criminals. When the corpuscles enter the body of the diviner, they affect his blood and cause some kind of fermentation, which results in spasms and contractions. These corpuscles will also fill the divining rod, but their outflow is hindered and this will cause the divining rod to twist and contract.

Bekker could have dismissed the story of Aymar as the result of a confidence trick by a rogue provincial farmer. After careful consideration, however, partly based on witness reports and partly guided by his own experience, he concluded that the phenomenon was real. In such a case, he needed a plausible natural explanation. Many philosophical models were available in order to propose such explanations, and Bekker did not want to exclude any, but he found the mechanical philosophy, or at least a theory of flowing small corpuscles, the most convincing option amenable to his goals. Bekker did not need to be a purist in this matter, and he was perfectly happy with an explanatory pluralism, as long as he knew there was a plausible natural explanation available. There were at least enough convincing natural explanations

[83] On analogical demonstrations, see Vermeir 2005.

[84] See e.g., Bekker 1693, Book 3, 178: "D'ervarentheid bewijst mijn seggen altemaal. De alderfijnste deeltjes en deselver vlugge uit en intrek van 't een lichaam in het ander doet den zeilsteen 't yser trekken; want dat malkander trekt dat raakt. Die rakinge bevestigt dese vleghtinge der deeltjes in malkander want dat niet in malkander haakt dat trekt ook niet." Note also that Bekker here follows the Aristotelian/Cartesian idea that action at a distance presupposes *contact* of invisible small particles.

[85] For a description of the cases in which Bekker was personally involved, see Bekker 1693, Book 4, Chaps. 7–8.

for divination available to chide Malebranche for ascribing this curious case of divination to the devil. It was the striking explanatory power of late seventeenth-century Cartesianism that allowed Bekker to explain any imaginable phenomenon as natural, and hence counter all empirical evidence of devil phenomena. This added to Bekker's confidence. The mechanical philosophy also had the advantage that it fitted seamlessly with the Cartesian metaphysics expounded in Book 2. It made it possible for the different parts of his work—historical, metaphysical, theological and natural philosophical—to cohere and mutually to buttress each other.

12.6 Conclusion

Book 4 of the *Betoverde Weereld* is seldom read. It contains heaps of crazy stories that seem inappropriate for serious historians of philosophy or theology. Nevertheless, it presents Bekker's empiricist response to the criticisms of his abstract and a priori philosophical and theological approaches. He regarded this empiricist project as an indispensable part of his general argument. His theoretical rationalist arguments would fall apart were it not for the empirical support given in book 4. Bekker was well known as a Cartesian. His early defence of Descartes as well as book 2 of the *Betoverde Weereld* show that he was steeped in Descartes' metaphysics. Not only Cartesian metaphysics, I have argued, but also the mechanical philosophy was central to his project. Bekker's self professed aim was to battle the commonly received view in which the devil was granted a lot of power over mankind. For Henry More, Joseph Glanvill and Robert Boyle, proving the existence of demonic activity was vital in order to combat atheism. Bekker thought it was exactly the other way around. His line of reasoning was not readily accepted, of course, and he was accused of atheism himself. Nevertheless, he was convinced he needed to take the veil of pagan prejudices away with which people read the bible. The importance of his work was not just to correct a widespread false belief, but to reform the reformation itself. In this way, he hoped to remove a great danger to the salvation and piety of his countrymen.

In the philosophical parts of the *Betoverde Weereld*, Bekker followed Cartesian metaphysics, and especially Descartes' dualism, to argue against devil beliefs. In the theological parts of his book, Bekker used a Cartesian style hermeneutics to explain the bible. His critics were not convinced, however. His hermeneutics could be questioned, and Cartesian philosophy did not necessarily contradict the power of the devil, since many Cartesians supported the possibility of demonic action with Cartesian arguments. Bekker was aware of these problems, and believed that his empiricism provided the necessary foundation on which his philosophical and theological work could stand. He collected many concrete stories of witchcraft and magic from written sources, credible witnesses or his personal experience. Most of these he could reject as false, but some crucial examples had to be explained by means of natural causes. Here, Descartes' natural philosophy came to the rescue. The mechanical philosophy, with its theories of flowing corpuscles, vortices and exhalations, made it possible for Bekker to naturalize even the most incredible phenomena. Nevertheless, Bekker was not dogmatic, and although he adhered to Cartesianism in metaphysics and exegetics, he followed a loosely Cartesian style in his natural philosophy. In fact, he proposed something like the greatest common divisor between older philosophies and the new mechanical philosophy, as far as the flux of subtle corpuscles was concerned. Bekker's empiricism in book 4 served to validate some of the reported extraordinary phenomena, but his natural philosophy allowed him to immediately reject any demonic explanation, by providing strong and probable natural causes.

Bekker's hybrid mechanical philosophy did not disenchant the world in the sense in which this is usually understood today. Strange phenomena like dowsing and some forms of magic were validated by Bekker, but he explained them naturalistically. Bekker did not weed out popular superstitions, as the Enlightenment would later try to do. Instead, he radicalised the reformation by rejecting all pagan remnants in Christianity. For Bekker, this would lead to a disenchantment of the world all the same. Indeed, there would be no magic or sorcery anymore, he exclaimed, if we only stopped believing in it.[86] The ghost with whom I started this paper was right after all. Indeed, the disembodied voice had correctly identified its enemies. These enemies comprised not only atheists, but also religious reformers who wanted to reform the reformation. His diatribe aptly represented the feelings of Bekker's critical and orthodox co-religionists when they were confronted with his work. Bekker's radicalisation of the reformation meant that he rejected as superstitious all the spirits, demons and devils that his contemporaries still believed in. This also meant that he accused his fellow Calvinists of superstition and false religion, for exactly the same reasons as these Calvinists condemned the Catholics. No wonder that his congregation reacted at least as furiously as the articulate ghost in the Saxon pamphlet, and that they expelled Bekker from their midst.

References

Ankarloo, Bengt, Stuart Clark, and William Monter. 2002. *Witchcraft and magic in Europe: The period of the Witch trials. Trials*, vol. 4. London: The Athlone Press.

[Anon.]. 1693. On publie qu'Aimar est un fourbe. Lettres à ce sujet. *Mercure Historique et Politique 14(Mai)*: 558.

Attfield, Robin. 1985. Balthasar Bekker and the decline of the Witch-Craze: The old demonology and the new philosophy. *Annals of Science* 42: 383–395.

Aubin, Nicolas. 1693. *Histoire des diables de Loudun, ou de la possession des religieuses Ursulines, et de la condamnation du supplice d'Urbain Grandier, curé de la meme ville.* Amsterdam: A. Wolfgang.

Bekker, Balthasar. 1668. *De Philosophia Cartesiana admonitio candida et sincera.* Wesel: Hoogenhuysens.

Bekker, Balthasar. 1670. *De vaste spyse der volmaakten.* Franeker: Karst Theunes Plekkenpoel.

Bekker, Balthasar. 1683. *Ondersoek van de Betekening der Kometen.* Leeuwarden: Hero Nauta.

Bekker, Balthasar. 1689. *Engelsch verhaal van ontdekte tovery.* Amsterdam: Daniel van den Dalen.

[86] Bekker 1693, Book 4, 269.

Bekker, Balthasar. 1693. *De Betoverde Weereld*. Amsterdam: Daniel Van Den Dalen.
Bekker, Balthasar. 1694. *Le Monde Enchanté*. Amsterdam: Pierre Rotterdam.
Bekker, Balthasar. 1739. *De Betoverde Weereld*. Deventer: De Vries.
Bever, Edward. 2009. Witchcraft prosecutions and the decline of magic. *Journal of Interdisciplinary History* 40: 263–293.
Brunsmands, Johann. 1693. *Energument Coagienses, sive Admirabilis historia, de horrenda Cacodaemonis tentatione*. Amsterdam: Daniel van den Dale.
Buisman, Jan Wim. 1998. Bekkers Wraak. Balthasar Bekker (1634–1698), de accommodatietheorie en Nederlandse protestantse theologen, 1750–1800. *De Achttiende Eeuw* 30: 97–111.
Buissière, Pierre. 1694. *Lettre à M. l'abbé D.L., sur les véritables effets de la baguette de Jacques Aymar*. Paris: Louis Lucas.
Chauvin, Pierre. 1692. *Lettre de M. Chauvin... a madame la marquise de Senozan, sur les moyens dont on s'est servy pour découvrir les Complices d'un assassinat commis à Lyon, le 5. de Juillet 1692*. Lyon: Chez J. Bapt. et Nicolas De Ville.
Clark, Stuart. 1997. *Thinking with Demons. The idea of witchcraft in early modern Europe*. Oxford: Oxford University Press.
Comiers, Claude. 1693. *La Baguette Justifiée*. s.l.
De Certeau, Michel. 1970. *La possession de Loudun*. Paris: Julliard.
de Vallemont, Pierre Lorrain. 1693. *La Physique Occulte, ou Traité de la baguette divinatoire*. Paris: Jean Anisson.
De Vet, Joannes J.V.M. 1980. *Pieter Rabus (1660–1702), een wegbereider van de Noordnederlandse verlichting*. Amsterdam: Holland University Press.
De Vleeschouwer, Herman Jean. 1939. Balthasar Bekker, avocat de Descartes. *Revue belge de philologie et d'histoire* 18: 63–84.
Della Porta, Giambattista. 1589. *Magiae naturalis libri XX in quibus scientiarum naturalium divitiae et deliciae demonstrantur*. Nápoles: Horatium Salvianum.
Deschales, Claude-François Milliet. 1674. *Cursus seu Mundus mathematicus*, vol. 2. Paris: Laurent Anisson.
Desmarets, Samuel. 1672. *Tractatus brevis de afflicto statu studii theologici in foederato belgio*. Groningen: Dominicus Lens.
Digby, Kenelm. 1658. *Late discourse made in a Solemne assembly of Nobles and learned men at Montpellier in France... Touching the cure of wounds by the powder of sympathy*. London: R. Lownes and T. Davies.
Dompnier, Bernard. 1998. *La Superstition à l'âge des Lumières*. Paris: Honoré Champion.
Duikerius, Johannes. 1691. *Het leven van Philopater, Opgewiegt in Voetiaensche Talmeryen en groot gemaeckt in de Verborgentheden der Coccejanen. Een Waere Historie*. Groningen: Siewert van der Brug.
Figuier, Louis. 1860. *Histoire du Merveilleux. Tome 2*. Paris: Hachette.
Fix, Andrew. 1989. Angels, devils, and evil spirits in seventeenth-century thought: Balthasar Bekker and the Collegiants. *Journal of the History of Ideas* 50: 527–547.
Fix, Andrew. 1993. Balthasar Bekker and the crisis of Cartesianism. *History of European Ideas* 17: 575–588.
Fix, Andrew. 1996. Hoe cartesiaans was Balthasar Bekker? *It Beaken* 58: 118–137.
Fix, Andrew. 1999. *Fallen Angels. Balthasar Bekker. Spirit belief and Confessionalism in the seventeenth-century Dutch Republic*. Dordrecht: Kluwer Academic Publishers.
Fix, Andrew. 2000. Bekker and Bayle on comets. *Geschiedenis van de wijsbegeerte in Nederland* 11: 81–96.
Fix, Andrew. 2013. What happened to Balthasar Bekker in England? A mystery in the history of publishing. *Church History and Religious Culture* 90: 609–631.
Frommann, Johann. 1675. *Tractatus de fascinatione novus et singularis*. Nürenberg: Endteri.
Glanvill, Joseph. 1681. *Saducismus triumphatus, or, Full and plain evidence concerning witches and apparitions*. London: J. Collins and S. Lownds.
Golden, Richard M. (ed.). 2006. *Encyclopedia of Witchcraft: The western tradition*. Santa-Barbara: ABC-CLIO.

Hazard, Paul. 1961. *La Crise de la conscience européenne (1680–1715)*. Paris: Fayard.
Henry, John. 1986. Occult qualities and the experimental philosophy: Active principles in pre-Newtonian matter theory. *History of Science* 24: 335–381.
Hilarides, Johannes. 1691. *Verklaaringe van't Sinnebeeld*. Amsterdam: Daniel van den Dalen.
Hutchison, Keith. 1982. What happened to occult qualities in the scientific revolution? *Isis* 73: 233–253.
Israel, Jonathan. 1996. The Bekker controversies as a turning point in the history of Dutch culture and thought. *Dutch Crossing: A Journal of Low Countries Studies* 20: 5–21.
Israel, Jonathan. 2001. *Radical enlightenment: Philosophy and the making of modernity, 1650–1750*. Oxford: Oxford University Press.
Israel, Jonathan. 2012. Spinoza and the religious radical enlightenment. In *The intellectual consequences of religious heterodoxy, 1600–1750*, ed. Sarah Mortimer and John Robertson, 181–203. Leiden: Brill.
Jorink, Eric. 2006. *Het "Boeck der Natuere". Nederlandse geleerden en de wonderen van Gods schepping 1575–1715*. Leiden: Primavera Pers.
Knuttel, Willem Pieter Cornelis. 1906. *Balthasar Bekker. De bestrijder van het bijgeloof*. 's-Gravenhage: Martinus Nijhoff.
Le Brun, Pierre. 1693. *Lettres qui découvrent l'illusion des philosophes sur la baguette*. Paris: Jean Boudot.
Le Brun, Pierre. 1702. *Histoire critique des pratiques superstitieuses*. Paris: J. de Nully.
Le Camus, Étienne. 1690. *Ordonnances synodales du diocèse de Grenoble*. Grenoble: Alexandre Giroud.
Levack, Brian P. 2006. *The Witch-Hunt in early modern Europe*, 3rd ed. Harlow: Pearson.
Malebranche, Nicolas. 1693. Réponse de l'Auteur de la Recherche de la Vérité; Réponse du même Auteur. *Mercure Galant (Janvier)*: 32–46.
Monfils, Lesley. 2004a. Een boek gaat door de wereld: verspreiding en vroege bezitters van Balthasar Bekkers "De betoverde weereld" (1691). *Jaarboek voor Nederlandse boekgeschiedenis* 11: 67–82.
Monfils, Lesley. 2004b. Een dominee en zijn "duivelse" boek. De eerste druk van Balthasar Bekkers De betoverde weereld. *De Boekenwerld* 21: 58–72.
Nicolson, Marjorie. 1929. The early stage of Cartesianism in England. *Studies in Philology* 26: 356–374.
Nooijen, Annemarie. 2009. *"Unserm grossen Bekker ein Denkmal"?: Balthasar Bekkers Betoverde Weereld in den deutschen Landen zwischen Orthodoxie und Aufklärung*. Münster: Waxmann.
Rabus, Pieter. 1693–1696. *Boekzaal van Europe*. Rotterdam: Pieter vander Slaart.
Schott, Gaspar. 1659. *Magia universalis naturae et artis*. Würzburg: Heirs of Johann Gottfried Schönwetter.
Simoni, Annae C. 1979. Balthasar Bekker and the Beckington Witch. *Quaerendo* 9: 135–142.
Thiers, Jean-Baptiste. 1697. *Traité des superstitions qui regardent les sacremens*, vol. 4. Paris: Dezallier.
van Bunge, Wiep. 1993. Balthasar Bekker's Cartesian hermeneutics and the challenge of Spinozism. *British Journal for the History of Philosophy* 1: 55–79.
van Bunge, Wiep. 1998. "Quelle extravagance": Balthasar Bekker (1634–1698) in Germany and France. *De achttiende eeuw* 30: 113–124.
[Vanini]. 1692. Deux lettres de Lyon relatives à la baguette de Jacques Aymar des 31 juillet et 6 septembre 1692. (Municipal Library of Grenoble).
van Ruler, Han. 2000. Minds, forms, and spirits: The nature of Cartesian Disenchantment. *Journal of the History of Ideas* 61: 381–395.
van Rusting, Salomon. 1694. *Duvels leven onder de duvelen, ofte duvels dispuyt om 't opper-duvelschap*. Amsterdam: Jan ten Hoorn.
van Sluis, Jacob (ed.). 1994. *Bekkeriana. Balthasar Bekker biografisch en bibliografisch*. Leeuwarden: Fryske Akademy.
Vermeir, Koen. 2004. The "physical prophet" and the powers of the imagination. Part I: A case-study on prophecy, vapours and the imagination (1685–1710). *Studies in History and*

Philosophy of Science Part C: Studies in History and Philosophy of Biological and Biomedical Sciences 35: 561–591. doi:10.1016/j.shpsc.2004.09.001.
Vermeir, Koen. 2005. The magic of the magic lantern (1660–1700): On analogical demonstration and the visualization of the invisible. *The British Journal for the History of Science* 38: 127–159.
Vermeir, Koen. 2011. Circulating knowledge or superstition? The Dutch debate on divination. In *Silent messengers: The circulation of material objects of knowledge in the early modern low countries*, ed. Sven Dupré and Christoph Herbert Lüthy, 293–328. Münster: LIT Verlag.
Vermeir, Koen. 2012. Castelli in aria: immaginazione e Spirito della Natura in Henry More. *LoSguardo* 10: 99–124.
Vermeir, Koen. 2013. Tout dépend des circonstances. Superstition, lois de la nature et dissolution du préternaturel au tournant du XVIIIe siècle. In *Marie mondialisée: penser le surnaturel à l'époque moderne*, ed. Olivier Christin, Fabrice Flückiger, and Naima Ghermani. Neuchâtel: Alphil.
Walker, Daniel Pickering. 1964. *The decline of Hell. Seventeenth-century discussions of eternal torment*. London: Routledge.
Wasser, Michael. 2008. The mechanical World-view and the decline of Witch beliefs in Scotland. In *Witchcraft and belief in early modern Scotland*, ed. Julian Goodare, Lauren Martin, and Joyce Miller, 206–226. New York: Palgrave Macmillan.

Bio-Bibliographical Appendix for Cartesians Discussed in Part II

Balthasar Bekker (1634–1698) was a Cartesian theologian and controversialist who became famous for his work *De Betoverde Weereld* (1691–1693). Born in Metslawier, a small Friesian village, Bekker studied philosophy and theology at the Universities of Groningen and Franeker, where he witnessed the burgeoning disputes about Cartesianism. Bekker openly took sides and published one of the first theological defenses of the Cartesian philosophy. He argued that it posed no risk to the Reformed Church, because philosophy and theology had their own domains. Bekker's anti-dogmatic theological views opposed Calvinist literalism and confessionalism, which provoked irate reactions from the religious establishment. After moving to Amsterdam, in search of a more tolerant climate, Bekker published a book on comets, using natural philosophical and exegetical arguments to show that these extraordinary phenomena were no portents. The arguments in this book would form the template of his magnum opus, *De Betoverde Weereld*, in which he denied that the devil could act in this world. By means of Cartesian hermeneutics, metaphysics and natural philosophy, Bekker argued that demonic activity and witchcraft could not occur in this world. Devil beliefs were nothing more than pagan remnants that had corrupted Christianity. In a reaction to this work, the church authorities condemned Bekker, and his book would spark the largest international controversy of the early modern period.

Primary Sources

Bekker, Balthasar. 1668. *De Philosophia Cartesiana admonitio candida et sincera*. Wesel: Hoogenhuysens.

Bekker, Balthasar. 1683. *Ondersoek van de Betekening der Kometen*. Leeuwarden: Hero Nauta.

Bekker, Balthasar. 1693. *De Betoverde Weereld*. Amsterdam: Daniel Van Den Dalen. (German translation 1693; French translation 1694; English translation 1695.)

Bekker, Balthasar. 1694. *Le Monde Enchanté*. Amsterdam: Pierre Rotterdam.

M. Dobre and T. Nyden (eds.), *Cartesian Empiricisms*, Studies in History and Philosophy of Science 31, DOI 10.1007/978-94-007-7690-6,
© Springer Science+Business Media Dordrecht 2013

Secondary Sources

Attfield, Robin. 1985. Balthasar Bekker and the decline of the Witch-Craze: The old demonology and the new philosophy. *Annals of Science* 42: 383–395.

De Vleeschouwer, and Herman Jean. 1939. Balthasar Bekker, avocat de Descartes. *Revue belge de philologie et d'histoire* 18: 63–84.

Fix, Andrew. 1993. Balthasar Bekker and the crisis of Cartesianism. *History of European Ideas* 17: 575–588.

Fix, Andrew. 1999. *Fallen Angels. Balthasar Bekker. Spirit belief and Confessionalism in the seventeenth-century Dutch Republic*. Dordrecht: Kluwer Academic Publishers.

Israel, Jonathan. 2001. *Radical enlightenment: Philosophy and the making of modernity, 1650–1750*. Oxford: Oxford University Press.

Nooijen, Annemarie. 2009. *"Unserm grossen Bekker ein Denkmal"?: Balthasar Bekkers Betoverde Weereld in den deutschen Landen zwischen Orthodoxie und Aufklärung*. Münster: Waxmann.

van Bunge, Wiep. 1993. Balthasar Bekker's Cartesian hermeneutics and the challenge of Spinozism. *British Journal for the History of Philosophy* 1: 55–79.

van Sluis, Jacob (ed.). 1994. *Bekkeriana. Balthasar Bekker biografisch en bibliografisch*. Leeuwarden: Fryske Akademy.

Vermeir, Koen. 2013. Mechanical philosophy in an enchanted world: Cartesian empiricism in Balthasar Bekker's radical reformation. In *Cartesian Empiricisms*, ed. Mihnea Dobre and Tammy Nyden, 275–306. Dordrecht: Springer.

Koen Vermeir

Robert Desgabets (1610–1678) was an early defender and teacher of the Cartesian philosophy in the region of Lorraine, France. Although he is little-known today, he played an important role in the development and transmission of the Cartesian philosophy, especially in Paris through his close association with Clerselier and in Toulouse through his student Pierre-Sylvain Régis. He is perhaps the most original of the Cartesian thinkers, although only one book (1675) and two small works (1668; 1671) were published during his lifetime. He is best known for his role in the theological controversy over the Cartesian explication of the Eucharist (Desgabets 1671), and for his defense of Nicolas Malebranche against the skeptic Simon Foucher (Desgabets 1675). In large part because of his role in the Eucharist controversy, his major philosophical writings only appeared in print in 1983. Desgabets' most important philosophical work, *Supplément à la philosophie de M. Descartes*, is a critical examination of Descartes' *Meditations*. In this work, Desgabets defends the Cartesian doctrines of sensible qualities, matter, mind-body dualism, mind-body union and interaction in man while criticizing Descartes' argument for the *cogito* as the first principle of knowledge, the claim that there is such a process as pure intellection, and that there are innate ideas. Desgabets' contributions in natural philosophy include pioneering work in the study of blood transfusion and mechanics, and a novel defence and development of the Cartesian philosophy. Desgabets' unusual marriage of Cartesianism and empiricism challenges many standard views of Descartes and the Cartesian philosophy.

Primary Sources

Desgabets, Robert. 1668. *Discourse de la communication ou transfusion du sang, published with, Lettre ècrite à M. Sorbière*. Paris: J. B. Denis.

Desgabets, Robert. 1671. *Considérations sur l'état présent de la controverse touchant le Très Saint-Sacrement de l'autel*, published anonymously, Holland.

Desgabets, Robert. 1675. *Critique de la Critique de la Recherche de la vérité*, Paris.

Desgabet, Robert. 1983. *Oeuvres philosophiques inédites*, Analecta Cartesiana 2, ed. J. Beaude with introduction by G. Rodis-Lewis, Amsterdam: Quadratures.

Secondary Sources

Allen-Hermanson, S. 2008. Desgabets: Rationalist or Cartesian Empiricist? In *Topics in early modern philosophy of mind*, ed. Jon Miller, 57–85. Dordrecht: Springer.

Armogathe, J.-R. 1977. *Theologia Cartesiana: L'Explication physique de l'Eucharisté chez Descartes et dom Desgabets*. The Hague: Martinus Nijhoff.

Beaude, Joseph. 1974. Desgabets et son oeuvre. *Revue de sythèse* 95: 7–17.

Cook, Monte. 2008. Desgabets as a Cartesian Empiricist. *Journal of the History of Philosophy* 46(4): 501–516.

Cousin, Victor. 1852. *Fragments de philosophie cartésienne*. Paris: Didier. (Reprinted 1970, Geneva: Slatkine Reprints.)

Easton, Patricia. Winter 2012 Edition. Robert Desgabets. *The Stanford encyclopedia of philosophy*, ed. Edward N. Zalta. http://plato.stanford.edu/archives/win2012/entries/desgabets/.

Easton, Patricia. 2013. Robert Desgabets on the physics and metaphysics of blood transfusion. In *Cartesian Empiricisms*, ed. Mihnea Dobre and Tammy Nyden, 185–202. Dordrecht: Springer.

Faye, Emmanuel. 2005. The Cartesianism of Desgabets and Arnauld and the problem of the eternal truths. In *Oxford studies in early modern philosophy*, vol. II, ed. Daniel Garber, 193–209. Oxford: Clarendon Press.

Lemaire, Paul. 1901. *Le Cartésianisme chez les Bénédictins: Dom Robert Desgabets, son système, son influence et son école*. Paris: Alcan.

Robinet, André. 1974. Dom Robert Desgabets, le conflit philosophique avec Malebranche et son l'oeuvre métaphysique. *Journée Desgabets, Revue de synthèse* 95: 65–83.

Rodis-Lewis, Généviève. 1974. L'écrit de Desgabets sur la transfusion du sang et sa place dans les polémiques contemporaines. *Journée de Desgabets Revue de synthèse* 95: 31–64.

Schmaltz, Tad M. 2002. *Radical Cartesianism: The French reception of Descartes*. Cambridge: Cambridge University Press.

Watson, Richard A. 1982. Transubstantiation among the Cartesians. In *Problems of Cartesianism*, ed. T.M. Lennon, J.M. Nicholas, and J.W. Davis, 127–148. Kingston/Montreal: McGill-Queens University Press.

Patricia Easton

Antoine Le Grand (1628/29–1699) was an important advocate of the Cartesian philosophy who wrote one of the earliest systematic textbooks of that philosophy in Latin. Le Grand was born in Douai (then part of the Spanish Netherlands) and trained there as a Franciscan with the second English Province, a group aiming to reconvert England to Catholicism. He began his mission to England in 1656, where he lived until his death, teaching philosophy and holding various church offices. After living primarily in London, in 1695 he moved to Oxford as a private tutor. As a Catholic priest in England he was in some danger and was briefly arrested in 1692. His teaching occurred in private homes or in rooms available to Franciscan missionaries. Le Grand's significant works, mainly in philosophy with some in Catholic theology, were published while he was in England. His first three books were ethical and not Cartesian. In 1671 he published a "digest" of Descartes' philosophy, expanded in 1672 into an *Institutio philosophiae*. These were intended as textbooks; the latter was revised through several editions. It was supplemented by two specialized books, the *Historia naturae, variis experimentis & ratiociniis elucidata* (1673) and *Dissertatio de carentia sensus & cognitionis in brutis* (1675). These three works were translated (reportedly by a member of the Royal Society) and published in 1694 in London by Richard Blome, as *An Entire Body of Philosophy, According to the Principles of the Famous Renate Des Cartes*. There were also theological works, works defending Descartes from attack, and an annotated edition of Rohault's physics in Latin.

Primary Sources

Le Grand, Antoine. 1672. *Institutio philosophiae, secundum principia domini Renati Descartes: novo methodo adornata & explicata: in usum juventutis academicae*. London: J. Martyn. Other editions: London 1675, 1677, 1680; Nuremberg 1679, 1683, 1695; Geneva 1694.

Le Grand, Antoine. 1679. *Apologia pro Renato Des-Cartes contra Samuelem Parkerum*. London: M. Clark. Another edition: Nuremberg 1681.

Le Grand, Antoine (ed.). 1682. *Jacobi Rohaulti Tractatus physicus*. Trans. Théophile Bonnet. London: G. Wells. Other editions: Amsterdam 1691; London 1692.

Le Grand, Antoine. 1694. *An entire body of philosophy according to the principles of the Famous Renate Des Cartes*, 2 vols., ed. Richard Blome. London: Samuel Roycroft.

Secondary Sources

Argomathe, Jean-Robert. 2003. The Roman Catholic Censure of the *Institutio Philosophiae* of Antoine Le Grand (1629–1699) according to unpublished documents from the archives of the Holy Office. In *Cartesian views: Papers presented to Richard A. Watson*, ed. Thomas M. Lennon, 193–203. Leiden: Brill.

Easton, Patricia. 2009. What is at Stake in the Cartesian Debates on the eternal truths? *Philosophy Compass* 4: 348–362.

Hatfield, Gary. 2005. Rationalist theories of sense perception and mind–body relation. In *Blackwell companion to rationalism*, ed. Alan Nelson, 31–60. Oxford: Blackwell.

Hatfield, Gary. 2013. The Cartesian psychology of Antoine Le Grand. In *Cartesian Empiricisms*, ed. Mihnea Dobre and Tammy Nyden, 251–274. Dordrecht: Springer.

Mautner, Thomas. 2000. From virtue to morality: Antoine Le Grand (1629–1699) and the new moral philosophy. *Jahrbuch für Recht und Ethik* 8: 209–232.

Ryan, John. 1935. Anthony Legrand (1629–1699): Franciscan and Cartesian. *New Scholasticism* 9:226–250. (Lists Le Grand's publications.)

Gary Hatfield

Henricus Regius or Hendrick de Roy (1598–1679) was one of the first followers of Descartes. He studied medicine in the Low Countries, Montpellier and Padua. In 1638, he was appointed professor *extra ordinem* of theoretical medicine and botany at the University of Utrecht. He had, by that point, read Descartes' *Discours de la méthode* and *Essais*, probably having been introduced to the works of the French philosopher through Reneri. He was subsequently put into contact with Descartes himself. In 1639, he became full professor at Utrecht. The year 1641 saw the start of the Utrecht dispute between Regius and the theologian Voetius. The dispute began because Regius, having been granted the right to lecture on physical problems, was taken to endorse the thesis that man was an accidental being. The dispute turned into a debate over the teaching of Cartesian philosophy. Descartes supported his friend Regius and advised him in his response to Voetius. This support ended when Regius decided, in 1646, to publish his *Fundamenta physices* with which Descartes disagreed on metaphysical grounds. In the 1647 letter-preface to the French edition of his *Principles of Philosophy*, Descartes deliberately distanced himself from his former disciple. Regius replied in a broadsheet titled *Explicatio Mentis Humanae* (1647) which Descartes commented upon in his *Notae in programma quoddam*, published in 1648. Regius responded by writing an expanded version of the broadsheet, titled *Brevis Explicatio Mentis Humanae* (1648).

Primary Sources

Bos, Erik-Jan (ed.). 2002. *The correspondence between Descartes and Henricus Regius*. PhD dissertation, Quaestiones Infinitae, Utrecht (includes Regius' 1641 *Physiologia, sive Cognitio Sanitatis.*)

Regius, Henricus. 1642. *Responsio, sive Notae in Appendicem…Voetii*. Utrecht: J. van Doorn.

Regius, Henricus. 1646. *Fundamenta physices*. Amsterdam: Elzevier.

Regius, Henricus. 1654. *Philosophia naturalis*. Amsterdam: Elzevier (second edition of Regius 1646; third edition with same title in 1661).

Secondary Sources

Bellis, Delphine. 2013. Empiricism without metaphysics: Regius' Cartesian natural philosophy. In *Cartesian Empiricisms*, ed. Mihnea Dobre and Tammy Nyden, 151–183. Dordrecht: Springer.

Clarke, Desmond. 2011. Henricus Regius. In *Stanford encyclopedia of philosophy*, ed. Edward N. Zalta. http://plato.stanford.edu/archives/sum2011/entries/henricus-regius/. Stanford: Stanford University.

de Vrijer, Marinus Johannes Antonie . 1917. *Henricus Regius. Een "cartesiaansch" hoogleraar aan de Utrechtsche Hoogeschool*. PhD dissertation, University of Groningen. The Hague: Martinus Nijhoff.

Farina, Paolo. 1977. Il corpuscularismo di Henricus Regius: Materialismo e medicina in un Cartesiano olandese del Seicento. In *Ricerche sull'atomismo del Seicento*, ed. Ugo Baldini, Giancarlo Zanier, Paolo Farina, and Francesco Trevisani, 119–178. Florence: La Nuova Italia Editrice.

Gariepy, Thomas Peter. 1990. *Mechanism without metaphysics. Henricus Regius and the establishment of Cartesian medicine*. PhD dissertation, Yale University.

Verbeek, Theo. 1992. *Descartes and the Dutch. Early reactions to Cartesian philosophy 1637–1650*. Carbondale/Edwardsville: Southern Illinois University Press.

Verbeek, Theo (ed.). 1993. *Descartes et Regius. Autour de l'Explication de l'esprit humain*. Amsterdam/Atlanta: Rodopi.

Verbeek, Theo. 1994. Regius's *Fundamenta Physices*. *Journal of the History of Ideas* 55(4): 533–551.

Delphine Bellis

Jacques Rohault (1618–1672) was one of the most important Cartesians in the second half of the seventeenth century. Trained as a mathematician and in close connection to the Parisian artisans, he quickly became known for his experimental approach to natural philosophical problems. In the late 1650s and early 1660s in Paris, he lectured in the conferences hosted at Habert de Montmor's house, but from 1660 onward he started his own Cartesian conferences at his own residence. Unofficial reports of his lectures circulated in manuscript, and in 1671 he published his *Traité de physique*, which quickly became the physics textbooks in numerous universities of Europe. English editions of this treatise circulated up to the mid-eighteenth century and they are of special interest, because Samuel Clarke appended Rohault's text with footnotes transforming it into a curious display of both Cartesian and Newtonian ideas. Another important aspect of Rohault is his involvement—together with his father in law, Claude Clerselier—in the dissemination of Cartesianism. In addition to his conferences, Rohault worked together with Clerselier in trying to convince theologians and philosophers that Descartes' views on the Eucharist should be accepted. On this matter he published a short treatise entitled *Entretiens sur la philosophie* (1671).

Primary Sources

Rohault, Jacques. 1671. *Traité de physique*. Paris. doi:dx.doi.org/10.3931/e-rara-4795.

Rohault, Jacques. 1682. *Oeuvres posthumes de M.Rohault*. Paris: Guillaume Desprez.

Rohault, Jacques. 1987. *System of natural philosophy, Illustrated with Dr. Samuel Clarke's notes, taken mostly out of Sir Isaac Newton's philosophy*. New York/ London: Garland Publishing. (Originally published: London: Printed for James and John Knapton at the Crown, 1723–1724.)

Rohault, Jacques. 2009. *Physique nouvelle (1667)*. Paris: Séha and Milan: Archè. (A manuscript reporting Rohault's lectures.)

Secondary Sources

Balz, Albert. 1930. Clerselier (1614–1684) and Rohault (1620–1675). *The Philosophical Review* 39(3): 445–458.

Clair, Pierre. 1978. *Jacques Rohault (1618–1672). Bio-Bibliographie*. Paris: Centre National de la Recherche Scientifique.

Des Chene, Dennis. 2002. Cartesian science: Régis and Rohault. In *Blackwell companions to philosophy: A companion to early modern philosophy*, ed. Steven Nadler, 183–196. Oxford: Blackwell.

Dobre, Mihnea. 2013. Rohault's Cartesian physics. In *Cartesian Empiricisms*, ed. Mihnea Dobre and Tammy Nyden, 203–226. Dordrecht: Springer.

Hoskin, Michael. 1961. 'Mining all within': Clarke's notes to Rohault's *Traité de Physique*. *The Tomist* 24: 353–363.

McClaughlin, Trevor. 1996. Was there an empirical movement in mid-seventeenth century France? Experiments in Jacques Rohault's *Traité De Physique*. *Revue d'histoire des sciences* 49(4): 459–481.

McClaughlin, Trevor. 2000. Descartes, experiments, and a first generation Cartesian, Jacques Rohault. In *Descartes' natural philosophy*, ed. Stephen Gaukroger, John Schuster, and John Sutton, 330–346. London: Routledge.

Mouy, Paul. 1934. *Le Développment de la physique cartésienne: 1646–1712*. Paris: J. Vrin.

Mihnea Dobre

Buchard de Volder (1643–1709) was a Cartesian philosophy professor at the University of Leiden from 1670 to 1705. Known today for his correspondence with Leibniz, his contemporaries knew him for his teaching, being the first to center a university physics course on the demonstration of experiments. In 1674 he traveled to England, met Boyle and Newton, and upon returning to Leiden successfully requested university funds to build the *Theatrum Physicum*, which he modeled on the University's famous anatomical theatre and recently built chemical theatre, both located in Leiden's Medical Faculty. Other Dutch and northern European universities patterned themselves on de Volder's success, acquiring equipment and facilities within Liberal Arts Faculties for teaching natural philosophy. De Volder partnered with Jan van Musschenbroek, a brass smith, to produce the initial equipment, including an air pump designed by de Volder, changing the course of the workshop, which went on to become an influential producer of scientific equipment. De Volder was a vehement defender of Cartesian metaphysics in his early career (when defending the teaching of Descartes against Voetian attacks in a 1676 pamphlet co-authored with his Cartesian colleagues Adriaan Heidanus and Christoph Wittichius) through to his retirement (as evidenced by his 1698–1706 correspondence with Leibniz).

Primary Sources

de Volder, Burchard. 1664. *Disputatio Medica, Inauguralis de Natura*. Leiden: Severinum Mattiae.
de Volder, Burchard. 1679. *Oratio funebris in obitum consultissimi vir Siberti Coeman, Philosophiae & U. Juris doctoris, hujusque in Acad. Lugduno Batava Professoris*. Leiden.
de Volder, Burchard. 1681a. *Disputationes philosophicae sive cogitations rationales de rerum naturalium principiis*. Middelburg: Remigii Schreverii.
de Volder, Burchard. 1681b. *Quastiones academicae De Aëris gravitate*. Middelburg: Remigii Schreverii.
de Volder, Burchard. 1682. *Oratio de conjugendis philosophicis et mathematicis disciplinis*. Leiden: Apud Jacobum Voorn.
de Volder, Burchard. 1685. *Disputationes philosophicae omnes contra atheos*. Middelburg: Johannem Lateranum.
de Volder, Burchard. 1689. *Oratio funebris in obitum Lucae Schacht*. Leiden.
de Volder, Burchard. 1695. *Exercitationes Academicae, quibus Ren. Cartesii Philosophia defenditur adversus Petri Danielis Heutii Censuram Philosophiae Cartesianae*. Amsterdam: A.v. Ravestein.
de Volder, Burchard. 1698. *Oratio de Rationis Viribus, Et Usu in Scientiis*. Leiden: Fridericum Haringium.
de Volder, Burchard. 1705. *Oratio Qua Consentientibus Illistr. Acad. Curatoribus, Urbisque Leidensis Coss. Sese Laboribus Academicis Abdicavit. Habita*. Leiden: Cornelium Boutestein.
de Volder, Burchard. 1709. *Oratio de Novis et Antiquis habita*. Leiden: Cornelium Boutestein. (Speech delivered by de Volder in 1699.)
[de Volder, Burchard, Christoph Wittichius] and Adriaan Heidanus. 1676. *Consideratien, over Eenige saecken onlanghs voorgevallen in de Universiteyt binnen Leyden. Den Derden Druck is achter by-gevoegt een Extract uyt de Resolutie van de Curateuren over de un iversiteyt en Burgermeesteren de Stadt Leyden*. Leiden: Aernoudt Doude.
Leibniz, Gottfried Wilhelm. 2013. *The Leibniz-De Volder correspondence: With selections from the correspondence between Leibniz and Johann Bernoulli*. Trans. Paul Lodge. New Haven: Yale University Press.
Vinson, Carolus. 1676. *Experimenta Anatomica Et Philosophica*, Sloane MS 1292, London: The British Library. (Student notes from de Volder's 1675–1676 physics lectures.)

Secondary Sources

de Cornelis, Pater. 1975. Experimental physics. In *Leiden University in the seventeenth century. An exchange in learning*, ed. Th.H. Lunsingh Scheurlee and G.H.M. Posthumus. Leiden: University of Leiden.
Klever, Wim. 1988. Burchard De Volder (1643–1709)—A Cryptospinozist on a Leiden Cathredra. *LIAS XV* 2: 191–241.

Le Clerc, Jean. 1709. Eloge de feu Mr. De Volder Professeur en Philosophie & aux Mathématiques, dans l'académie de Leide. *Bibliothèque Choisie* 18: 346–401.

Lodge, Paul. 1998. The failure of Leibniz's correspondence with De Volder. *Leibniz Society Review* 8: 47–67.

Lodge, Paul. 2001. The debate over extended substance in Leibniz's correspondence with De Volder. *International Studies in the Philosophy of Science* 15(2): 155–165.

Lodge, Paul. 2004. Leibniz' close encounter with Cartesianism in the correspondence with De Volder. In *Leibniz and his correspondents*, ed. Paul Lodge, 162–192. Cambridge: Cambridge University Press.

Lodge, Paul. 2005. Burchard De Volder: Crypto-Spinozist or disenchanted Cartesian? In *Receptions of Descartes: Cartesianism and anti-Cartesianism in early modern Europe*, ed. Tad M. Schmaltz, 128–145. London: Routledge.

McGahagan, T.A. 1976. *Cartesianism in The Netherlands, 1639–1676. The new science and the Calvinist reformation.* Unpublished PhD, University of Pennsylvania, Pennsylvania.

Nyden, Tammy. 2013. De Volder's Cartesian physics and experimental pedagogy. In *Cartesian Empiricisms*, ed. Mihnea Dobre and Tammy Nyden, 227–249. Dordrecht: Springer.

Nyden, Tammy. In press. Living force at Leiden: De Volder, 's Gravesande and the reception of Newtonianism. In *Newton and Empiricism*, ed. Zvi Biener and Eric Schliesser. Oxford: Oxford University Press.

Otterspeer, Willem. 2001. The University of Leiden: An eclectic institution. *Early Science and Medicine* 6(4): 324–333.

Otterspeer, Willem. 2008. *The bastion of liberty: Leiden University today and yesterday*. Leiden: Leiden University Press.

Ruestow, Edward G. 1973. *Physics at seventeenth and eighteenth-century Leiden: Philosophy and the new science in the University*. The Hague: Martinus Nijhoff.

Schuurman, Paul. 2004. *Ideas, mental faculties and method: The logic of ideas of Descartes and Locke and its reception in the Dutch Republic, 1630–1750*. Brill: Leiden/Boston.

Wiesenfeldt, Gerhard. 2002. *Leerer Raum in Minervas Haus: Experimentelle Naturlehre an der Universität Leiden, 1675–1715*. Amsterdam: Royal Netherlands Academy of Arts and Science.

Tammy Nyden

Index

M. Dobre and T. Nyden (eds.), *Cartesian Empiricisms*, Studies in History and Philosophy of Science 31, DOI 10.1007/978-94-007-7690-6,
© Springer Science+Business Media Dordrecht 2013

T

The manufacturer's authorised representative in the EU is Springer Nature Customer Service Centre GmbH, Europaplatz 3, 69115 Heidelberg, Germany. If you have any concerns regarding our products, please contact ProductSafety@springernature.com

Printed and bound by CPI Group (UK) Ltd, Croydon, CR0 4YY
22/09/2026
02220958-0003